The Works of Archimedes

Archimedes was the greatest scientist of antiquity and one of the greatest of all time. This book is Volume I of the first fully fledged translation of his works into English. It is also the first publication of a major ancient Greek mathematician to include a critical edition of the diagrams, and the first translation into English of Eutocius' ancient commentary on Archimedes. Furthermore, it is the first work to offer recent evidence based on the Archimedes Palimpsest, the major source for Archimedes, lost between 1915 and 1998. A commentary on the translated text studies the cognitive practice assumed in writing and reading the work, and it is Reviel Netz's aim to recover the original function of the text as an act of communication. Particular attention is paid to the aesthetic dimension of Archimedes' writings. Taken as a whole, the commentary offers a groundbreaking approach to the study of mathematical texts.

REVIEL NETZ is Associate Professor of Classics at Stanford University. His first book, *The Shaping of Deduction in Greek Mathematics: A Study in Cognitive History* (1999), was a joint winner of the Runciman Award for 2000. He has also published many scholarly articles, especially in the history of ancient science, and a volume of Hebrew poetry, *Adayin Bahuc* (1999). He is currently editing The Archimedes Palimpsest and has another book forthcoming with Cambridge University Press, *From Problems to Equations: A Study in the Transformation of Early Mediterranean Mathematics.*

THE WORKS OF
ARCHIMEDES

Translated into English, together with
Eutocius' commentaries, with commentary,
and critical edition of the diagrams

REVIEL NETZ

Associate Professor of Classics, Stanford University

Volume I
The Two Books *On the
Sphere and the Cylinder*

CAMBRIDGE
UNIVERSITY PRESS

CAMBRIDGE UNIVERSITY PRESS
Cambridge, New York, Melbourne, Madrid, Cape Town, Singapore, São Paulo, Delhi

Cambridge University Press
The Edinburgh Building, Cambridge CB2 8RU, UK

Published in the United States of America by Cambridge University Press, New York

www.cambridge.org
Information on this title: www.cambridge.org/9780521117982

First published 2004
This digitally printed version 2009

A catalogue record for this publication is available from the British Library

ISBN 978-0-521-66160-7 hardback
ISBN 978-0-521-11798-2 paperback

To Maya

CONTENTS

Acknowledgments   *page* ix

Introduction   1
1   Goal of the translation   1
2   Preliminary notes: conventions   5
3   Preliminary notes: Archimedes' works   10

Translation and Commentary   29
On the Sphere and the Cylinder, Book I   31
On the Sphere and the Cylinder, Book II   185
Eutocius' Commentary to *On the Sphere and*
    the Cylinder I   243
Eutocius' Commentary to *On the Sphere and*
    the Cylinder II   270

Bibliography   369
Index   371

ACKNOWLEDGMENTS

Work on this volume was begun as I was a Research Fellow at Gonville and Caius College, Cambridge, continued as a Fellow at the Dibner Institute for the History of Science and Technology, MIT, and completed as an Assistant Professor at the Classics Department at Stanford University. I am grateful to all these institutions for their faith in the importance of this long-term project.

Perhaps the greatest pleasure in working on this book was the study of the manuscripts of Archimedes kept in several libraries: the National Library in Paris, the Marcian Library in Venice, the Laurentian Library in Florence and the Vatican Library in Rome. The librarians at these institutions were all very kind and patient (not easy, when your reader bends over diagrams, ruler and compass in hand!). I wish to thank them all for their help.

Special words of thanks go to the Walters Art Museum in Baltimore, where the Archimedes Palimpsest has recently been entrusted for conservation. I am deeply grateful to the Curator of Manuscripts there, William Noel, to the conservator of manuscripts, Abigail Quandt, to the imagers of the manuscript, especially Bill Christens-Barry, Roger Easton, and Keith Knox and finally, and most importantly, to the anonymous owner of the manuscript, for allowing study of this unique document.

My most emphatic words of thanks, perhaps, should go to Cambridge University Press, for undertaking this complicated project, and for patience when, with the Archimedes Palimpsest rediscovered, delay – of the most welcome kind – was suddenly imposed upon us. I thank Pauline Hire, the Classics Editor in the Press at the time this work was begun, and Michael Sharp, current Classics Editor, for invaluable advice, criticism and friendliness. Special words of thanks go to my student, Alexander Lee, for his help in proofreading the manuscript.

To mention by name all those whose kind words and good advice have sustained this study would amount to a publication of my private list of addresses. Let it be said instead that this work is a product of many intersecting research communities – in the History of Greek Mathematics, in Classics, in the History and Philosophy of Science, as well as in other fields – from whom I continue to learn, and for whom I have produced this work, as a contribution to an ongoing common study – and as a token of my gratitude.

INTRODUCTION

I GOAL OF THE TRANSLATION

The extraordinary influence of Archimedes over the scientific revolution was due in the main to Latin and Greek–Latin versions handwritten and then printed from the thirteenth to the seventeenth centuries.[1] Translations into modern European languages came later, some languages served better than others. There are, for instance, three useful French translations of the works of Archimedes,[2] of which the most recent, by C. Mugler – based on the best text known to the twentieth century – is still easily available. A strange turn of events prevented the English language from possessing until now any full-blown translation of Archimedes. As explained by T. L. Heath in his important book, *The Works of Archimedes*, he had set out there to make Archimedes accessible to contemporary mathematicians to whom – so he had thought – the mathematical contents of Archimedes' works might still be of practical (rather than historical) interest. He therefore produced a paraphrase of the Archimedean text, using modern symbolism, introducing consistency where the original is full of tensions, amplifying where the text is brief, abbreviating where it is verbose, clarifying where it is ambiguous: almost as if he was preparing an undergraduate textbook of "Archimedean Mathematics." All this was done in good faith, with Heath signalling his practices very clearly, so that the book is still greatly useful as a mathematical gloss to Archimedes. (For such a mathematical gloss, however, the best work is likely to remain Dijksterhuis' masterpiece from 1938 (1987), *Archimedes*.) As it turned out, Heath had acquired in the twentieth century a special position in the English-speaking world. Thanks to his good English style, his careful and highly scholarly translation of Euclid's *Elements*, and, most important, thanks to the sheer volume of his activity, his works acquired the reputation of finality. Such reputations are always

---

[1] See in particular Clagett (1964–84), Rose (1974), Hoyrup (1994).
[2] Peyrard (1807), Ver Eecke (1921), Mugler (1970–74).

1

deceptive, nor would I assume the volumes, of which you now hold the first, are more than another transient tool, made for its time. Still, you now hold the first translation of the works of Archimedes into English.

The very text of Archimedes, even aside from its translation, has undergone strange fortunes. I shall return below to describe this question in somewhat greater detail, but let us note briefly the basic circumstances. None of the three major medieval sources for the writings of Archimedes survives intact. Using Renaissance copies made only of one of those medieval sources, the great Danish scholar J. L. Heiberg produced the first important edition of Archimedes in the years 1880–81 (he was twenty-six at the time the first volume appeared). In quick succession thereafter – a warning to all graduate students – two major sources were then discovered. The first was a thirteenth-century translation into Latin, made by William of Moerbeke, found in Rome and described in 1884,[3] and then, in 1906, a tenth-century Palimpsest was discovered in Istanbul.[4] This was a fabulous find indeed, a remarkably important text of Archimedes – albeit rewritten and covered in the thirteenth century by a prayer book (which is why this manuscript is now known as a *Palimpsest*). Moerbeke's translation provided a much better text for the treatise *On Floating Bodies*, and allowed some corrections on the other remaining works; the Palimpsest offered a better text still for *On Floating Bodies* – in Greek, this time – provided the bulk of a totally new treatise, the *Method*, and a fragment of another, the *Stomachion*. Heiberg went on to provide a new edition (1910–15) reading the Palimpsest as best he could. We imagine him, through the years 1906 to 1915, poring in Copenhagen over black-and-white photographs, the magnifying glass at hand – a Sherlock Holmes on the Sound. A fine detective work he did, deciphering much (though, now we know, far from all) of Archimedes' text. Indeed, one wishes it was Holmes himself on the case; for the Palimpsest was meanwhile gone, Heiberg probably never even realizing this. Rumored to be in private hands in Paris yet considered effectively lost for most of the twentieth century, the manuscript suddenly reappeared in 1998, considerably damaged, in a sale at New York, where it fetched the price of two million dollars. At the time of writing, the mystery of its disappearance is still far from being solved. The manuscript is now being edited in full, for the first time, using modern imaging techniques. Information from this new edition is incorporated into this translation. (It should be noted, incidentally, that Heath's version was based solely on Heiberg's first edition of Archimedes, badly dated already in the twentieth century.) Work on this first volume of translation had started even before the Palimpsest resurfaced. Fortunately, a work was chosen – the books *On the Sphere and the Cylinder*, together with Eutocius' ancient commentary – that is largely independent from the Palimpsest. (Eutocius is not represented in the Palimpsest, while Archimedes' text of this work is largely unaffected by the readings of the Palimpsest.) Thus I can move on to publishing this volume even before the complete re-edition of the Palimpsest has been made, basing myself on Heiberg's edition together with a partial consultation of the Palimpsest. The

---

[3] Rose (1884).   [4] Heiberg (1907).

translations of *On Floating Bodies*, the *Method* and the *Stomachion* will be published in later volumes, when the Palimpsest has been fully deciphered. It is already clear that the new version shall be fundamentally different from the one currently available.

The need for a faithful, complete translation of Archimedes into English, based on the best sources, is obvious. Archimedes was not only an outstanding mathematician and scientist (clearly the greatest of antiquity) but also a very influential one. Throughout antiquity and the middle ages, down to the scientific revolution and even beyond, Archimedes was a living presence for practicing scientists, an authority to emulate and a presence to compete with. While several distinguished studies of Archimedes had appeared in the English language, he can still be said to be the least studied of the truly great scientists. Clearly, the history of science requires a reliable translation that may serve as basis for scholarly comment. This is the basic purpose of this new translation.

There are many possible barriers to the reading of a text written in a foreign language, and the purpose of a scholarly translation as I understand it is to remove all barriers having to do with the foreign language itself, leaving all other barriers intact. The Archimedean text approaches mathematics in a way radically different from ours. To take a central example, this text does not use algebraic symbolism of any kind, relying, instead, upon a certain formulaic use of language. To get habituated to this use of language is a necessary part of understanding how Archimedes thought and wrote. I thus offer the most faithful translation possible. Differences between Greek and English make it impossible, of course, to provide a strict one-to-one translation (where each Greek word gets translated constantly by the same English word) and thus the translation, while faithful, is not literal. It aims, however, at something close to literality, and, in some important intersections, the English had to give way to the Greek. This is not only to make sure that specialist scholars will not be misled, but also because whoever wishes to read Archimedes, should be able to read *Archimedes*. Style and mode of presentation are not incidental to a mathematical proof: they constitute its soul, and it is this soul that I try, to the best of my ability, to bring back to life.

The text resulting from such a faithful translation is difficult. I therefore surround it with several layers of interpretation.

- I intervene in the body of the text, in clearly marked ways. Glosses added within the standard pointed-brackets notation (<. . .>) are inserted wherever required, the steps of proofs are distinguished and numbered, etc. I give below a list of all such conventions of intervention in the text. The aim of such interventions is to make it easier to construe the text as a sequence of meaningful assertions, correctly parsing the logical structure of these assertions.
- Footnotes add a brief and elementary mathematical commentary, explaining the grounds for the particular claims made. Often, these take the form of references to the tool-box of known results used by Archimedes. Sometimes, I refer to Eutocius' commentary to Archimedes (see below). The aim of these footnotes, then, is to help the readers in checking the validity of the argument.

- A two-part set of comments follows each proposition (or, in some cases, units of text other than propositions):
  - The first are *textual comments*. Generally speaking, I follow Heiberg's (1910–15) edition, which seems to remain nearly unchanged, for the books *On the Sphere and the Cylinder*, even with the new readings of the Palimpsest. In some cases I deviate from Heiberg's text, and such deviations (excepting some trivial cases) are argued for in the textual comments. In other cases – which are very common – I follow Heiberg's text, while doubting Heiberg's judgment concerning the following question. Which parts of the text are genuine and which are interpolated? Heiberg marked what he considered interpolated, by square brackets ([. . .]). I print Heiberg's square brackets in my translation, but I very often question them within the textual comments.
  - The second are *general comments*. My purpose there is to develop an interpretation of certain features of Archimedes' writing. The comments have the character not of a reference work, but of a monograph. This translation differs from other versions in its close proximity to the original; it maps, as it were, a space very near the original writing. It is on this space that I tend to focus in my general comments. Thus I choose to say relatively little on wider mathematical issues (which could be equally accessed through a very distant translation), only briefly supply biographical and bibliographical discussions, and often focus instead on narrower cognitive or even linguistic issues. I offer three apologies for this choice. First, such comments on cognitive and linguistic detail are frequently necessary for understanding the basic meaning of the text. Second, I believe such details offer, taken as a whole, a central perspective on Greek mathematical practices in general, as well as on Archimedes' individual character as an author. Third and most important, having now read many comments made in the past by earlier authors, I can no longer see such comments as "definitive." Mine are "comments," not "commentary," and I choose to concentrate on what I perceive to be of relevance to contemporary scholarship, based on my own interest and expertise. Other comments, of many different kinds, will certainly be made by future readers of Archimedes. Readers interested in more mathematical commentary should use Eutocius as well as Dijksterhuis (1987), those interested in more biographical and historical detail on the mathematicians mentioned should use Knorr (1986), (1989), and those looking for more bibliographic references should use Knorr (1987) (which remains, sixteen years later nearly complete). (Indeed, as mentioned above, Archimedes is not intensively studied.)
- Following the translation of Archimedes' work, I add a translation of Eutocius' commentary to Archimedes. This is a competent commentary and the only one of its kind to survive from antiquity. Often, it offers a very useful commentary on the mathematical detail, and in many cases it has unique historical significance for Archimedes and for Greek mathematics in general. The translation of Eutocius follows the conventions of the translation of Archimedes, but I do not add comments to his text, instead supplying, where necessary, fuller footnotes.

A special feature of this work is a critical edition of the diagrams. Instead of drawing my own diagrams to fit the text, I produce a reconstruction based on the independent extant manuscripts, adding a critical apparatus with the variations between the manuscripts. As I have argued elsewhere (Netz [forthcoming]), I believe that this reconstruction may represent the diagrams as available in late antiquity and, possibly, at least in some cases, as produced by Archimedes himself. Thus they offer another, vital clue to our main question, how Archimedes thought and wrote. I shall return below to explain briefly the purpose and practices of this critical edition.

Before the translation itself I now add a few brief preliminary notes.

2 PRELIMINARY NOTES: CONVENTIONS

2.1 Some special conventions of Greek mathematics

In the following I note certain practices to be found in Archimedes' text that a modern reader might find, at first, confusing.

1. Greek word order is much freer than English word order and so, selecting from among the wider set of options, Greek authors can choose one word order over another to emphasize a certain idea. Thus, for instance, instead of writing "A is equal to B," Greek authors might write "to B is equal A." This would stress that the main information concerns B, not A – word order would make B, not A, the focus. (For instance, we may have been told something about B, and now we are being told the extra property of B, that it is equal to A.) Generally speaking, such word order cannot be kept in the English, but I try to note it when it is of special significance, usually in a footnote.

2. The summation of objects is often done in Greek through ordinary conjunction. Thus "the squares ABΓΔ and EZHΘ" will often stand for what we may call "the square ABΓΔ plus the square EZHΘ." As an extension of this, the ordinary plural form can serve, as well, to represent summation: "the squares ABΓΔ, EZHΘ" (even without the "and" connector!) will then mean "the square ABΓΔ plus the square EZHΘ." In such cases, the sense of the expression is in itself ambiguous (the following predicate may apply to the sum of the objects, or it may apply to each individually), but such expressions are generally speaking easily disambiguated in context. Note also that while such "implicit" summations are very frequent, summation is often more explicit and may be represented by connectors such as "together with," "taken together," or simply "with."

3. The main expression of Greek mathematics is that of proportion:

As A is to B, so is C to D.

(A, B, C, and D being some mathematical objects.) This expression is often represented symbolically, in modern texts, by

A:B::C:D

and I will use such symbolism in my footnotes and commentary. In the main text I shall translate, of course, the original non-symbolic form. Note especially that this expression may become even more concise, e.g.:

As A is to B, C to D, As A to B, C to D.

And that it may have more complex syntax, especially:

> A has to B the same ratio as C has to D, A has to B a greater ratio than C has to D.

The last example involves an obvious extension of proportion, to ratio-inequalities, i.e. A:B>C:D. More concisely, this may be expressed by:

> A has to B a greater ratio than C to D.

4. Greek mathematical propositions have, in many cases, the following six parts:

- *Enunciation*, in which the claim of the proposition is made, in general terms, without reference to the diagram. It is important to note that, generally speaking, the enunciation is equivalent to a conditional statement that *if x* is the case, *then* so is *y*.
- *Setting-out*, in which the antecedent of the claim is re-stated, in particular terms referring to the diagram (with the example above, *x* is re-stated in particular reference to the diagram).
- *Definition of goal*, in which the consequent of the claim is re-stated, as an exhortation addressed by the author to himself: "I say that . . . ," "it is required to prove that . . . ," again in the particular terms of the diagram (with the same example, we can say that *y* is re-stated in particular reference to the diagram).
- *Construction*, in which added mathematical objects (beyond those required by the setting-out) may be introduced.
- *Proof*, in which the particular claim is proved.
- *Conclusion*, in which the conclusion is reiterated for the general claim from the enunciation.

Some of these parts will be missing in most Archimedean propositions, but the scheme remains a useful analytical tool, and I shall use it as such in my commentary. The reader should be prepared in particular for the following difficulty. It is often very difficult to follow the enunciations as they are presented. Since they do not refer to the particular diagram, they use completely general terms, and since they aspire to great precision, they may have complex qualifications and combinations of terms. I wish to exonerate myself: this is not a problem of my translation, but of Greek mathematics. Most modern readers find that they can best understand such enunciations by reading, first, the setting-out and the definition of goal, with the aid of the diagram. Having read this, a better sense of the *dramatis personae* is gained, and the enunciation may be deciphered. In all probability the ancients did the same.

2.2 Special conventions adopted in this translation

1. The main "<. . .>" policy:

Greek mathematical proofs always refer to concrete objects, realized in the diagram. Because Greek has a definite article with a rich morphology, it can elide the reference to the objects, leaving the definite article alone. Thus the Greek may contain expressions such as

> "The by the AB, BΓ"

whose reference is

"The <rectangle contained> by the <lines> AB, BΓ"

(the morphology of the word "the" determines, in the original Greek, the identity of the elided expressions, given of course the expectations created by the genre).

In this translation, most such elided expressions are usually added inside pointed brackets, so as to make it possible for the reader to appreciate the radical concision of the original formulation, and the concreteness of reference – while allowing me to represent the considerable variability of elision (very often, expressions have only partial elision). This variability, of course, will be seen in the fluctuating positions of pointed brackets:

"The <rectangle contained> by the <lines> AB, BΓ," as against, e.g.

"The <rectangle> contained by the <lines> AB, BΓ."

(Notice that I do not at all strive at consistency *inside* pointed brackets. Inside pointed brackets I put whatever seems to me, in context, most useful to the reader; the duties of consistency are limited to the translation proper, outside pointed brackets.)

The main exception to my general pointed-brackets policy concerns points and lines. These are so frequently referred to in the text that to insist, always, upon a strict representation of the original, with expressions such as

"The <point> **A**," "The <line> **AB**"

would be tedious, while serving little purpose. I thus usually write, simply:

A, AB

and, in the less common cases of a non-elliptic form:

"The point A," "The line AB"

The price paid for this is that (relatively rarely) it is necessary to stress that the objects in question are points or lines, and while the elliptic Greek expresses this through the definite article, my elliptic "A," "AB" does not. Hence I need to introduce, here and there, the expressions:

"The <point> **A**," "The <line> **AB**"

but notice that these stand for precisely the same as

A, AB.

2. The "<=. .>" sign is also used, in an obvious way, to mean essentially the same as the "[*Scilicet.* . . .]" abbreviation. Most often, the expression following the "=" will disambiguate pronouns which are ambiguous in the English (but which, in the Greek, were unambiguous thanks to their morphology).

3. Square brackets in the translation ("[. . .]") represent the square brackets in Heiberg's (1910–15) edition. They signify units of text which according to Heiberg were interpolated.

4. Two sequences of numbering appear inside standard brackets. The Latin alphabet sequence "(a) . . . (b) . . ." is used to mark the sequence of constructions: as each new item is added to the construction of the geometrical configuration (following the setting-out) I mark this with a letter in the sequence of the Latin alphabet. Similarly, the Arabic number sequence "(1) . . . (2) . . ." is used to mark the sequence of assertions made in the course of the proof: as each new assertion is made (what may be called "a step in the argument"), I mark this with a number. This is meant for ease of reference: the footnotes and the

commentary refer to constructions and to claims according to their letters or numbers. **Note that this is purely my editorial intervention, and that the original text had nothing corresponding to such brackets.** (The same is true for punctuation in general, for which see point 6 below.) Also note that these sequences refer only to construction and proof: enunciation, setting-out, and definition of goal are not marked in similar ways.

5. The "/. . ./" symbolism: for ease of reference, I find it useful to add in titles for elements of the text of Archimedes, whether general titles such as "introduction" or numbers referring to propositions. I suspect Archimedes' original text had neither, and such titles and numbers are therefore mere aids for the reader in navigating the text.

6. Ancient Greek texts were written without spacing or punctuation: they were simply a continuous stream of letters. Thus punctuation as used in modern editions reflects, at best, the judgments of late antiquity and the middle ages, more often the judgments of the modern editor. I thus use punctuation freely, as another editorial tool designed to help the reader, but in general I try to keep Heiberg's punctuation, in deference to his superb grasp of the Greek mathematical language, and in order to facilitate simultaneous use of my translation and Heiberg's edition.

7. Greek diagrams can be characterized as "qualitative" rather than "quantitative." This is very difficult to define precisely, and is best understood as a warning: **do not assume that relations of size in the diagram represent relations of size in the depicted geometrical objects**. Thus, two geometrical lines may be assumed equal, while their diagrammatic representation is of two unequal lines and, even more confusingly, two geometrical lines may be assumed *un*equal, while their diagrammatic representation is of two *equal* lines. Similar considerations apply to angles etc. What the diagram most clearly *does* represent are relations of connection between the geometrical constituents of the configuration (what might be loosely termed "topological properties"). Thus, in an extreme case, the diagram may concentrate on representing the fact that two lines touch at a single point, ignoring another geometrical fact, that one of the lines is straight while the other is curved. This happens in a series of propositions from 21 onwards, in which a dodecagon is represented by twelve *arcs*; but this is an extreme case, and generally the diagram may be relied upon for such basic qualitative distinctions as straight/curved. See the following note on purpose and practices of the critical edition of diagrams.

2.3   Purpose and practices of the critical edition of diagrams

The main purpose of the critical edition of diagrams is to reconstruct the earliest form of diagrams recoverable from the manuscript evidence. It should be stressed that the diagrams across the manuscript tradition are strikingly similar to each other, often in quite trivial detail, so that there is hardly a question that they derive from a common archetype. For most of the text translated here, diagrams are preserved only for one Byzantine tradition, that of codex A (see below, note on the text of Archimedes). However, for most of the diagrams from

SC I.32 to *SC* II.6, diagrams are preserved from the Archimedes Palimpsest (codex C, see below note on the text of Archimedes). Once again, the two Byzantine codices agree so closely that a late ancient archetype becomes a likely hypothesis. I shall not dwell on the question, how closely the diagrams of late antiquity resembled those of Archimedes himself: to a certain extent, the same question can be asked, with equal futility, for the text itself. Clearly, however, the diagrams reconstructed are genuinely "ancient," and provide us with important information on visual practices in ancient mathematics.

Since the main purpose of the edition is the recovery of an ancient form, I do not discuss as fully the issues – very interesting in themselves – of the various processes of transmission and transformation. Moerbeke's Latin translation (codex B) is especially frustrating in this respect. In the thirteenth century, Moerbeke clearly used his source as inspiration for his own diagram, often copying it faithfully. However, he transformed the basic layout of the writing, so that his diagrams occupied primarily not the space of writing itself but the margins. This resulted in various transformations of arrangement and proportion. To compound the difficulty, 250 years later the same manuscript was very carefully read by Andreas Coner, a Renaissance humanist. Coner erased many of Moerbeke's diagrams, covering them with his own diagrams that he considered more "correct." This would form a fascinating subject for a different kind of study. In this edition, I refer to the codex only where, in its present state, some indications can be made for the appearance of Moerbeke's source. When I am silent about this codex, readers should assume that the manuscript, at least in its present state, has a diagram quite different from all other manuscripts, as well as from that printed by me.

Since the purpose of the edition is to recover the ancient form of Archimedes' diagrams, "correctness" is judged according to ancient standards. Obvious scribal errors, in particular in the assigning of letters to the figure, are corrected in the printed diagram and noted in the apparatus. However, as already noted above, I do not consider diagrams as false when they do not "appear right." The question of the principles of representation used by ancient diagrams requires research. Thus one purpose of the edition is simply to provide scholars with the basic information on this question. Furthermore, it is my view, based on my study of diagrams in Archimedes and in other Greek textual traditions of mathematics, that the logic of representation is in fact simple and coherent. Diagrams, largely speaking, provide a schematic representation of the pattern of configuration holding in the geometrical case studied. This pattern of configuration is what can be reliably "read" off the diagram and used as part of the logic of the argument, since it is independent of metrical values. Ancient diagrams are taken to represent precisely that which can be exactly represented and are therefore, unlike their modern counterparts, taken as tools for the logic of the argument itself.

This is difficult for modern readers, who assume that diagrams represent in a more pictorial way. Thus, for instance, a chord that appears like a diameter could automatically be read by modern readers to signify a diameter, with a possible clash between text and diagram. Indeed, if you have not studied Greek mathematics before, you may find the text just as perplexing. For both text and diagram we must always bear in mind that the reading of a document produced in a foreign culture requires an effort of imagination. (For the same reason, I also follow the ancient practice of putting diagrams immediately following their text.)

The edition of diagrams cannot have the neat logic of textual editions, where critical apparatuses can pick up clearly demarcated units of text and note the varia to each. Being continuous, diagrams do not possess clear demarcations. Thus a more discursive text is called for (I write it in English, not Latin) and, in some cases, a small "thumbnail" figure best captures the varia. There are generally speaking two types of issues involved: the shape of the figure, and the assignment of the letters. I try to discuss first the shape of the figure, starting from the more general features and moving on gradually to the details of the shape, and, following that, discuss the assignment of the letters. Obviously, in a few cases the distinction can not be clearly made. For both the shape of the figure and the letters, I start with varia that are widespread and are more likely to represent the form of the archetype, and move on to more isolated varia that are likely to be late scribal adjustments or mistakes. It shall become obvious to the reader that while codices BDG tend to adjust, i.e. deliberately to change the diagrams (usually rather minimally) for various mathematical or practical reasons, codices EH4 seem to aim at precise copying, so that varia tend to consist of mistakes alone – from which, of course, BDG are not free either (for the identity of the codices, see note on the text of Archimedes below).

While discussing the shape of the figure, I need to use some labels, and I use those of the printed diagram. It will sometimes happen that a text has some noteworthy varia on a detail of the shape, compounded by a varia on the letter labelling that detail. When referring to the shape itself, I use the label of the printed, "correct" diagram, regardless of which label the codex itself may have.

3   PRELIMINARY NOTES: ARCHIMEDES´ WORKS

3.1   Archimedes and his works

This is not the place to attempt to write a biography of Archimedes and perhaps this should not be attempted at all. Our knowledge of Archimedes derives from two radically different lines of tradition. One is his works, for whose transmission see the following note. Another is the ancient biographical and historical tradition, usually combining the factual with the legendary. The earliest source is Polybius,[5] the serious-minded and competent historian writing a couple of generations after Archimedes' death: an author one cannot dismiss. It can thus be said with certainty that Archimedes was a leading figure in the defense of Syracuse from the Romans, dying as the city finally fell in 212 BC, in one of the defining moments of the Second Punic War – the great World War of the classical Mediterranean. It is probably this special role of a scientist, in such a pivotal moment of history, which gave Archimedes his fame. Details of his

---

[5] Polybius VIII.5 ff.

engineering feats, of his age in death, and of the various circumstances of his life and death, are all dependent on later sources and are much more doubtful. It does seem likely that he was not young at the time of his death, and the name of his father – Pheidias – suggests an origin, at least some generations back, in an artistic, that is artisanal, background.

Perhaps alone among ancient mathematicians, a clearly defined character seems to emerge from the writings themselves: imaginative to the point of playfulness, capable of great precision but always preferring the substance to the form. It is easy to find this character greatly attractive, though one should add that the playfulness, in the typical Greek way, seems to be antagonistic and polemic, while the attention to substance over form sometimes verges into carelessness. (On the whole, however, the logical soundness of the argument is only extremely rarely in doubt.) One of my main hopes is that this translation may do justice to Archimedes' personality: I often comment on it in the course of my general comments.

Even the attribution of works to Archimedes is a difficult historical question. The corpus surviving in Greek – where I count Eutocius' commentaries as well – includes the following works (with the abbreviations to be used later in the translation):

SC I	The first book *On the Sphere and the Cylinder*
Eut. *SC* I	Eutocius' commentary to the above
SC II	The second book *On the Sphere and the Cylinder*
Eut. *SC* II	Eutocius' commentary to the above
SL	*Spiral Lines*
CS	*Conoids and Spheroids*
DC	*Measurement of the Circle* (*Dimensio Circuli*)
Eut. *DC*	Eutocius' commentary to the above
Aren.	*The Sand Reckoner* (*Arenarius*)
PE I, II	*Planes in Equilibrium*[6]
Eut. *PE* I, II	Eutocius' commentary to the above
QP	*Quadrature of the Parabola*
Meth.	The *Method*
CF I	The first book *On Floating Bodies* (*de Corporibus Fluitantibus*)
CF II	The second book *On Floating Bodies* (*de Corporibus Fluitantibus*)
Bov.	The *Cattle Problem* (*Problema Bovinum*)
Stom.	*Stomachion*

Some works may be ascribed to Archimedes because they start with a letter by Archimedes, introducing the work by placing it in context: assuming these are not forgeries (and their sober style suggests authenticity), they are the best evidence for ascription. These are *SC* I, II, *SL*, *CS*, *Aren.*, *QP*, *Meth.*

---

[6] The traditional division of *PE* into two books is not very strongly motivated; we shall return to discuss this in the translation of *PE* itself.

Even more useful to us, the introductory letters often connect the works introduced to other, previous works by Archimedes. Thus the author of the Archimedean introductions claims authorship to what appears to be *SC* I (referred to in introductions to *SC* II, *SL*), *SC* II (referred to in introduction to *SL*), *CS* (referred to in introduction to *SL*), *QP* (referred to in introduction to *SC* I). In the course of the texts themselves, the author refers to further works no longer extant: a study in numbers addressed *To Zeuxippus* (mentioned in *Aren.*), and a study *Of Balances*, which seems to go beyond our extant *PE* (mentioned in the *Method*). A special problem concerns an appendix Archimedes promised to attach to *SC* II in the course of the main text: now lost from the manuscript tradition of *SC* II, it was apparently rediscovered by Eutocius, who includes it in his commentary to that work.

Other works, while not explicitly introduced by an Archimedean letter, belong to areas where, based on ancient references, we believe Archimedes had an interest, and generally speaking show a mathematical sophistication consistent with the works mentioned above: these are *DC*, *PE*, *CF* I, II. Thus the fact that they are ascribed to Archimedes by the manuscript tradition carries a certain conviction.

Furthermore, several works show a certain presence of Doric dialect – that is, the dialect used in Archimedes' Syracuse. As it differs from the main literary prose dialect of Hellenistic times, Koine, only in relatively trivial points (mainly those of pronounciation), it is natural that the dialect was gradually eroded from the manuscript tradition, disappearing completely from some works. Still, larger or smaller traces of it can be found in *SL*, *CS*, *Aren.*, *PE*, *QP*, *CF* I, II.

Questions may be raised regarding the precise authorship of Archimedes, based on logical and other difficulties in those texts. *The Measurement of the Circle*, in particular, seems to have been greatly modified in its transmission (see the magisterial study of this problem in Knorr (1989), part 3). Doubts have been cast on the authenticity of *Planes in Equilibrium* I, as its logical standards seem to be lower than those of many other works (Berggren 1976): I tend to think this somewhat overestimates Archimedean standards elsewhere, and underestimates *PE* I. I shall return to this question in the translation of *PE*.

It thus seems very probable that, even if sometimes modified by their transmission, all the works in the Greek corpus are by Archimedes, with the possible exceptions of the *Cattle Problem* and of the *Stomachion* – two brief jeux d'esprit whose meaning is difficult to tell, especially given the fragmentary state of the *Stomachion*.

The Arabo-Hebraic tradition of Archimedes is large, still not completely charted and of much more complicated relation to Archimedes the historical figure. It now seems that, of thirteen works ascribed to Archimedes by Arabic sources, five are paraphrases or extracts of *SC* I, II, *DC*, *CF* I and *Stom.*, four are either no longer extant or, when extant, can be proved to have no relation to Archimedes, while four may have some roots in an Archimedean original. These four are:

Construction of the Regular Heptagon
On Tangent Circles
On Lemmas
On Assumptions

None of these works seems to be in such textual shape that we can consider them, as they stand, as works by Archimedes, even though some of the results there may have been discovered by him. (In a sense, the same may be said of *DC*, extant in the Greek.) I thus shall not include here a translation of works surviving only in Arabic.[7]

Finally, several works by Archimedes are mentioned in ancient sources but are no longer extant. These are listed by Heiberg as "fragments," collected at the end of the second volume of the second edition:

On Polyhedra
On the Measure of a Circle
On Plynths and Cylinders
On Surfaces and Irregular Bodies
Mechanics
Catoptrics
On Sphere-Making
On the Length of the Year

Some of those references may be based on confusions with other, extant works, while others may be pure legend. The reference to the work *On Polyhedra*, however, made by Pappus in his *Mathematical Collection*,[8] is very detailed and convincing.

In the most expansive sense, bringing in the Arabic tradition in its entirety, we can speak of thirty-one works ascribed to Archimedes. Limiting ourselves to extant works whose present state seems to be essentially that intended by Archimedes, we can mention in great probability ten works: *SC* I, *SC* II, *SL*, *CS*, *Aren.*, *PE*, *QP*, *Meth.*, *CF* I, *CF* II. It is from these ten works that we should build our interpretation of Archimedes as a person and a scientist.

I shall translate here all these works, adding in *DC*, *Bov.*, and *Stom.* Brief works, the first clearly not in the form Archimedes intended it, the two last perhaps not by him, they are still of historical interest for their place in the reception of Archimedes: by including them, I make this translation agree with Heiberg's second edition.

This first volume is dedicated to the longest self-contained sequence in this corpus: *SC* I, II. Division of the remaining works between volumes II and III will be determined by the progress of the reading of the Palimpsest.

3.2   The text of Archimedes

Writing was crucial to Archimedes' intellectual life who, living in Syracuse, seems to have had his contacts further east in the Mediterranean, in Samos (where his admired friend Conon lived) and especially in Alexandria (where Dositheus and Eratosthenes were addressees of his works). Most of the treatises, as explained above, are set out as letters to individuals, and while this is essentially a literary trope, it gains significance from Archimedes' practice of

---

[7] For further discussions of the Arabic traditions, see Lorch (1989), Sesiano (1991).

[8] Pappus V, Hultsch (1876–78) I.352–58.

sending out enunciations without proofs – puzzles preceding the works themselves. Thus in the third century BC, to have known the works of Archimedes
would mean to have been privy to a complex web of correspondence between
Mediterranean intellectuals. How and when this web of correspondence got
transformed into collections of "treatises by Archimedes" we do not know.
Late authors often reveal an acquaintance with many works by Archimedes,
but the reference is more often to results than to works, and if to works, it is
to individual works rather than to any collection of works. No one in antiquity
seems to have known the works in the precise form or arrangement of any of
the surviving manuscripts.

Indeed the evidence of the surviving manuscripts is very indirect – as it usually is for ancient authors. Late antiquity was a time of rearrangement, not least
of ancient books. Most important, books were transformed from papyrus rolls
(typically holding a single treatise in a roll) into parchment codices (typically
holding a collection of treatises). Books from late antiquity very rarely survive, and we can only guess that, during the fifth and sixth centuries – during
Byzantium's first period of glory – several such collections containing works
by Archimedes were made. In particular, it appears that an important collection
was made by Isidore of Miletus – no less than the architect of Hagia Sophia.
The evidence for this is translated in this volume, and is found at the very end
of both of Eutocius' commentaries.

As for most ancient authors, our evidence begins to be surer at around the
ninth century AD. It was then that, following a long period of decline, Byzantine
culture began one of its several renaissances, producing a substantial number
of copies of ancient works in the relatively recent, minuscule script. At least
three codices containing works by Archimedes were produced during the ninth
and tenth centuries. The same tradition where we see the evidence for the
presence of Isidore of Miletus, also has evidence for the presence of Leo the
geometer,[9] a leading Byzantine intellectual of the ninth century. It thus appears
that a book collecting several treatises by Archimedes was prepared, by Isidore
of Miletus or his associates, in Constantinople in the sixth century AD, and
that this book was copied, by Leo the geometer or his associates, once again in
Constantinople, in the ninth century AD. Lost now, enough is known about this
book (as to be explained below) to give it a name. This is Heiberg's codex A.
It contained, in this sequence, the works: *SC* I, II, *DC*, *CS*, *SL*, *PE* I, II, *Aren.*,
QP; Eutoc. *In SC* I, II, *In DC*, *In PE* I, II, as well as, following that, a work by
Hero.

This volume then was essentially a *collected works of Archimedes*. This is
not typical of codices for ancient science, where one usually has collections
that are defined by subject matter. (For instance, we may have a collection of
various works dedicated to astronomy, or when we have a collection from a
single author, such as Euclid, the author himself provides us with an introduction to a field.) That collected works of Archimedes were put together in

---

[9] The evidence survives in a scribal note made, in codex A, at the end of *QP*, to be
translated and discussed in a later volume of this translation.

late antiquity, and then again by Byzantine scholars, is a mark of the esteem by which Archimedes was held. Even more remarkable, then, that Byzantium (and hence, probably, late antiquity as well) had not one, but at least two versions of collected works of Archimedes. At around 975 (judging from the nature of the script), another such codex was made, once again, probably, a copy of a late ancient book. This codex seems to have contained the following works in sequence: *PE* (I?), II, *CF* I, II, *Meth.*, *SL*, *SC*, *DC*, *Stom.* This was called by Heiberg codex C, and is extant as the Archimedes Palimpsest.

Indirect evidence, to be explained below, leads us to believe that a third Byzantine codex included works by Archimedes, though here in the more common context of a codex setting out a field – that of mechanics and optics. This codex included at least the following works by Archimedes (we do not know in what order): *PE* I, II, *CF* I, II, *QP*. It was called by Heiberg codex 𝕭.

At the turn of the millennium, then, the Byzantine world had access to all the works of Archimedes we know today, often in more than one form. Two centuries later, all this was gone.

Codex C, the Archimedes Palimpsest was, obviously, palimpsested. By the twelfth or thirteenth centuries, the value of this collection was sufficiently reduced to suggest that it could better serve as scrap parchment for the production of a new book – obviously, not a highly valuable one. Thus a run-of-the-mill Greek prayer book was written over this collection of the works of Archimedes, so that the collection and its fabulous contents remained unknown for seven centuries. However, this is the only surviving Byzantine manuscript of Archimedes and, ironically, it is probably the prayer book that protected the works from destruction.

No longer extant today, codices A and 𝕭 had a very important role to play in the history of Western science. It was in Western Europe, indeed, that they performed their historical service, removed there following Western Europe's first colonizing push. In the Crusades, Western Europe was trying to assert its authority over the Eastern Mediterranean. The culmination of this push was reached in 1204, when Constantinople itself was sacked by Venice and its allies, its old territories parceled out to western knights, many of its treasures looted. Codices A and 𝕭, among such looted works, soon made their way to Europe, and by 1269 were in the papal library in Viterbo, where William of Moerbeke used them for his own choice of collected works by Archimedes, translated into Latin. The autograph for this translation is extant, and was called by Heiberg *codex B*. Conforming to Moerbeke's practice elsewhere, the translation is faithful to the point where Latin is no longer treated as Latin, but as Greek transposed to a different vocabulary. Thus this codex B is almost as useful a source for Archimedes' text as a Greek manuscript would be. The works translated from the Archimedean corpus are, in order: *SL*, *PE* I, II, *QP*, *DC*, *SC* I, II, Eutoc. *In SC* I, II, *CS*, Eutoc. *In PE* I, II, *CF* I, II. Thus Moerbeke has translated all the Archimedean content of codex A, though not in order, excepting *Aren.*, Eutoc. *In DC*, comparing A to 𝕭 for *PE*, *QP*, and using 𝕭 as his source for *CF*.

Moerbeke's translation was not unknown at the time, but it was the mathematical renaissance of the fifteenth and sixteenth centuries that brought

Archimedes into prominence.[10] All of a sudden, works by Archimedes were considered of the highest value. In 1491, Poliziano writes back from Venice to Florence – "I have found [here] certain mathematical books by Archimedes . . . that we miss" – no doubt referring to codex A – and straight away a copy is made. We can follow the rapid sequence of new copies, translations, and finally printed editions: a copy of codex A, made in the mid-fifteenth century in Venice, Heiberg's codex E; another, the Florentine copy made by Poliziano's efforts, Heiberg's codex D; Francois I, never one to be left behind, has another copy made in 1544 for his library at Fontainebleau, now in the French National Library – Heiberg's codex H; the same library has another sixteenth-century copy of the same works, which Heiberg called codex G. Finally, the Vatican Library has another copy, which Heiberg called simply codex 4. These codices – D, E, H, G, 4 – are all independent copies made of the same codex A, as Heiberg meticulously proved by studying the pattern of recurring and non-recurring errors in all manuscripts. Many other manuscripts were prepared at the same time, of great importance for the diffusion of Archimedes' works in Europe (though not for the reconstruction of Archimedes' text, as all those manuscripts were derived not directly from codex A, but indirectly from the copies mentioned above, so that they add nothing new to what we know already from the five copies D, E, H, G, and 4). Heiberg lists thirteen such further copies, and doubtless others were made as well, most during the sixteenth century.

As Europe was gaining in manuscripts of Archimedes, it was also losing some. Codex 𝕭 apparently could have been lost as early as the fourteenth century; codex A certainly disappeared towards the end of the sixteenth century. Objects of value, and greed, such codices rapidly transfer from hand to hand, laying themselves open to the ravages of fortune. Codex C, meanwhile, survived hidden in its mask of anonymity.

Together with the growing number of Greek manuscripts, and even more important for the history of western science, Latin manuscripts were accumulating. In the middle of the fifteenth century – exactly when Greek copies begin to be made of codex A – Jacob of Cremona had once again translated Archimedes' works into Latin. This translation no longer had access to codex 𝕭 and thus did not contain *CF*. It was nearly as frequently copied as codex A itself was and, written in Latin, its copies were more frequently consulted, one of them, famously, by Leonardo.[11] Europe, flush with works of Archimedes in Greek and Latin, soon had them represented in print, to begin with numerous publications with brief extracts from works of Archimedes or with a few treatises translated into Latin, reaching finally the First Edition of Archimedes in Basel, 1544. Those printed versions relied on many separate lines of transmission, often inferior, the Basel edition using a derivative copy of codex A for the Greek, and Cremona's translation for the Latin. The two later editions of the works of Archimedes, made by Rivault in Paris, 1615, and then by Torelli in Oxford, 1792, were a bit better, relying on codices G and E, respectively. It was only Heiberg and his generation that brought to light all the extant manuscripts and

---

[10] See Rose (1974), especially chapter 10.

[11] For the Latin tradition of Archimedes – with antecedents prior to Moerbeke, and the complex Renaissance history – see the magisterial study, Clagett (1964–84).

discovered their order, a process ending in Heiberg's second edition of 1910–15. Thus "the text of Archimedes" – an authoritative setting out of the best available evidence on the writings of Archimedes – is a very recent phenomenon.

Before that, Europe had not a "text of Archimedes" but many of them: various versions available in both Greek and Latin. Going beyond copies and translations in the narrow sense, Europe also had more and more works produced to comment upon or recast the Archimedean corpus as known to various authors – by the sixteenth century authors such as Tartagla, Commandino and Maurolico, and leading on to the famous works of the seventeenth century by Galileo, Huygens, and others.[12] In a word, we can say that Archimedes' methods of measurement formed the basis for reflections leading on to the calculus, while Archimedes' statics and hydrostatics formed the basis for reflections leading on to mathematical physics. In this sense, the text of Archimedes is with us, simply, as modern science.

To sum up the discussion, I now offer the tree setting out the order of transmission of the manuscripts of Archimedes referred to in this book (especially in the apparatus to the diagrams). I follow Heiberg's sigla, which call for a word of explanation.

Heiberg had studied the manuscript tradition of Archimedes for over thirty-five years, starting with his dissertation, Quaestiones Archimedeae (1879), going to his First Edition (1880–81) and leading, through numerous articles detailing new discoveries and observations, to the Second Edition (1910–15). He considerably refined his views throughout the process, and the final position reached in 1915 seems to be solidly proven. Still, his final choice of sigla reflects the circuitous path leading there, and is somewhat confusing.

Heiberg uses symbols of different kinds A, $\mathfrak{B}$ and C, for codices that are similar in nature: mutually independent Byzantine manuscripts, from the ninth to tenth centuries.

Heiberg gave the siglum B to Moerbeke's autograph translation. This may be misleading, as it might make us think of this codex as having an authority comparable to that of A and C. In fact, codex B is partly a copy of A, partly a copy of $\mathfrak{B}$. It has special value for the works of $\mathfrak{B}$, for whom B is our only surviving witness. However, there are many other copies based on codex A, and for such works codex B has no special status as a witness. The works translated in this volume are of this nature. Heiberg's apparatus frequently refers to "AB" – the consensus of the manuscripts A and B. This is misleading, in creating the impression that the authority of two separate lines of tradition support the reading. "AB," essentially, is the same as A, and it is only codex C – the Palimpsest – that provides extra information.

Heiberg gave the further Greek manuscripts – all copies of codex A – either lettered names, or numerals. On the whole, lettered codices are direct copies of codex A (and, codex A being missing, are thus witnesses to be summoned to the critical apparatus), while numbered codices are copies made out of extant, lettered codices (and thus do not need to feature in the critical apparatus). Heiberg has gradually shifted his views but not his sigla, so that this division,

---

[12] For "Archimedism" as a force in the scientific revolution, see Hoyrup (1994).

as well, is imperfect. Codex F was not copied directly from codex A, but from the derivative codex E; while codex 4 was not derived from codex D, as Heiberg thought at first, but directly from codex A. (The same is true for a very small fragment of a copy made of codex A which survives as codex 13.)

Finally, it should be noted that Heiberg made a decision, in using Moerbeke as an authority but not Jacob of Cremona. It seems likely that the codex of the Marciana Library in Venice, Marc. Lat. 327, is an authograph by Jacob of Cremona, containing a translation made directly from codex A (and also relying on codex B). Jacob's translation is much less faithful than Moerbeke's, and the diagrams (which we attempt to edit here) were clearly largely re-made rather than copied. Thus we shall not use this codex ourselves. But in setting out the tree of transmission of the works of Archimedes, Marc. Lat. 327, lacking a siglum from Heiberg, is in fact parallel to codex B.[13]

We are now finally in a position to set out the sigla and tree for Archimedes' works in the Greco-Latin tradition (ignoring derivative manuscripts):

A	Lost archetype for B, D–H, 4, ninth–tenth centuries?
ℬ	Lost archetype for parts of codex B, ninth–tenth centuries?
B	Ottobon. lat. 1850, autograph of Moerbeke, 1269.
C	The Archimedes Palimpsest, tenth century.
D	Laurent. XXVIII 4, fifteenth century.
E	Marc. Gr. 305, fifteenth century.
G	Paris. Gr. 2360, sixteenth century.
H	Paris. Gr. 2361, 1544.
4	Vatican Gr. Pii. II nr. 16, sixteenth century.
13	Monac. 492, fifteenth century.
Marc. Lat. 327	Autograph of Jacob of Cremona, fifteenth century.

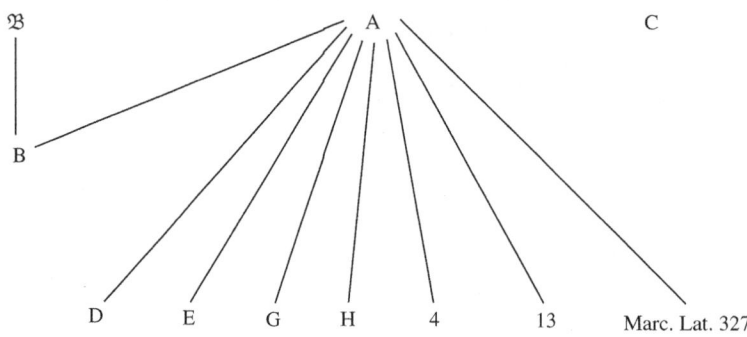

3.3   The two books *On the Sphere and the Cylinder*

The two books translated here by Archimedes, together with the two commentaries on them by Eutocius, constitute four works very different from each other.

---

[13] For the nature of Marc. Lat. 327 see Clagett (1978) 326–7, 334–8.

The relation between Archimedes' *SC* I and *SC* II is definitely not the simple one of two parts of the same work, while Eutocius seems to have had very different aims in his two commentaries. *SC* I is a self-contained essay deriving the central metrical properties of the sphere – the surface and volume both of itself and of its segments. *SC* II is another self-contained essay, setting out a collection of problems in producing and cutting spheres according to different given parameters. Eutocius' *Commentary on SC* I is mostly a collection of minimal glosses, selectively explicating mathematical details in the argument. His *Commentary on SC* II is a very thorough work, commenting upon a very substantial proportion of the assertions made by Archimedes, sometimes proceeding further into separate mathematical and historical discussions with a less direct bearing on Archimedes' text. I shall say no more on Eutocius' works, besides noting that his *Commentary on SC II* is among the most interesting works produced by late ancient mathematical commentators.

Archimedes' two books, *SC* I and II, are, of course, exceptional masterpieces. According to a testimony by Cicero,[14] whom there is no reason to doubt, Archimedes' tomb had inscribed a sphere circumscribed inside a cylinder, recalling the major measurement of volume obtained in *SC* I: if so, either Archimedes or those close to him considered *SC* I to be somehow the peak of his achievement. The reason is not difficult to find. Archimedes' works are almost all motivated by the problem of measuring curvilinear figures, all of course indirectly related to the problem of measuring the circle. Archimedes had attacked this problem directly in *DC*, obtaining, however, no more than a boundary on the measurement of the circle. Measuring the sphere is the closest Archimedes, or mathematics in general, has ever got to measuring the circle. The sphere is measured by being reduced to other *curvilinear* figures (otherwise, this would have been equivalent to measuring the circle itself). Still, the main results obtained – that the sphere as a solid is two thirds the cylinder circumscribing it, its surface four times its great circle – are remarkable in simplifying curvilinear, three-dimensional objects, that arise very naturally. The Spiral Lines, the Conoids and Spheroids, the Parabolic Segments and all the other figures that Archimedes repeatedly invented and measured, all fall short of the sphere in their inherent complexity and, indeed, artificiality. Yet the sphere is as simple as the circle – merely going a dimension further – and as natural.

The structure of *SC* I is anything but simple. The work can be seen to consist of two main sections, further divided into eight parts (the titles are, of course, mine):

Section 1: Introduction

Introduction   Covering letter, "Axiomatic" introduction, and so-called Proposition 1 (which is essentially a brief argument for a claim being part of the "Axiomatic" Introduction).

Chapter 1   Propositions 2–6, problems for the construction of geometrical proportion inequalities.

---

[14] *Tusculan Disputations*, V.23.

Chapter 2   Propositions 7–12, measuring the surfaces of various pyramidical figures.

Chapter 3   Propositions 13–16, measuring and finding ratios involving conical surfaces.

Chapter 4   Propositions 17–20, measuring conical volumes.

Interlude   Propositions 21–2, finding proportions holding with a circle and an inscribed polygon.

Section 2: Main treatise

Chapter 5   Propositions 23–34, reaching the major measurements of the work: surface and volume of a sphere.

Chapter 6   Propositions 35–44, extending the same measurements to sectors of spheres.

The defining features of the book as a whole are intricacy, surprise – and inherent simplicity.

Intricacy and surprise are created by Archimedes' way of reaching the main results. He starts the treatise in an explicit introduction, stating immediately the main results. The surface of the sphere is four times its great circle; the surface of a spherical segment is equal to another well-defined circle; the volume of an enclosing cylinder is half as much again as the enclosed sphere. This introduction immediately leads to a sequence of four chapters whose relevance to the sphere is never explained: problems of proportion inequality, measurements of surfaces and volumes other than the sphere. The interlude, finally, moves into a seemingly totally unrelated subject, of the circle and a polygon inscribed within it.

The introductory section is difficult to entangle, in that it moves from theme to theme, in a non-linear direction (typically, while the various chapters do rely on previous ones – the interlude is an exception – they are mostly self-contained and require little background for their arguments). It is also difficult to entangle in that it moves between modes: problems and theorems, proportional and direct relations, equalities and inequalities. The Axiomatic Introduction mainly provides us with criteria for judging inequalities. The first chapter has problems of proportion inequality. Chapter 2 moves from equality (Propositions 7–8) to inequality (Propositions 9–12). Chapter 3 moves from equalities (Propositions 13–14) to proportions (Propositions 15–16). While chapters 4 and the interlude are simpler in this sense (equalities in chapter 4, proportions in the interlude), they also deal with some very contrived and strange objects.

In short, the introduction sets out a clear goal: theorems on equalities of simple objects. Archimedes moves through problems, inequalities, and very complex objects.

Proposition 23, introducing the second section with the main work, effects a dramatic transformation. A circle with a polygon inscribed within it is imagined rotated in space, yielding a sphere with a figure inscribed within it. The inscribed figure is made of truncated cones, measured through the results of chapters 3–4. Furthermore, with the same idea extended to a circumscribed polygon yielding a circumscribed figure made of truncated cones, proportion inequalities come about involving the circumscribed and inscribed figures.

Combining chapter 1 and the interlude, such proportion inequalities can be manipulated to combine with the measurements of the inscribed and circumscribed figures, reaching, indirectly, a measurement of the sphere itself. Thus the simple idea of Proposition 23 immediately suggests how order can emerge out of the chaotic sequence of the introductory section. The following intricate structure of chapter 5 unpacks this suggestion, going through the connections between the previous parts. Unlike the previous parts, chapter 5 assumes a complex logical structure of dependencies, and many previous results are required for each statement. A similar structure, finally, is then obtained for spherical segments, in chapter 6. (See the trees of logical dependencies of chapters 5 and 6.)

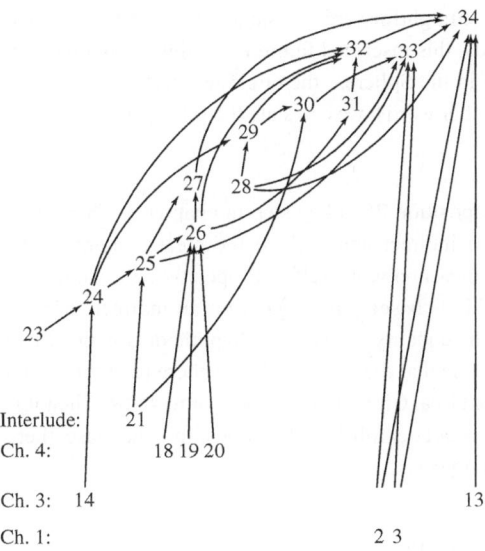

Chapter 5

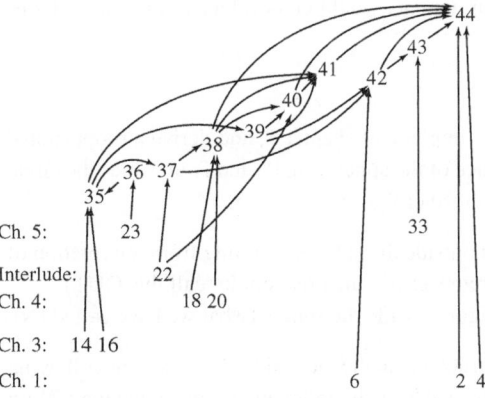

Chapter 6

Intricacy and surprise govern the arrangement of the text (thus, for instance, it was Archimedes' *decision*, to postpone Proposition 23 till after the introductory discussion was completed). Intricacy and surprise are, indeed, the mathematical keynotes of the work. The main idea is precisely that, through inequalities and unequal proportions holding between complex figures, equalities of simple figures can be derived. How? Through an indirect argument. We want to prove the equality between, for example, the surface of the sphere, and four times its great circle (Proposition 33). We attack the problem indirectly, through two separate routes. First, we approach what is not an equality involving simple objects, but an inequality involving complex objects. We consider the complex object made of the rotation of a regular polygon inscribed in a circle, leading to a figure inscribed inside a sphere. Let us call this inscribed figure IN. Following Proposition 23, we know that the object IN is composed of truncated cones, which we can measure through chapter 3; we can also correlate it with the circle and regular polygon from which it derives and apply the results of the interlude. A straightforward measurement can then show that the following inequality holds: this inscribed figure IN is always smaller than Four Times the Great Circle in the Sphere – the circle we started from in the rotation (call this 4CIR). In other words, we can state the following inequality:

(1) IN<4CIR

This is the conclusion of Proposition 25, a key step in chapter 5. This forms the first line of attack in the indirect approach to the measurement of the surface. Now, getting to the measurement itself in Proposition 33, we pursue the second line of attack. This is based, once again, on an indirect strategy. Instead of showing directly the equality, we assume, hypothetically, that there is an inequality. Thus we assume that the surface of the sphere (call this SUR) is not equal to Four Times the Great Circle in the Sphere or to 4CIR. Then it is either greater or smaller – we shall demolish both options. In either case, there is an *inequality* between two objects:

- The surface (SUR).
- Four times the great circle (4CIR).

For instance (analogous arguments would be developed either way) let us take the inequality:

(2) SUR>4CIR

An inequality allows us to implement chapter 1, and derive a proportional inequality involving the surface of the sphere (SUR) and Four Times the Great Circle (4CIR), as well as two further figures:

- The figure circumscribed outside the sphere, resulting from the rotation of a regular polygon that circumscribes the great circle (call this OUT).
- The similarly inscribed figure, inside the sphere (what we have called IN).

For example, we can construct, given the inequality (1) above, the following two figures OUT and IN so that we have the following proportional inequality:

(3) OUT:IN<SUR:4CIR.

(Ultimately, this proportion inequality is also based on the measurements due to chapter 3, as well as the interlude).

This proportion inequality implies another:

(4) OUT:SUR<IN:4CIR.

Now, from the Axiomatic Introduction we can also derive the direct inequality that does not involve proportions:

(5) OUT>SUR.

which, together with the inequality (3), gives rise to another direct inequality:

(6) IN>4CIR.

which contradicts the independently proved inequality of proposition 25, (1) above:

(1) IN<4CIR.

Hence the inequality (2) is impossible:
it is not the case that (2) SUR>4CIR.
Arguing analogously against the alternative inequality (SUR<4CIR) only the equality remains, the QED:

(7) SUR=4CIR.

In sum, we look for equalities holding between simple objects. We go indirectly, in three different ways:

- We throw in complex objects that we compare with the simple objects,
- We start with proving inequalities (direct as well as proportion inequalities) rather than equalities,
- We assume that the desired equality does not hold.

All of which converge on the desired equality.

With inequalities and complex objects, it is easier to make progress. Indeed, we can derive certain inequalities that follow in general, no matter what our special assumptions – general inequalities – and we can also derive other inequalities that follow from the specific hypothetical assumption that the desired equality does not hold – special inequalities. The general and special inequalities are incompatible, and this is what proves the desired equality.

The intellectual connection between *SC* I and the calculus is as follows. Archimedes measures curvilinear objects, and since measurement is ultimately a reduction to rectilinear objects, what he is trying to do is to equate the curvilinear with the rectilinear. Now, a sphere is not made of rectilinear figures, unless we are willing to see it as made of infinitely many such figures. Effectively, then, measuring a sphere must always have something to do, however indirectly, with a certain act of imagination where the sphere is considered as made of infinitely many objects, giving rise to an operation we may always compare, if we so wish, to the calculus. Archimedes, we may say, discusses the sphere as if it were composed of infinitely many truncated cones of zero width. Instead of moving directly into infinity, however, Archimedes uses the rigorous approach of the mature calculus, bounding the sphere by external and internal bounding figures composed of *indefinitely many* truncated cones. In this, of course, Archimedes extends the approach taken in our extant Book XII of Euclid's *Elements*, whose contents may have been due to Eudoxus. In the introduction, Archimedes specifically praises Eudoxus' measurement of the cone that, at least in Euclid's version, is indeed comparable to his own.

This is not the place to discuss in detail the position of Archimedes in the pre-history of the calculus. This much should be said, not so much in terms of Archimedes' approach compared to that of the modern calculus, but in terms of Archimedes' approach compared to the models and options available to Archimedes himself. In *Elements* XII, as in *SC* I, the bounding of the measured figure is made in a "natural" way. The cone is bounded in *Elements* XII by pyramids whose base is an equilateral polygon inscribed inside the base; the sphere is bounded by Archimedes by truncated cones arising from an equilateral polygon outside and inside the great circle. In other words, both measurements rely on the most natural reduction of a circle to a simpler figure – the reduction of a circle to an equilateral polygon with indefinitely many sides. This polygon is the natural rectilinear correlate, in geometrical terms, to the circle, since it gradually approaches the shape of the circle. There are other possible rectilinear reductions of the circle, that do not take such a natural geometrical approach. For instance, one may envisage the circle as (i) composed of all its radii; or (ii) of all the chords parallel to a given diameter. When this is translated from the language of infinitely many line-segments into the language of indefinitely many slices, the circle is then bounded by (i) sectors, in the first case, or by (ii) rectangles, in the second case. An approach analogous to (ii), bounding the circle by rectangles, was taken by Archimedes in *CS*. An approach analogous to (i), bounding the circle by sectors, was taken by Archimedes in *SL*. Thus *CS* and *SL* both differ in their character from *SC* I, which is more like *Elements* XII. *SC* I deals throughout with natural geometrical objects that arise directly from the shape of the sphere. It is typical that the center of the sphere is always an important point in the figures, and that the defining property of the equidistance of points on the surface from the center is always relevant to the arguments. The geometrical conception is, in this sense, simple. On the other hand, the price paid for simplicity is that of indirectness. *CS* and *SL* deal with their respective objects in a more brutal fashion, as it were, cutting them down almost into purely quantitative objects, with a somewhat less clear geometrical significance. Those objects, however, can be more directly summed up, by quantitative principles.

Further, and related to this geometrical "naturalness," it is typical to *SC* I that relatively little is required as mathematical background – essentially, nothing beyond Euclid's *Elements*. Archimedes' preliminary propositions, in the First Section, are all quite simple, almost direct consequences of the *Elements*. The one strange preliminary set of results – propositions 21–2 of the interlude – is indeed strange in calling up relations whose import is not immediately obvious, but it is also very easy to prove. There is thus nothing here like the highly complex special quantitative results proved by Archimedes for the sake of *CS*, *SL*. Finally, no use is made of special curves, and the objects are all made of straight lines and circles alone. Elsewhere in the Archimedean corpus, the very objects studied arise from special curves and, in principle, the circle can always be treated alongside other conic sections to derive useful relations. None of this is done in *SC* I, whose mathematical universe is identical to that of the *Elements*. It truly could have been *Elements* XIV. This perhaps can be said of no other extant work by Archimedes.

We shall follow this comparison between *SC* I and other Archimedean works (especially *CS*, *SL*) in greater detail, in later volumes. Let us stress here, finally, the inherent relation between simplicity and indirectness in *SC* I. The work is intricate and surprising – which Archimedes clearly values – but it is also simple, in that we finally realize that no objects extraneous to the nature of the sphere are required. The intricacy arises, in a sense, from the simplicity: to bound the sphere meaningfully between truncated cones, special results about cones and polygons are required. The mathematical elegance of the work goes hand-in-hand with its surprise and suspense.

SC II has an entirely different character. Instead of the elegance of surprise and simplicity, it goes directly to the spectacular effect of the tour-de-force.

Among the extant works by Archimedes, *SC* II is the only one whose main theme is not theorems, but problems. For instance, whereas *SC* I has several problems, in propositions 2–6 ("chapter 1"), they are there as part of the prepara-tory material to the main results of Propositions 23 and following ("chapters 5–6"). *SC* II, on the other hand, has a few theorems (Propositions 2, 8, and 9) but at least 2, and possibly 8 and 9 as well, are there for the sake of proving problems. Thus *SC* I foregrounds theorems over problems, whereas *SC* II fore-grounds problems over theorems. The logical distinction between theorems and problems is difficult to specify, and it seems to do mostly with the different emphasis on the task set to the geometer. In a theorem, the task is to judge the truth of a result, while in a problem the task is to obtain a way for a result. Thus the theorem puts the emphasis on the result itself, while the problem puts the emphasis on the way to obtaining the result. One has the sense that, perhaps for the reason explained just now, Greek mathematicians had more of a propri-etary sense towards problems than towards theorems. A problem represented your own way of reaching a result, whereas a theorem belonged, in a sense, to all mathematicians. A typical example of this is the catalogue of solutions to the problem of finding two mean proportionals, translated here in Eutocius' commentary to *SC* II 2.

Thus the sense of the tour-de-force. Here are problems that, Archimedes claims, are now soluble for the first time – thanks to the theorems of *SC* I. In a turn-about, the theorems of *SC* I, foregrounded inside *SC* I itself, are now seen as background to the problems of *SC* II:

1   To find a plane equal to the surface of a given sphere (introduction).
2   To find a sphere equal to a given cone or cylinder (Proposition 1).
3   To cut a sphere so that the surfaces of the segments have to each other a given ratio (Proposition 3).
4   To cut a sphere so that the segments have to each other a given ratio (Proposition 4).
5   To find a segment of a sphere similar to a given segment, and equal to another given segment (Proposition 5).

6   To find a segment of a sphere similar to a given segment, its surface equal to a surface of a given segment (Proposition 6).

7   To cut a sphere so that the segment has to the cone enclosed within it a given ratio (Proposition 7).

8   Finally, an implied problem is: to find the greatest spherical segment with a given surface. (This is shown by Proposition 9 to be the hemisphere.)[15]

Proposition 2 is a theorem, transforming the relation for segments of sphere, shown in *SC* I, into a property more useful for the problems of this book. This is used in several propositions 4–5, 7–8, while proposition 8, another theorem, is used in proposition 9. Other than this, the propositions are largely self-enclosed, with little overall structure binding the book. It is typical that, unlike the problems of *SC* I 2–6, no problem is ever applied: Archimedes never requires to produce a cut, using a previous problem, as part of the construction of a new result. Each problem is thus clearly marked as an end in itself.

Poor in any structure binding propositions to each other, the work is rich in the internal structure of the propositions taken separately. Many of the problems are proved using the analysis and synthesis mode, where each proposition encompasses two separate proofs: first, assuming the problem as solved, a concomitant construction is shown in the analysis. Then, in the synthesis, the found construction is used as a basis for solving the problem. Further, both preparatory theorems – 2 and 8 – also have a bipartite structure, proving the same result twice. This may represent later accretions into the text, or it may represent Archimedes' explicit decision, to make the texture of the work as rich as possible. Finally, proposition 4 required a preliminary problem which formed a mini-treatise on its own right. Archimedes postponed that mini-treatise to the end of the work, and it got lost from the main line of transmission of his work. Fortunately, Eutocius was able to retrieve that work and to preserve it in his commentary. This, once again, has a complex internal structure, consisting of an analysis, a synthesis, and a special theorem showing the limits of solubility.

The sense of tour-de-force is mostly sustained by the sheer complexity of the results shown. The work combines linear, surface, and solid measurements. Through Proposition 2, segments of sphere can be equated with cones and thus their ratios can be equated with ratios of heights of cones (with a common base), thus simplifying solid to linear measurements; many problems combine such linear measurements with conditions that specify both volume and surface. Thus the work comes close to being a study in complicated cubic

---

[15] It appears from Archimedes' introduction to *SL* that he had a further motivation to some of the theorems in the book, namely, to imply the *falsity* of certain claims whose truth he had earlier asserted, as a stratagem made to attract *false* proofs. Thus, for instance, the last theorem implies the falsity of the following statement: the greatest segment of the sphere is obtained with the plane orthogonal to the diameter passing at the point at the diameter where the square on its greater segment is three times the square on its smaller segment (!). Archimedes had patiently waited for someone to fall into this trap and finally, when nobody did, he had sent out *SC* II. We shall return to discuss this in the translation of *SL* itself.

equations. Nothing like the "elementary" nature of *SC* I, then. The ancillary objects, constructed for the sake of the proofs, result from complex proportion manipulations, not from any direct geometric significance. Greek geometrical tools are stretched to the limits and beyond: the mini-treatise at the end of the work relies essentially upon conic sections; the notion of the exponent is adumbrated in Proposition 8. Both treat geometrical objects in a semi-algebraic way, as objects of manipulation in calculation.

The difference in character between *SC* I and *SC* II should remind us, finally, of how misleading their understanding as a single work is. Archimedes did not write a work *On the Sphere and the Cylinder*, consisting of two parts. He wrote two separate books, of which the second relied, to a large extent, on results proved in the first. Each book was published separately – whatever "publish" exactly means – and had different goals. The creation of *Sphere and Cylinder*, a single work by Archimedes, is the product of late readers who, unlike Dositheus, the original recipient, read the two works simultaneously and lumped them together. It may well be that this process of unification reached its final form only following the work of Eutocius. By commenting upon *SC* I and *SC* II in sequence, Eutocius created, potentially, the work by Archimedes, the *On the Sphere and the Cylinder*. All that was left was for the Byzantine schools to keep these two works, as well as Eutocius' commentaries, together. It is thus fitting that we translate here all four works together, an organic unity composed, in the sixth century AD, out of four separate entities.

TRANSLATION AND COMMENTARY

ON THE SPHERE AND THE CYLINDER, BOOK I

/Introduction: general/

Archimedes to Dositheus:[1] greetings.

Earlier, I have sent you some of what we had already investigated then, writing it with a proof: that every segment contained by a straight line and by a section of the right-angled cone[2] is a third again as much as a triangle having the same base as the segment and an equal height.[3] Later, theorems worthy of mention suggested themselves to us, and we took the trouble of preparing their proofs. They are these: first, that the surface of every sphere is four times the greatest circle of the <circles> in it.[4] Further, that the surface of every segment of a sphere is equal to a circle whose radius is equal to the line drawn from the vertex of the segment to the circumference of the circle which is the base of the segment.[5] Next to these, that, in every sphere, the cylinder having a

---

[1] The later reference is to *QP*, so this work – *SC* I – turns out to be the second in the Archimedes–Dositheus correspondence. Our knowledge of Dositheus derives mostly from introductions by Archimedes such as this one (he is also the addressee of *SC* II, *CS*, *SL*, besides of course *QP*): he seems to have been a scientist, though perhaps not much of one by Archimedes' own standards (more on this below). See Netz (1998) for further references and for the curious fact that, judging from his name, Dositheus probably was Jewish.

[2] "Section of the right-angled cone:" what we call today a "parabola." The development of the Greek terminology for conic sections was discussed by both ancient and modern scholars: for recent discussions referring to much of the ancient evidence, see Toomer (1976) 9–15, Jones (1986) 400.

[3] A reference to the contents of *QP* 17, 24.   [4] *SC* I.33.

[5] Greek: "that to the surface . . . is equal a circle . . ." The reference is to *SC* I.42–3.

base equal to the greatest circle of the <circles> in the sphere, and a height equal to the diameter of the sphere, is, itself,[6] half as large again as the sphere; and its surface is <half as large again> as the surface of the sphere.[7]

In nature, these properties always held for the figures mentioned above. But these <properties> were unknown to those who have engaged in geometry before us – none of them realizing that there is a common measure to those figures. Therefore I would not hesitate to compare them to the properties investigated by any other geometer, indeed to those which are considered to be by far the best among Eudoxus' investigations concerning solids: that every pyramid is a third part of a prism having the same base as the pyramid and an equal height,[8] and that every cone is a third part of the cylinder having the base the same as the cylinder and an equal height.[9] For even though these properties, too, always held, naturally, for those figures, and even though there were many geometers worthy of mention before Eudoxus, they all did not know it; none perceived it.

But now it shall become possible – for those who will be able – to examine those <theorems>.

They should have come out while Conon was still alive.[10] For we suppose that he was probably the one most able to understand them and to pass the appropriate judgment. But we think it is the right thing, to share with those who are friendly towards mathematics, and so, having composed the proofs, we send them to you, and it shall be possible – for those who are engaged in mathematics – to examine them. Farewell.

---

[6] The word "itself" distinguishes this clause, on the relation between the volumes, from the next one, on the relation between the surfaces. In other words, the cylinder "itself" is what we call "the volume of the cylinder." This is worth stressing straight away, since it is an example of an important feature of Greek mathematics: relations are primarily between geometrical objects, not between quantitative functions on objects. It is not as if there is a cylinder and two quantitative functions: "volume" and "surface." Instead, there are two geometrical objects discussed directly: a cylinder, and its surface.

[7] *SC* I.34.

[8] *Elements* XII.7 Cor. Eudoxus was certainly a great mathematician, active probably in the first half of the fourth century. The most important piece of evidence is this passage (together with a cognate one in Archimedes' *Method*: see general comments). Aside for this, there are many testimonies on Eudoxus, but almost all of them are very late or have little real information on his mathematics, and most are also very unreliable. Thus the real historical figure of Eudoxus is practically unknown. For indications of the evidence on Eudoxus, see Lasserre (1966), Merlan (1960).

[9] *Elements* XII.10.   [10] See general comments.

TEXTUAL COMMENTS

The first page of codex A was crumbling already by 1269 (when its first extant witness, codex B, was prepared), and the page was practically lost by the fifteenth century (when the Renaissance codices began to be copied). Heiberg's first edition (1880–81), based only on A's Greek Renaissance copies, was very much a matter of guesswork as far as that page was concerned, so that this page was thoroughly revised in the second edition in light of the codices B and (the totally independent) C. I translate Heiberg's text as it stands in the second edition (1910). It is interesting that Heath (1897), based on Heiberg's first edition, was never revised: at any rate, this is the reason why my text here has to be so different from Heath's, even though this is one of the cases where Heath attempts a genuine translation rather than a paraphrase. Otherwise this general introduction is textually unproblematic.

GENERAL COMMENTS

Introduction: the genre

Introductory letters to mathematical works could conceivably have been a genre pioneered by Archimedes (of course, this is difficult to judge since we have very few mathematical works surviving from before Archimedes in their original form). At any rate, they are found in other Greek Hellenistic mathematical works, e.g. in several books of Apollonius' *Conics*, Hypsicles' *Elements* XIV, and Diocles' *On Burning Mirrors*. The main object of such introductions seems to set out the relation of the text to previous works, by the author (in this case, Archimedes relates the work to *QP*), and by others (in this case, Archimedes relates the work to that of Eudoxus). Correlated with the external setting-out – how the work relates to works external to it – is an internal setting-out – how the work is internally structured, and especially what are its main results.

For the internal setting-out, it is interesting that Archimedes orders his results as I.33, I.42–3, I.34, i.e. not the order in which they are set out in the text itself. Sequence, in fact, is not an important consideration of the work. Once the groundwork is laid, in Propositions 1–22, the second half of the work is less constrained by strong deductive relations, one result leading to the next: the main results of the second part are mainly independent of each other. Archimedes stresses then the nature of the discoveries, not their order. The main theme for those discoveries is that of the "common measure" (which is a theme of both his new results on the sphere, and his old results on the parabola). The Greek for "common measure" is *summetria*, which, translated into Latin, is a cognate of "commensurability." *Summetria* is indeed a technical term in Greek mathematics, meaning "commensurability" in the sense of the theory of irrationals (Euclid's *Elements* X Def. 1). In Greek, however, it has the overtone of "good measure," something like "harmony." What is so remarkable, then: the very fact that curvilinear and rectilinear figures have a common measure, or the fact that their ratio is so simple and pleasing? (It is even possibly relevant that, in Greek mathematical musical theory – well known to Archimedes and his audience – 4:3 and 3:2 are, respectively, the ratios of the fourth and the fifth.)

To return to the external setting-out: this is especially rich in historical detail, and should be compared with Archimedes' *Method*, 430.1–9, which is the only other sustained historical excursus made by Archimedes. The comparison is worrying in two ways. First, the *Method* passage concerns, once again, the same relation between cone and cylinder, i.e. it seems as if Archimedes kept recycling the same story. Second, the *Method* version seems to contradict this passage (*SC*: no knowledge prior to Eudoxus. *Method*: no proof prior to Eudoxus, however known already to Democritus).

Was Archimedes an old gossip then? A liar? More to the point: we see Archimedes constantly comparing himself to Eudoxus, arguing for his own superiority over him. This is the best proof we have of Eudoxus' greatness. And as for the facts, Archimedes was no historian.

Archimedes' audience: conon and dositheus

Conon keeps being dead in Archimedes' works: in the introductions to *SL* (2.2 ff.) and *QP* (262.3 ff.), also *SC* II (168.5). Born in Samos, dead well before Archimedes' own death in 212 BC, he must have been a rare person as far as Archimedes was concerned: a mathematician. That he was a mathematician, and that this was so rare, is signaled by Archimedes' shrill tone of despair: the death of Conon left him very much alone. (A little more – no more – is known of Conon from other sources, and he appears, indeed, to have been an accomplished mathematician and astronomer: the main indications are Apollonius' *Conics*, introduction to Book IV, Diocles' *On Burning Mirrors*, introduction, and Catullus' poem 66.)

Archimedes shows less admiration towards Dositheus. The letter is curt, somewhat arrogant, almost dismissive – though note that the first person plural would be normal and therefore less jarring for the ancient reader. The concluding words, with the refrain "but now it shall become possible – for those who will be able – to examine those <theorems>," "... and it shall be possible – for those who are engaged in mathematics – to examine them" stress that only one readership may *examine* the results – "those who are engaged in mathematics." There is another, much more peripheral readership: "... those who are friendly towards mathematics," and it is with them that Archimedes says that he had decided to "share." In other words, Dositheus is one of the "friends." He is no mathematician according to Archimedes' standards. Archimedes' hope is that, through Dositheus, the work will become public and may reach some genuine mathematicians (the one he had known – Conon – being dead).

It seems, to judge by the remaining introductions to his works, that Archimedes never did find another mathematician.

/"Axiomatic" introduction/

First are written the principles and assumptions required for the proofs of those properties.

/Definitions/

Eut. 244

Eut. 245

/1/ There are in a plane some limited[11] curved lines, which are either wholly on the same side as the straight <lines>[12] joining their limits or have nothing on the other side.[13] /2/ So[14] I call "concave in the same direction" such a line, in which, if any two points whatever being taken, the straight <lines> between the <two> points either all fall on the same side of the line, or some fall on the same side, and some on the line itself, but none on the other side. /3/ Next, similarly, there are also some limited surfaces, which, while not themselves in a plane, do have the limits in a plane; and they shall either be wholly on the same side of the plane in which they have the limits, or have nothing on the other side. /4/ So I call "concave in the same direction" such surfaces, in which, suppose two points being taken, the straight <lines> between the points either all fall on the same side of the surface, or some on the same side, and some on <the surface> itself, but none on the other side.

/5/ And, when a cone cuts a sphere, having a vertex at the center of the sphere, I call the figure internally contained by the surface of the cone, and by the surface of the sphere inside the cone, a "solid sector." /6/ And when two cones having the same base have the vertices on each of the sides of the plane of the base, so that their axes lie on a line, I call the solid figure composed of both cones a "solid rhombus."

And I assume these:

---

[11] The adjective "limited," throughout, is meant to exclude not only infinitely long lines (which may not be envisaged at all), but also closed lines (e.g. the circumference of a circle), which do not have "limits."

[12] The words "straight <line>" represent precisely the Greek text, *eutheia*: "straight" is written and "line" is left to be completed. This is the opposite of modern practice, where often the word "line" is used as an abbreviation of "straight line." Outside this axiomatic introduction, whenever the sense will be clear, I shall translate *eutheia* (literally meaning "straight") by "line."

[13] See Eutocius for the important observation that "curved lines" include, effectively, any one-dimensional, non-straight objects, such as "zigzag" lines. See also general comments on Postulate 2.

[14] Here and later in the book I translate the Greek particle δή with the English word 'so'. The Greek particle has in general an emphatic sense underlining the significance of the words it follows. In the mathematical context, it most often serves to underline the significance of a transitional moment in an argument. It serves to emphasize that, a conclusion having been reached, a new statement can finally be made or added. The English word "so" is a mere approximation to that meaning.

/Postulates/

Eut. 245
Eut. 246

/1/ That among lines which have the same limits, the straight <line> is the smallest. /2/ And, among the other lines (if, being in a plane, they have the same limits): that such <lines> are unequal, when they are both concave in the same direction and either one of them is wholly contained by the other and by the straight <line> having the same limits as itself, or some is contained, and some it has <as> common; and the contained is smaller.

/3/ And similarly, that among surfaces, too, which have the same limits (if they have the limits in a plane) the plane is the smallest. /4/ And that among the other surfaces that also have the same limits (if the limits are in a plane): such <surfaces> are unequal, when they are both concave in the same direction, and either one is wholly contained by the other surface and by the plane which has the same limits as itself, or some is contained, and some it has <as> common; and the contained is smaller.

/5/ Further, that among unequal lines, as well as unequal surfaces and unequal solids, the greater exceeds the smaller by such <a difference> that is capable, added itself to itself, of exceeding everything set forth (of those which are in a ratio to one another).

Assuming these it is manifest that if a polygon is inscribed inside a circle, the perimeter of the inscribed polygon is smaller than the circumference of the circle; for each of the sides of the polygon is smaller than the circumference of the circle which is cut by it.

TEXTUAL COMMENTS

It is customary in modern editions to structure Greek axiomatic material by titles and numbers. These do not appear in the manuscripts. They are convenient for later reference, and so I add numbers and titles within obliques (//). Paragraphs, as well, are an editorial intervention. The structure is much less clearly defined in the original and, probably, no clear visual distinction was originally made between the introduction (in its two parts) and the following propositions. This is significant, for instance, for understanding the final sentence, which is neither a postulate nor a proposition. Archimedes does not set a series of definitions and postulates, but simply makes observations on his linguistic habits and assumptions.

GENERAL COMMENTS

Definitions 1–4

Following Archimedes, we start with Definition 1. Imagine a "curved line," and the straight line joining its two limits. For instance, let the "curved line" be the railroad from Cambridge to London as it is in reality (let this be called *real railroad*); the straight line is what you wish this railroad to be like: ideally straight (let this be called *ideal railroad*). Now, as we take the train from Cambridge to London, we compare the two railroads, the real and the ideal. Surprisingly perhaps, the two do have to coincide on at least two points (namely, the start and end points). Other than this, the real veers from the ideal. If the real sometimes coincides with the ideal, sometimes veers to the east, but never veers to the west, then it falls under this definition. If the real sometimes coincides with the ideal, sometimes veers to the west, but never veers to the east, once again it falls under this definition. But if – as I guess is the case – the real sometimes veers to the east of the ideal, sometimes to the west, then (and only then) it does not fall under this definition. In other words, this definition singles out a family of lines which, even if not always straight, are at least consistent in their direction of non-straightness, always to the same side of the straight. It is only this family which is being discussed in the following Definition 2 (a similar family, this time for planes, is singled out in Definition 3, and is discussed in Definition 4: whatever I say for Definitions 1–2 applies *mutatis mutandis* for Definitions 3–4).

Definition 2, effectively, returns to the property of Definition 1, and makes it global. That is, if Definition 1 demands that the line be consistent in its non-straightness relative to its start and end points only, Definition 2 demands that the line be consistent in its non-straightness relative to any two points taken on it (the obvious example would be the arc of a circle). It follows immediately that whatever line fulfils the property of Definition 2, must also fulfil the property of Definition 1 (the end and start points are certainly some points on the line). Thus, the lines of Definition 2 form a subset of the lines of Definition 1. This is strange, since the only function of Definition 1 is to introduce Definition 2 (indeed, since originally the definitions were not numbered or divided, we should think of them as two clauses of a single statement). But, in fact, Definition 1 adds nothing to Definition 2: Definition 2 defines the same set of points, with or without the previous addition of Definition 1. That is, to say that the property of Definition 2 is meant to apply only to the family singled out in Definition 1 is an empty claim: the property can apply to no other lines. It seems to me that the clause of Definition 1 is meant to introduce the main idea of Definition 2 with a simple case – which is what I did above. In other words, the function of Definition 1 may be pedagogic in nature.

Postulates 1–2: about what?

The wording of the translation of Postulate 1 gives rise to a question of translation of significant logical consequences. My translation has ". . . among lines which have the same limits, the straight <line> is the smallest . . ." Heiberg's Latin translation, as well as Heath's English (but not Dijksterhuis') follow Eutocius' own quotation of this postulate, and read an "all" into the text, translating as if it had "among *all* lines having the same limits . . ."

The situation is in fact somewhat confusing. To begin with, there is no unique set of "lines having the same limits," simply because there are many couples of limits in the world, each with its own lines. So, to make some sense of the postulate, we could, possibly, imagine a Platonic paradise, in it a single straight line, a sort of Adam-line; and an infinite number of curved lines produced between the two limits of this line – a harem of Eves produced from this Adam's rib. And then the postulate would be a statement about this Platonic, uniquely given "straight line." This is Heiberg's and Heath's reading, which make Postulate 1 into a general statement about the straight line as such. The temptation to adopt this reading is considerable. But I believe the temptation should be avoided. The postulates do not relate to a Platonic heaven, but are firmly situated in this world of ours where there are infinitely many straight lines. (The postulates will be employed in different propositions, with different geometrical configurations, different sets of lines.) The way to understand the point of the postulates, is, I suggest, the following:

There are many possible clusters of lines, such that: all the lines in the cluster share the same limits. Within any such cluster, certain relations of size may obtain. Postulate 2 gives a rule that holds between any two curved lines in such a given cluster (assuming the two lie in a single plane). Why do we have Postulate 1? This is because Postulate 2 cannot be generalized to cover the case of straight lines. (This is because the straight line is not contained, even partly, by "the other line and the line having the same limits as itself." See my explanation of the second postulate below.) So a special remark – hardly a postulate – is required, stating that, in any such cluster, the smallest line will be (if present in the cluster) the straight line. Thus, nothing like "a definition of the straight line" may be read into Postulate 1.

Unpacking Postulate 2

Take a limited curved line, and close it – transform it into a closed figure – by attaching a straight line between the two limits, or start and end points, of the line. This is, as it were, "sealing" the curved line with a straight line. So any curved line defines a "sealed figure" associated with it. (In the case of lines that are concave to the same direction, they even define a continuous sealed figure, i.e. one that never tapers to a point: a zigzagging line, veering in this and that direction would define a sequence of figures each attached to the next by the joint of a single point – whenever the line happened to cross the straight line between its two extremes).

Now take any such two curved lines. Assume they both have the *same* limits, and that they both lie in a single plane. Now let us have firmly before our mind's eye the sealed figure of one of those lines; and while we contemplate it, we look at the other curved line. It may fall into several parts: some that are inside the sealed figure, some that are outside the sealed figure, and some that coincide with the circumference of the sealed figure. If it has at least one part that is inside the sealed figure, and no part that is outside the sealed figure, then it has the property of the postulate. Note then that a straight line can never have this property: it will be all *on* the circumference of the sealed figure, none of it ever *inside* it (hence the need for Postulate 1).

Unpacking Postulate 3

The caveat, "when they have the limits in a plane," is slightly difficult to visualize. The point is that a couple of three-dimensional surfaces may share the same limit; yet that limit may still fail to be contained by a single plane (so this latter possibility must be ruled out explicitly). Imagine two balloons, one inside the other, somehow stitched together so that their mouths precisely coincide. Thus they have "the same limit," but the limit – the mouth – need not necessarily lie on a plane. Imagine for instance that you want to block the air from getting out of the balloons – you want a surface to block the mouth; you put the mouth next to the wall, but it just will not be blocked: the wall is a perfect plane, and the mouth does not lie on a single plane: some of it is further out than the rest. This, then, is what we do *not* want in this postulate.

The overall structure of Definitions 1–4, Postulates 1–4

This combination of definitions and postulates forms a very detailed analysis of the conditions for stating equalities between lines and surfaces. So many ideas are necessary!

1   "The same side," requiring the following considerations:
    – A generalization of "curved" to include "zigzag" lines.
    – What I call "real and ideal railroads" (Definition 1).
    – A disjunctive analysis (the real either wholly on one side of the ideal, or partly on it, but none on the other side).
2   "Concave," requiring the following considerations:
    – The idea of "lines joining any two points whatsoever."
    – The same disjunctive analysis as above.
3   "Contain," requiring the following considerations:
    – Having the same limits.
    – What I call the "sealed figure" (Postulate 2).
    – A disjunctive analysis (Whether wholly inside, or part inside and none outside).
4   Finally one must see:
    – The independence of the special case of the straight line – which requires a caveat in Postulate 1.
    – Also there is the special problem with the special case of the plane – which requires the caveat mentioned above, in Postulate 3.

There was probably no rich historical process leading to this conceptual elucidation. The only seed of the entire analysis is *Elements* I.20, that any two lines in a triangle are greater than the third. But the argument there (relying on considerations of angles in triangles) does not yield any obvious generalizations. So how did this analysis come about? A simple answer, apparently: Archimedes thought the matter through.

He is not perfectly explicit. The sense of "curved lines" must have been clear to him, but as it stands in the text it is completely misleading, and requires Eutocius' explication with his explanation of what I call "zigzag" lines. My own explications, too, with their "real and ideal" and "sealed figures," were also left by Archimedes for the reader to fill in. The use of disjunctive properties serves to make the claims even less intuitive.

Most curiously, this entire analysis of concavity will *never* be taken up in the treatise. No application of the postulates relies on a verification of its applicability, through the definitions; there is not even the slightest gesture towards such a verification.

This masterpiece had no antecedents, and no real implementation, even by Archimedes himself. A logical, conceptual *tour-de-force*, an indication of the kind of mathematical *tour-de-force* to follow. Archimedes portrayed himself as the one who sees through what others before him did not even suspect, and he gave us now a first example.

Postulate 5

This postulate, often referred to as "Archimedes' axiom," recurs, in somewhat different forms, elsewhere in the Archimedean corpus: in the introduction to the *SL* (12.7–11) [this may be a quotation of our own text], and in the introduction to the *QP* (264.9–12). As the modern appellation implies, the postulate has great significance in modern mathematics, with its foundational interests in the structure of continuity, so that one often refers to "Archimedean" or various "non-Archimedean" structures, depending on whether or not they fulfil this postulate. This is not the place to discuss the philosophical issues involved, but something ought to be said about the problem of historically situating this postulate.

Two presuppositions, I suggest, ought to be questioned, if not rejected outright:

1 "Archimedes is engaged here in axiomatics." We just saw Archimedes offering an axiomatic study (clearing up notions such as "concavity") almost for its own sake. This should not be immediately assumed to hold for this postulate as well. The postulate might also be here in order to do a specific job – as a tool for a particular geometrical purpose. In this case, it need not be seen as a contribution to axiomatic analysis as such. For instance, it is conceivable that Archimedes thought this postulate could be proved (I do not say he did; I just point out how wide the possibilities are). Nor do we need to assume Archimedes was particularly interested in this postulate; he need not necessarily have considered it "his own."

2 "Archimedes extends Euclid/Eudoxus." The significance of the postulate, assuming that it was a new discovery made by Archimedes, would depend on its precise difference from other early statements on the issue of size, ratio and excess. Indeed, the postulate relates in some ways to texts known to us through the medieval tradition of Euclid's Elements (*Elements* V Def. 4, X.1), often associated by some scholars, once again (perhaps rightly) with the name of Eudoxus. It is not known who produced those texts, and when but, even more importantly, it is absolutely unknown in what form, if any, such texts were known to Archimedes himself. (Archimedes makes clear, in both *SL* and *QP*, that the postulate – in some version – *was* known to him from earlier geometers; but we do not know *which* version). It is even less clear which texts Dositheus (or any other intended reader) was expected to know.

So nothing can be taken for granted. The text must be read and understood in the light of what it says, how it is used, and the related material in the Archimedean corpus. This calls for a separate study, which I shall not pursue here.

/ 1 /

If a polygon is circumscribed around a circle, the perimeter of the circumscribed polygon is greater than the perimeter of the circle.

For let a polygon – the one set down[15] – be circumscribed around a circle. I say that the perimeter of the polygon is greater than the perimeter of the circle.

(1) For since BAΛ[16] taken together is greater than the circumference BΛ (2) through its <=BAΛ> containing the circumference <=BΛ> while having the same limits,[17] (3) similarly, ΔΓ, ΓB taken together <are greater> than ΔB, as well; (4) and ΛK, KΘ taken together <are greater> than ΛΘ; (5) and ZHΘ taken together <is greater> than ZΘ; (6) and once more, ΔE, EZ taken together <are greater> than ΔZ; (7) therefore the whole perimeter of the polygon is greater than the circumference of the circle.

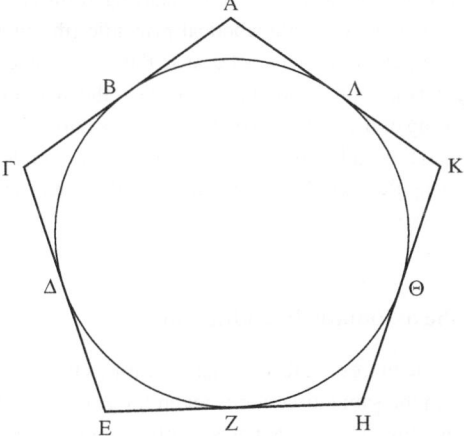

I.1
In most Codices EH is parallel to base of page. Codices BG, however, both have E rather lower than H. I suspect codex A had a slight slope, ignored in most copies and exaggerated in BG.

---

[15] "Set down" = "in the diagram." See general comments for this strange expression.

[16] BAΛ: an alternative way of referring to the sequence of two lines BA, AΛ. (This might be related to Archimedes' generalized notion of "line," including "zigzag" line, which was implicit in the axiomatic introduction).

[17] Post. 2; see general comments for the limited use of the axiomatic introduction.

TEXTUAL COMMENTS

The manuscripts agree that this should be numbered as the "first" proposition (i.e. the preceding passage is still "introductory"). Discrepancies between the numbering in the various manuscripts begin later (with proposition "6"). I print Heiberg's numbering, mainly for ease of reference, but it is possible that Archimedes' text had no numbers for propositions. If so, there was no break, originally, between the "introduction" and this passage. The only mark that a new type of text had begun (and the reason why all manuscripts chose to place the first number here) was the first occurrence of a diagram.

GENERAL COMMENTS

The use of the diagram

Archimedes is impatient here, and employs all sorts of shortcuts. The sentence "let the polygon – the one set down – be circumscribed around a circle" is the setting-out: the only statement translating the general enunciation in particular terms. Instead of guiding in detail the precise production of the diagram, then – as is the norm in Euclid's *Elements* – Archimedes gives a general directive. He is an architect here, not a mason. As a side-result of this, all the letters of this proposition rely, for their identification, on the diagram alone. Without looking at the diagram, there is no way you could know what the letters stand for: the text says nothing explicit about that, and instead totally assumes the diagram. Thus, the principle according to which letters are assigned to points is spatial (a counter-clockwise tour around the polygon, starting from A at the hour 12). This is instead of the standard alphabetical principle (the first mentioned letter: A, the second: B, etc. . . .). This is because there is not even the make-believe of producing the diagram through the text – as if the diagram were constructed during the reading of the text. This make-believe occurs in the standard Euclidean proposition: as the readers follow the alphabetical principle, they imagine the diagram gathering flesh gradually, as it were, as the letters are assigned to their objects.

The use of the axiomatic introduction

The use of Postulate 2 is remarkable in its deficiency. That the various lines and circumferences are all concave in the same direction is taken for granted. Not only is this concavity property not proven – it is not even explicitly mentioned. So why have the careful exposition of the concept of "concave in the same direction" in Definition 2? There, a precise test for such concavity was formulated – *not* to be applied here. Perhaps, Archimedes' goal is not axiomatic perfection (where every axiom, and every application of an axiom, must be made explicit), but *truth*. He has discovered what he is certain is true – Postulate 2, based on Definition 2. When applying the postulate, Archimedes is much more relaxed: as long as the applicability of the postulate is sufficiently clear, there is no need to mention it explicitly.

Generality of the proof

This proposition raises the problem of mathematical generality and the complex way in which it is achieved in Greek mathematics. First, consider the choice of object for discussion. Any polygon would do, and a triangle would have been the simplest, yet Archimedes chose a pentagon. Why? Perhaps, because choosing a more complex case makes the proof appear more general. At least, this is not the simplest case (which, just because it is "simplest," is in some sense "special").

But, still, how to generalize from the pentagon to any-gon? Archimedes does not even make a gesture towards such a generalization. For instance, the selection of lines along the polygon does not follow any definite principle (e.g. clockwise or anti-clockwise). Such a definite principle could have suggested a principle of generalization ("and go on if there are more . . ."). But Archimedes suggests none, erratically jumping from line to line. Even more: Archimedes does not pause to generalize *inside* the particular proof. There are no "three dots" in this proof. He goes on and on, exhausting the polygon (instead of saying "and so on" at some stage). While there is an effort to make the particular case "as general as possible," there is no gesture towards making the generalization explicit.

/2/

Given two unequal magnitudes, it is possible to find two unequal lines so that the greater line has to the smaller a ratio smaller than the greater magnitude to the smaller.

Let there be two unequal magnitudes, AB, Δ, and let AB be greater. I say that it is possible to find two unequal lines producing the said task.

(a) Let BΓ be set out equal to Δ (1) through the second <proposition> of the first <book> of Euclid <=*Elements*>, (b) and let there be set out some straight line, ZH; (2) so, ΓA being added onto itself will exceed Δ.[18] (c) So let it be multiplied,[19] and let it <=the result of multiplication> be AΘ, (d) and as many times AΘ is of AΓ, that many let ZH be of HE. (3) Therefore it is: as ΘA to AΓ, so ZH to HE;[20] (4) and inversely, it is: as EH to HZ, so AΓ to AΘ.[21] (5) And

Eut. 250

---

[18] Post. 5. Note that the elided word is here often "magnitude" and not "line."

[19] "Multiplied" is taken to mean the same as "being added onto itself." It is implicit that ΓA is multiplied until it exceeds Δ.

[20] The derivation from Step 2 to Step 3 ("A is the same multiple of B as D is of C; therefore A:B::C:D") is too simple to be *proved* by Euclid. It is part of the *definition* of proportion, but only in the case of numbers (*Elements* VII. Def. 21).

[21] *Elements* V.7 Cor. Note that changing the sequence of sides, i.e. a change such as A:B::C:D → C:D::A:B is not considered as a move at all, and requires no word of the "alternately" family. The symmetry of proportion is seen as a *notational* freedom.

since AΘ is greater than Δ, (6) that is than ΓB, (7) therefore ΓA has to AΘ a ratio smaller than ΓA to ΓB.[22] (8) But as ΓA to AΘ, so EH to HZ; (9) therefore EH has to HZ a smaller ratio than ΓA to ΓB. (10) And

Eut. 251   compoundly;[23] (11) [therefore] EZ has to ZH a smaller ratio than AB to BΓ [(12) through lemma]. (13) But BΓ is equal to Δ; (14) therefore EZ has to ZH a smaller ratio than AB to Δ.

(15) Therefore two unequal lines have been found, producing the said task [(16) namely the greater has to the smaller a smaller ratio than the greater magnitude to the smaller].

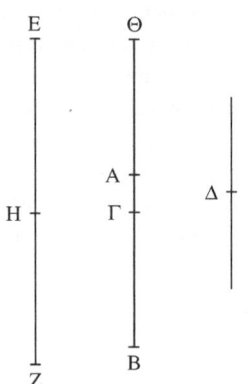

I.2
Codices DH: EZ=ΘB.
Codex G: H permuted with E, m.2 introduced E next to both points E, Z.   Codex H: E (?) instead of Γ.
Heiberg permutes A/B. See general comments.

TEXTUAL COMMENTS

Step 1 is an interpolation, unbracketed by Heiberg for the bad reason that Proclus had already read a reference of Archimedes to Euclid (Proclus, *In Eucl.* 68.12) – which shows merely that the interpolation antedated Proclus. From our knowledge of Euclid, the reference should be to I.3, not to I.2, but even so, this reference is only speciously relevant. I.3 shows how to cut off, from a given line, a line equal to some other given line. There is – there can be – no generalization for magnitudes in general, even if by "magnitudes" geometrical objects alone are meant. Even if Archimedes could commit such a blunder, it remains a fact that such references are the most common scholia. Hence, most likely, this is indeed an interpolation.

This was a *sui generis* textual problem. The next three are all typical of many others we shall come across later on.

First, in Step 11, Heiberg brackets the word "therefore" because of its absence from Eutocius' quotation. This is not a valid argument, as Eutocius does not aim to copy the text faithfully. Why should he copy such words as "therefore," which have no meaning outside their context?

---

[22] *Elements* V.8.
[23] *Elements* V.18: A:B::C:D → (A+B):B::(C+D):D (Archimedes, however, assumes an extension of the *Elements* to inequalities of ratios. This extension is supplied by Eutocius' commentary).

Second, Heiberg must be right about Step 12. Had it been Archimedes' we would certainly have a lemma, following this proposition, by Archimedes himself. This is a scholion, referring to Eutocius' own commentary.

Finally, Step 16 belongs to an important class: pieces of text which may be authentic (and then must shape accordingly our understanding of Archimedes' practices) or may be interpolated. How to tell? Only by our general understanding of Archimedes' practice – an understanding which is itself dependent upon such textual decisions! Heiberg imagined a purist, minimalist Archimedes. In this, he may have been right: my sense, too, is that Step 16 is by a later scholiast. But we should keep our minds open.

GENERAL COMMENTS

Existence and realism

The proposition is a problem: not showing the truth of an assertion (as theorems do), but performing a task. However, it is in a sense akin to a theorem. In the Euclidean norm, problems are formulated as "given X . . . to do Y." Archimedes often uses, as here, the format "given X . . . it is possible to do Y." This turns the problem into a truth-claim, more akin to a theorem.

A problem, which is rather like a truth-claim, may strike a modern reader as a *proof of existence*. This has been the subject of a modern controversy: Zeuthen (1886) had suggested that ancient problems, in general, are existence proofs, while Knorr (1983) has argued that, within geometry itself, questions of mathematical existence were often of less importance. Even when the issue of mathematical existence arose, it was handled through techniques other than those of problems. What about the present proposition, then? I would side with Knorr, and suggest that the problem does not aim to show the existence of an object, but to furnish a tool. Postulates 1–4 (followed by their quick, un-numbered sequel, and by the first proposition) furnished tools for obtaining inequalities between geometrical *objects*. We now move on to develop tools (based on the fifth Postulate) for obtaining inequalities between geometrical *ratios*. Both types of *in*equalities will then be used to prove the geometrical *equalities* of this treatise. This is what the proposition does. On the philosophical side, it does not deal at all with the question of mathematical existence. The question of "existence" is basically that: do you assume that mathematical objects exist, or do you prove their existence? Archimedes reveals here what may be considered to be the usual realism of Greek mathematics, where objects are simply taken for granted.

First, let us assume that Step 1 is an interpolation. It follows then that the proposition requires an unstated postulate ("to take away a magnitude from a magnitude, so as to have left a magnitude equal to a given magnitude"). This is a strong tacit existence assumption. Further, in Step d, we need to know how many times ΓA was multiplied (in Step c) before exceeding Δ. This is because we define ZH – as so many times HE as ΓA was multiplied. But we do not *know* how many times ΓA was multiplied. On the basis of Postulate 5, we are promised that ΓA *may* exceed Δ. But there is no algorithm for finding a *specific* number of times required for exceeding Δ. Once again, we assume that we can obtain an object (the number of times X was multiplied to exceed Y), without specifying a procedure for obtaining it: its existence, once again, assures its being obtainable.

In both cases, Archimedes reveals his realism. No algorithm is required: the relevant magnitudes and ratios exist, and there is no need to spell out how exactly you get them.

Shortcuts used in exposition

Archimedes displays a certain "laziness;" it manifested itself in the preceding proposition in the setting-out of the particular case (the diagram was simply assumed), it is here manifested in the setting out of the particular *demonstrandum*: instead of saying *what* is to be possible in the particular case, Archimedes says that "it is possible to find two unequal lines *producing the said task*" (leaving it to the reader to supply just what *is* the task: hence perhaps the interpolated Step 16?).

Schematic nature and the intended generality of the diagram

I have suggested in the introduction that Greek mathematical diagrams are more "schematic" than their modern counterparts, and that they serve to display the logical structure of the geometrical configuration, rather than to provide a metrically correct picture of the geometrical objects. This, I suggest, is a strength of ancient diagrams. Here we see a remarkable example of this strength. The general issue is that, if a diagram is taken to be a metrically correct picture, then it must specify a single range of metrical values. If in the diagram one line appears greater, equal, or smaller than another, this is because, in the geometrical situation depicted, the one line is indeed, respectively, greater, equal, or smaller than the other. In a diagram that is understood to be metrically correct, there is no such thing as an indefinite relation of size. In a schematic diagram, however, the relation of size between non-overlapping lines is indefinite. Whether the one appears greater than the other, or whether they appear equal, is just irrelevant, as long as they are indeed non-overlapping. Now let us compare Archimedes' diagram with Heiberg's. Heiberg permutes the letters A/B, so that he makes a choice: AΘ is greater not only than Δ, but also than AΓ. In geometrical reality, the situation admits of a certain generality or indefiniteness: AΘ can stand in any relation to AΓ. Archimedes allows AΘ to be non-overlapping with AΓ, in this way signaling this crucial indefiniteness. For Heiberg – who took his diagrams to be metrical – indefiniteness was ruled out from the outset, hence he failed to notice the loss of generality that resulted from his transformation of the diagram.

/ 3 /

Given two unequal magnitudes and a circle, it is possible to inscribe a polygon inside the circle and to circumscribe another, so that the side of the circumscribed polygon has to the side of the inscribed polygon a smaller ratio than the greater magnitude to the smaller.

Let the given two magnitudes be A, B, and the given circle the one set down.[24] Now, I say that it is possible to produce the task.

(a) For let there be found two lines, Θ, KΛ, (b) of which let the greater be Θ, so that Θ has to KΛ a smaller ratio than the greater magnitude to the smaller, (c) and let ΛM be drawn from Λ at right <angles> to ΛK, (d) and let KM be drawn down from K, equal to Θ [(1) for this is possible],[25] (e) and let two diameters of the circle be drawn, at right <angles> to each other, ΓB, ΔZ.[26] (f) Now, bisecting the angle <contained> by ΔHΓ, and bisecting its half, and doing the same ever again, (2) we will have left some angle smaller than twice the <angle contained> by ΛKM.[27] (g) Let it be left, and let it be NHΓ, (h) and let NΓ be joined. (3) Therefore NΓ is a side of an equilateral polygon [(4) Since in fact the angle <contained by> NHΓ measures the <angle contained> by ΔHΓ,[28] (5) which is right, (6) and therefore the circumference NΓ measures the <circumference> ΓΔ (7) which is a quarter of a circle; (8) so that it <=NΓ> measures the circle, too, (9) therefore it is a side of an equilateral polygon.[29] (10) For this is obvious]. (i) And let the angle <contained by> ΓHN be bisected by the line HΞ, (j) and, from Ξ, let ΟΞΠ touch the circle, (k) and let HNΠ, HΓΟ be produced; (11) so that ΠΟ, too, is a side of the polygon circumscribed around the circle, <which is> also equilateral[30] [(12) it is obvious that it is also similar to the inscribed, whose side is NΓ].[31] (13) And

Eut. 252

Eut. 253

Eut. 253

---

[24] I.e. in the diagram.

[25] See Eutocius. Also see Steps 2–3 in the following proposition and the footnote there.

[26] Confusingly, the letter B is reduplicated in this proposition, serving once as a given magnitude and once as a point on the circle. See textual comments.

[27] An extension of *Elements* X.1.

[28] "To measure" is to have the ratio of a unit to an integer. Step e: NHΓ has been produced by bisecting ΔHΓ, recursively; hence their ratio is that of a unit to an integer (we will say it is $1:2^n$).

[29] If this circumference measures the circle, the circle can be divided into a whole number of such circumferences. Dividing it in this way, and drawing the chords for each circumference, we will get a polygon inscribed inside the circle. All its sides are chords subtending equal circumferences, hence through *Elements* III.29 they are all equal: an equilateral polygon.

[30] See Eutocius.

[31] We want to show that NΓ is parallel to ΠΟ. (1) The angle at Ξ is right (*Elements* III.18). (2) The angle NHT is equal to the angle ΓHT (Step i). (3) NH and ΓH are equal (both radii, *Elements* I Def. 15). (4) And TH is common to the triangles NHT, ΓHT; (5) which are therefore congruent (2–4 in this argument, *Elements* I.4), (6) so the angle at T is right (5 in this argument, *Elements* I.13), (7) and so NΓ is parallel to ΠΟ (1, 6 in this argument, *Elements* I.28).

since the <angle contained> by NHΓ is smaller than twice the <angle contained> by ΛKM, (14) but it is twice the <angle contained> by THΓ, (15) therefore the <angle contained> by THΓ is smaller than the <angle contained> by ΛKM. (16) And the <angles> at Λ, T are right;[32] (17) therefore MK has to ΛK a greater ratio than ΓH to HT.[33] (18) And ΓH is equal to HΞ; (19) so that HΞ has to HT a smaller ratio (that is ΠO to NΓ)[34] (20) than MK to KΛ; (21) further, MK has to KΛ a smaller ratio than A to B.[35] (22) And ΠO is a side of the circumscribed polygon, (23) while ΓN <is a side> of the inscribed; (24) which it was put forward to find.

Eut. 254

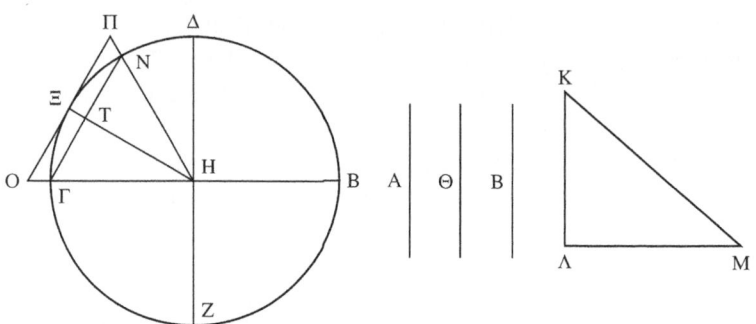

TEXTUAL COMMENTS

In Step e, the text refers to lines ΓB, ΔZ, a labeling that agrees with the diagram and which I follow. Heiberg has changed the letter B to E, in both text and diagram, first, because the letter B, in the manuscripts' reading, is reduplicated (used once for a given magnitude and another time for the end of a line), and second, because the letter E, in the manuscripts' reading, is not used at all, creating a gap in the alphabetical sequence. (Otherwise, the only gap is the missing letters Ρ, Σ prior to the final T). The argument for Heiberg's correction is almost compelling, yet it does require making two separate transformations, in text and diagram. Generally speaking, there are enough scribal errors in the letters of both diagram and text to suggest that neither was systematically corrected to agree with the other, so that it is not very probable that a mistake in one could influence the other (though, of course, this remains a possibility). Finally, our overall judgment that letters in Greek diagrams are not reduplicated is based precisely on such textual decisions (see also Proposition I.44 below). With little certainty either way, I keep the manuscripts' reading.

---

[32] Λ right: Step c. T right: see note to Step 12.   [33] See Eutocius.

[34] We would expect the word order "so that HΞ has to HT (that is ΠO to NΓ) a smaller ratio . . ." (see general comments). Then the content of Step 19 would have been clearer: it asserts that HΞ:HT::ΠO:NΓ (*Elements* VI.2).

[35] Steps b (Θ:KΛ<A:B), d (KM=Θ).

I.3

All codices except B have B twice, on a line and on the circle. Thus certainly A. Codex B, and Heiberg following him, has changed the B on the circle to E. See textual comments. Codices DG: Θ somewhat smaller than A, B. Codex H: Θ somewhat greater than A, B.   Codex B exchanges the positions of B, Θ, and makes Θ considerably greater than A, B.   Codices E, 4, have A=Θ=B. My conjecture is that, in codex A, the three lines were drawn rather freely, with small size differences (which, in truth, we now cannot reconstruct). Codices BD have the side KΛ a little longer than the side ΛM, but this clearly represents bad judgment of the margins, as in all other manuscripts the triangle is as in the figure printed. A strange mistake in Heiberg: he claims mistakenly that he has added a Π which is missing from the codices' diagrams (there is some confusion with I.4).

Steps 4–10 are probably rightly bracketed. Had they been in Eutocius' text, he would not have given his own commentary (besides, the subjective judgment that this piece of mathematics is of low quality, seems particularly strong here). Step 1 seems strange, but could be Archimedean (see also my footnotes on Eutocius' commentary on this step). Step 12 is not directly useful, but it is the *kind* of thing required by many assumptions of the proof, and does not have the look of a scholion (a scholion would prove, or gesture at a proof of such a claim). Finally, the strange word order of Step 19 may represent a textual problem. An interesting option (no more than an option) is that Archimedes completely left out the words "that is ΠΟ to ΝΓ" (accentuating the "hide-and-seek" aspect of this stage of the proof),[36] and then an honest interpolator inserted them at a *strange* location – signaling, perhaps, the interpolation *as such* by inserting it in the "wrong" position? – But this is *sheer* guesswork.

GENERAL COMMENTS

The scholiast's regress

The scholiast of 4–10 offers a good illustration of the scholiast's paradoxical position. This is the paradox of Carol (1895): you can never prove anything. You are arguing from P to Q; but you really need an extra premise, that P entails Q; and then you discover the extra premise, that P and "P entails Q" entail Q; and so on. So where to stop arguing? Mathematicians stop when they are satisfied (or when they think their audience will be) that the result is convincing enough. But scholiasts – for instance, a translator who offers also a brief commentary – face a tougher task. They should explain everything. Exasperating – and we sympathize with the author of Step 10. Having given the explanation, the scholiast wrings his hands in despair, realizing that this is not yet *quite* a *final*, decisive *proof*, and exclaims: "for this is obvious!"

/4/

Again, there being two unequal magnitudes and a sector, it is possible to circumscribe a polygon around the sector and to inscribe another, so that the side of the circumscribed has to the side of the inscribed a smaller ratio than the greater magnitude to the smaller.

---

[36] I refer to these features of the ending: Step 21 takes for granted Steps b and d, made much earlier (so that their tacit assumption is somewhat tricky); Step 24 asserts that the task has been produced, but to see this we actually need to piece together all of the Steps 19–23 (of which, Step 19 is doubly buried, in this "that is" clause which in turn is awkwardly placed).

For let there be again two unequal magnitudes, E, Z, of which let the greater be E, and some circle ABΓ having Δ <as> a center, and let a sector be set up at Δ, <namely> AΔB; so it is required to circumscribe and inscribe a polygon, around the sector ABΔ, having the sides equal except BΔA, so that the task will be produced.

(a) For let there be found two unequal lines H, ΘK, the greater H, so that H has to ΘK a smaller ratio than the greater magnitude to the smaller [(1) for this is possible],[37] (b) and similarly, after a line is drawn from Θ at right <angles> to KΘ, (c) let KΛ be produced equal to H [(2) for <this is> possible, (3) since H is greater than ΘK].[38] (d) Now, the angle <contained> by AΔB being bisected, and the half bisected, and the same being made forever, (4) there will be left a certain angle, which is smaller than twice the <angle contained> by ΛKΘ.[39] (e) So let it be left <as> AΔM; (5) so AM is then a side of a polygon inscribed inside the circle.[40] (f) And if we bisect the angle <contained> by AΔM by ΔN (g) and, from N, draw ΞNO, tangent to the circle, (6) that <tangent> will be a side of the polygon circumscribed around the same circle, similar to the one mentioned;[41] (7) and similarly to what was said above (8) ΞO has to AM a smaller ratio than the magnitude E to Z.

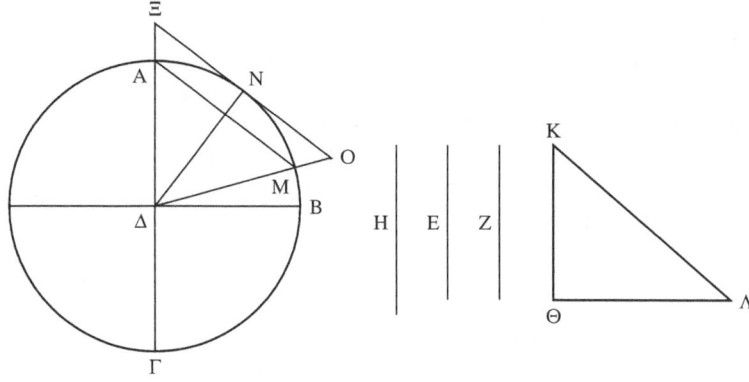

I.4
Here is the first diagram where we begin to see codex B having a radically different lay-out – see unlabelled thumbnail. This has no consequence for reconstructing codex A. Moerbeke has changed the basic page layout, so that his diagrams were in the margins (instead of inside the columns of writing) forcing very different economies of space. I shall mostly ignore codex B's lay-out in the following. Codex G has a different arrangement for the circle, for which see labelled thumbnail; codex D rotates the circle slightly counterclockwise (probably for space reasons). Codices DGH have H extend a little lower than E, Z, which I follow, but codices BE4 have E=Z=H. In codex D, Z>E as well. Codex D has ΘK>ΘΛ. Codex G has Θ instead of E.   Codex H has omitted Γ, has both K and H (!) instead of M. Heiberg has introduced, strangely, the letter Π at the intersection of NΔ/MA.

---

[37] *SC* I.2.

[38] *Elements* I.32: the sum of angles in a triangle is two right angles (so a right angle must be the greatest angle). *Elements* I.19: the greater angle is subtended by the greater side (so the right angle must subtend the greater side). Since the angle at Θ is right, KΛ must be greater than ΘK, which is guaranteed, indeed, by the relations KΛ=H (Step c), H>ΘK (Step a).

[39] *Elements* X.1 Cor.

[40] See Step 3 in *SC* I.3 (and the following Steps 4–10 there).

[41] See Step 11 in *SC* I.3, and Eutocius on that step.

TEXTUAL COMMENTS

Step 1 is reminiscent of Step 1 in I.3 above. Both assert the possibility of an action. In Proposition 3, the possibility was guaranteed by facts external to this work. Here, the possibility is guaranteed by Archimedes' own Proposition 2. If Step 1 in this proposition were to be considered genuine, this would throw an interesting light on Archimedes' references to his own proximate results – but we can not say that this is genuine.

Steps 2–3 are even more problematic. Here, again, the steps assert the possibility of an action – the very same action whose possibility is asserted in Step 1 of the preceding proposition. There the text was no more than "for this is possible." Here, there is some elaboration, explaining *why* this is possible. The elaboration is the right sort of elaboration – better in fact than Eutocius' commentary on Step 1 of the preceding proposition. Everything makes sense – except for the fact that this elaboration comes only *the second time* that this action is needed. Why not give it earlier? There is something arbitrary about giving it here. But who says Archimedes was not arbitrary? In fact, he is perhaps more likely to be arbitrary than a commentator; but once again, we simply cannot decide.

GENERAL COMMENTS

Repetition of text, and virtual mathematical actions

Here starts the important theme of *repetition*. Many propositions in this book contain partial repetitions of earlier propositions. This proposition partially repeats Proposition 3.

Repetitions arise because the same argument is applied to more than a single object. In this case, the argument for circles is repeated for sectors. Modern mathematicians will often "generalize" – look for the genus to which the argument applies ("circles or sectors," for instance), and argue in general for this genus. This is not what is commonly done in Greek mathematics, whose system of classification to genera and species is taken to be "natural" – objects are what they are, a circle is a circle and a sector is a sector, and if a proof is needed for both, one tends to have a separate argument for each.

There are many possible ways of dealing with repetitions. One extreme is to pretend it is not there: to have precisely the same argument, without the slightest hint that it was given earlier in another context. This is then *repetition simpliciter*. Less extreme is a full repetition of the same argument, which is at least honest about it, i.e. giving signals such as "similarly," "again," etc. This is *explicit repetition*. Or repetitions may involve an abbreviation of the argument (on the assumption that the reader can now fill in the gaps): this is *abbreviated repetition*. And finally the entire argument may be abbreviated away, by e.g. "similarly, we can show . . . ," the readers are left to see for themselves that the same argument can be applied in this new case. This may be called the *minimal repetition*.

Usually what we have is some combination of all these approaches – which is strange. Once the possibility of a minimal repetition is granted, anything else is redundant. And yet the Greek mathematician labors through many boring repetitions, goes again and again through the same motions, and then airily remarks "and then the same can be shown similarly" – so why did you go through all those motions? Consider this proposition. First, there are many signals of repetition: "again" at the very start of both enunciation and setting-out; "similarly" at Steps b, 7. Also much is simply repeated: the basic construction phase (i.e. the construction up to and excluding the construction of the polygons themselves). (This may even be *more* elaborate here than in the preceding proposition – see textual comments.) The main deductive action, on the other hand, is completely abbreviated away: Steps 13–24 of the preceding proposition are abbreviated here into the "similarly" of Step 7.

The most interesting part is sandwiched between the full repetition and the full abbreviation: the construction of the polygons. In Step d the angle is bisected. The equivalent in Proposition 3 is Step f, where *we bisect it*. The difference is that of passive and active voice, and it is meaningful. The active voice of Proposition 3 signifies the real action of bisecting. The passive voice of Proposition 4 signifies the virtual action of contemplating the *possibility* of an action. Going further in the same direction is the following: "(f) And if we bisect the angle <contained> by AΔM by ΔN (g) and, from N, draw ΞNO, tangent to the circle, (6) that <tangent> will be a side of the polygon . . ." The equivalent in the preceding proposition (Steps i–k, 11) has nothing conditional about it. Instead, it is the usual sequence of an action being done and its results asserted. The conditional of Proposition 4, Steps f–g, 6, is very different. Instead of doing the mathematical action, we argue through its *possibility* – through its virtual equivalent. So these two examples together (passive voice instead of active voice, conditional instead of assertion) point to yet another way in which the mathematical action can be "abbreviated:" not by chopping off bits of the text, but by standing one step removed from it, contemplating it from a greater distance – by substituting the virtual for the actual. This substitution, I would suggest, is essential to mathematics: the quintessentially mathematical way of abbreviating the infinite repetition of particular cases through a general argument, virtually extendible *ad infinitum.*

/ 5 /

Given a circle and two unequal magnitudes, to circumscribe a polygon around the circle and to inscribe another, so that the circumscribed has to the inscribed a smaller ratio than the greater magnitude to the smaller.

Let a circle be set out, A, and two unequal magnitudes, E, Z, and <let> E <be> greater; so it is required to inscribe a polygon inside the polygon and to circumscribe another, so that the task will be produced.

(a) For I take two unequal lines, Γ, Δ, of which let the greater be Γ, so that Γ has to Δ a smaller ratio than E to Z;[42] (b) and, taking H as a mean proportional of Γ, Δ,[43] (1) therefore Γ is greater than H, as well. (c) So let a polygon be circumscribed around the circle, (d) and another inscribed, (e) so that the side of the circumscribed polygon has to that of the inscribed a smaller ratio than Γ to H [(2) as we learned];[44] (3) so, through this, (4) the duplicate ratio, too, is smaller than the duplicate.[45] (5) And the <ratio> of the polygon to the polygon is duplicate that of the side to the side [(6) for <the polygons are> similar],[46] (7) and <the ratio> of Γ to Δ is of Γ to H;[47] (8) therefore the circumscribed polygon, too, has to the inscribed a smaller ratio than Γ to Δ; (9) much more, therefore, the circumscribed has to the inscribed a smaller ratio than E to Z.

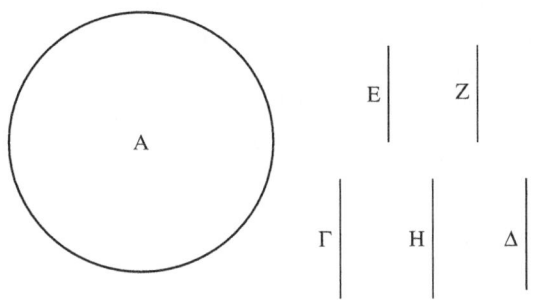

I.5
The diagram follows the consensus of codices EH4. Codex D has all five lines in a single row (E, Z to the left of G, H, D), while codex G has the three line Γ, H, Δ more to the left (so that H is to the left of E, Δ is between E, Z), while codex B, of course, has a different layout altogether. In codex D, H=Γ=Δ; in codex G, Γ>Δ>H; in Codex B, Γ>H>Δ (so Heiberg).

TEXTUAL COMMENTS

Step 2 is an obvious interpolation (the verb "learn" must come from a scholiast, not from Archimedes). There is no compelling reason, however, to suspect Step 6. Heiberg tended to doubt any backwards-looking argument (everything starting with a "for"), as if they were all notes by scholia, whereas Archimedes himself only used forward-looking, "therefore" arguments. Heiberg *could* have been right: once again, our view of Archimedes' practice on this matter will have to depend on our reconstruction of Archimedes' text.

---

[42] *SC* I.2.

[43] "Mean proportional:" X is mean proportional between A and B when A:X::X:B; here, Γ:H::H:Δ. *Elements* VI.13.

[44] *SC* I.3.

[45] "Duplicate ratio:" in our terms, if the original ratio is a:b, then the duplicate ratio is a²:b².

[46] *Elements* VI.20.   [47] Step b.

GENERAL COMMENTS

Impatience revealed in exposition

The proof proper starts with a very remarkable first person. Are we to imagine extreme impatience: "now listen, I just do this, and then that, and that's all; clear now?" – impatience is a constant feature of the style throughout this introductory sequence, and this proposition, in particular, is a variation on the preceding ones, no more. The proof, once again, is very abbreviated. Possibly, Step 6 is by Archimedes, and if so, it would be an interesting case of abbreviation, as the text is literally "for similar" – no more than an indication of the kind of argument used; almost a footnote to *Elements* VI. 20.

What makes this preliminary sequence of problems important is not their inherent challenge, but their being required, later, in the treatise. These are mere stepping-stones. Briefly: later in the treatise, Archimedes will rely upon "compressing" circular objects between polygons, and these interim results are required to secure that the "compression" can be as close as we wish. Effectively, this sequence unpacks Postulate 5 to derive the specific results about different kinds of compressions. It is natural that a work of this kind shall start with such "stepping-stones," but the natural impatience with this stage of the argument favors a certain kind of informality that will remain typical of the work as a whole.

/ 6 /

So similarly we shall prove that given two unequal magnitudes and a sector it is possible to circumscribe a polygon around the sector and to inscribe another similar to it, so that the circumscribed has to the inscribed a smaller ratio than the greater magnitude to the smaller. And this is obvious, too: that if a circle or a sector are given, and some area, it is possible, by inscribing equilateral polygons inside the circle or the sector, and ever again inside the remaining segments, to have as remainders some segments of the circle or the sector, which are smaller than the area set out. For these are given in the *Elements*.[48]

But it is to be proved also that, given a circle or a sector, and an area, it is possible to circumscribe a polygon around the circle or the sector, so that the remaining segments of the circumscription[49] are smaller

---

[48] *Elements* X.1; or this may be a wider reference to the "method of exhaustion," where polygons are inscribed in this way, first used in the *Elements* in XII.2. (This assumes – which need not necessarily be true – that this reference is by Archimedes, and that the reference is to something largely akin to the *Elements* as we have them.)

[49] "Circumscription" stands for περιγραφή, a deviation from the standard περιγραφέν, "the circumscribed <polygon>."

"The remaining segments of the circumscription" are the polygon minus the circle.

than the given area. (For, after proving for the circle, it will be possible to transfer a similar argument to the sector, as well.)

Let there be given a circle, A, and some area, B. So it is possible to circumscribe a polygon around the circle, so that the segments left between the circle and the polygon are smaller than the area B; (1) for <this>, too: <that>, there being two unequal magnitudes – the greater being the area and the circle taken together, the smaller being the circle – (a) let a polygon be circumscribed around the circle and another inscribed, so that the circumscribed has to the inscribed a smaller ratio than the said greater magnitude to the smaller.[50] (2) Now, this is the circumscribed polygon whose remaining <segments> will be smaller than the area set forth, B.

(3) For if the circumscribed has to the inscribed a smaller ratio than: both the circle and the area B taken together, to the circle itself, (4) the circle being greater than the inscribed, (5) much more will the circumscribed have to the circle a smaller ratio than: both the circle and the area B taken together, to the circle itself; (6) and therefore, dividedly, the remaining <segments> of the circumscribed polygon have to the circle a smaller ratio than the area B to the circle;[51] (7) therefore the remaining <segments> of the circumscribed polygon are smaller than the area B.[52]

(8) Or like this: since the circumscribed has to the circle a smaller ratio than: both the circle and the area B taken together, to the circle, (9) through this, then, the circumscribed will be smaller than <them> taken together;[53] (10) and so the whole of the remaining <segments> will be smaller than the area B.

(11) And similarly for the sector, too.

Eut. 254

I.6
Codex G has the pentagons upside-down, as in the thumbnail. The codices, except for codices DG, introduce a letter E at the top vertex of the circumscribing pentagon. Codex D has the height of the rectangle greater than its base.

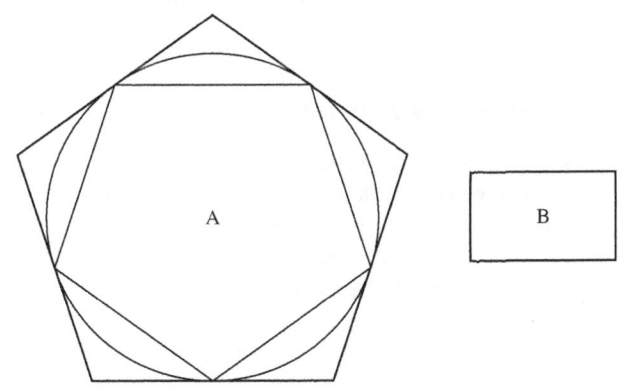

A   B

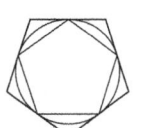

[50] *SC* I.5.

[51] *Elements* V.17 shows A:B::C:D → (A–B):B::(C–D):D which is called "division." This is extended here to cover inequalities (as can be supplied also from Eutocius' commentary to Proposition 2).

[52] *Elements* V.10.   [53] *Elements* V.10.

TEXTUAL COMMENTS

The number "6" for the proposition probably appeared at the top of this proposition, in the lost codex A. It appears there in all the manuscripts dependent upon A – with the exception of Moerbecke's translation. There is a gap here in the Palimpsest. At any rate, the manuscripts begin to diverge in their numbering; I will not report the further divergences, which are considerable. More important, we begin to see why they diverge: the text really does not come in clear units. The text makes digressions, repetitions, alternations: it is not the simple sequence of claim and proof, as in the Euclidean norm.

We have here another rare reference to the *Elements* (cf. Prop. 2, Step 1), at the end of the second paragraph. It is not to be dismissed straight away, since it is *functional*. The reference to the *Elements* is this time meant to explain why one thing is obvious (it's in the *Elements*!) while the other requires proof (it isn't in the *Elements*!). It thus may perhaps be authorial.

GENERAL COMMENTS

Lack of pedagogic concerns

The second paragraph has the strange word "circumscription," which, as explained in the footnote, is the same as "circumscribed." This shift in vocabulary is insignificant except for betraying a certain looseness – a looseness which can be seen with more serious logical points. Most important, the text is vague: what are the "remaining segments" mentioned again and again? Remaining from *what*? Archimedes just takes it for granted that his meaning is understood (namely that they are what is left after we take away the circle). Possibly, such cases may show that Archimedes did not seriously try to put himself in the place of the prospective reader. Archimedes betrays a certain haughtiness, even, towards such a reader: the "impatience" towards the argument easily becomes an impatience towards the reader. Rigor and clarity are sacrificed for the sake of brevity.

Meta-mathematical interests displayed in the text

The second paragraph ends remarkably, with a meta-mathematical statement "after proving for the circle, it will be possible to transfer a similar argument to the sector as well." A corollary following the proof is normal (as in fact we get in Step 11), but to anticipate the corollary in such a way is a remarkable intrusion of the second-order discourse inside the main, first-order discourse: even before getting down to the proof itself, Archimedes notes its possible extendability. The same is true, of course, for the whole of the beginning of this proposition.

Further, consider the alternative proof in Steps 8–10. It may of course be a later scholion, but it may also be authentic – and was already known to Eutocius. Assume then that this is by Archimedes: why should he have it? Now, the ratio-manipulation with the "dividedly" move in Step 6 is a rigorous way to derive the inequality there – based on the tools of proportion with which the Greek mathematician is always at home. But it is also an artificial move and one strongly feels that many other manipulations could do as well. The alternative offered in Steps 8–10 follows, I would suggest, a more direct visual intuition. The decision to keep both proofs is interesting, and may reveal Archimedes wavering between two ideals of proof.

The impression is this. The first few propositions are less interesting in their own right; they are obviously anticipatory. Archimedes gradually resorts to abbreviation, to expressions of impatience. In this proposition, it is as if he finally moves away from the sequence of propositions, looking at them from a certain distance, discoursing not so much *through* them as *about* them. As we shall have occasion to see further below, such modulations of the authorial voice are used by Archimedes to guide us through the text; in this case, this final modulation of voice, from first order to second order, serves to signal the conclusion of this preliminary sequence.

/ 7 /

If a pyramid having an equilateral base is inscribed in an equilateral cone, its <=the pyramid's> surface without the base is equal to a triangle having a base equal to the perimeter of the base <of the pyramid> and, <as> height, the perpendicular drawn from the vertex on one of the sides of the base <of the pyramid>.

Let there be an isosceles cone, whose base is the circle ABΓ, and let an equilateral pyramid be inscribed inside it, having ABΓ <as> base; I say that its <=the pyramid's> surface without the base is equal to the said triangle.

(1) For since the cone is isosceles, (2) and the base of the pyramid is equilateral, (3) the heights of the triangles containing the pyramid[54] are equal to each other.[55] (4) And the triangles have <as> base the lines AB, BΓ, ΓA, (5) and, <as> height, the said; (6) so that the triangles are equal to a triangle having a base equal to AB, BΓ, ΓA, and, <as> height, the said line[56] [(7) that is the surface of the pyramid without the triangle ABΓ].[57]

---

[54] By "triangles containing the pyramid" are meant the faces of the pyramid (excluding the base). By "their heights" are meant those drawn from the vertex of the cone to the sides of the base.

[55] The move from Steps 1–2 to 3 can be obtained in several ways; see general comments.

[56] *Elements* VI.1.

[57] Step 7 refers by the words "that is" to "the triangles" mentioned at the beginning of Step 6 (and not to the "triangle having a base equal to AB, BΓ, ΓA" mentioned later in Step 6). This is as confusing in the original Greek as it is in the translation.

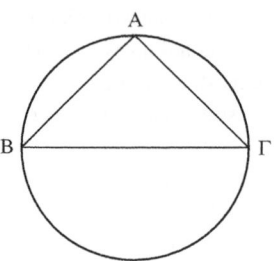

I.7
Codex D has the line
BΓ lower than a
diameter, so that ABΓ
becomes equilateral.

[The proof is clearer in another way.

Let there be an isosceles cone, whose base is the circle ABΓ, <its> vertex the point Δ, and let a pyramid be inscribed inside the cone, having <as> base the equilateral triangle ABΓ, and let ΔA, ΔΓ, ΔB be joined. I say that the triangles AΔB, AΔΓ, BΔΓ are equal to a triangle, whose base is equal to the perimeter of the triangle ABΓ, and whose perpendicular <drawn> from the vertex on the base is equal to the perpendicular drawn from Δ on BΓ.

(a) For let there be drawn perpendiculars, ΔK, ΔΛ, ΔM; (1) therefore these are equal to each other.[58] (b) And let a triangle, EZH, be set out, having: the base EZ equal to the perimeter of the triangle ABΓ and the perpendicular HΘ equal to ΔΛ. (2) Now, since the <rectangle contained> by BΓ, ΔΛ is twice the triangle ΔBΓ, (3) and the <rectangle contained> by AB, ΔK is also twice the triangle ABΔ, (4) and the <rectangle contained> by AΓ, ΔM is twice the triangle AΔΓ, (5) therefore the <rectangle contained> by the perimeter of the triangle ABΓ, that is EZ, (6) and by ΔΛ, that is HΘ, (7) is twice the triangles AΔB, BΔΓ, AΔΓ. (8) But the <rectangle contained> by EZ, HΘ is also twice the triangle EZH;[59] (9) therefore the triangle EZH is equal to the triangles AΔB, BΔΓ, AΔΓ].

I.7
Alternative proof
Codex D, once again,
has an equilateral
triangle with a very
different figure as a
result, see thumbnail
(did the scribe try to
achieve a
three-dimensional
image?).   In codex
G, line EZ is a little
higher than line AΓ.

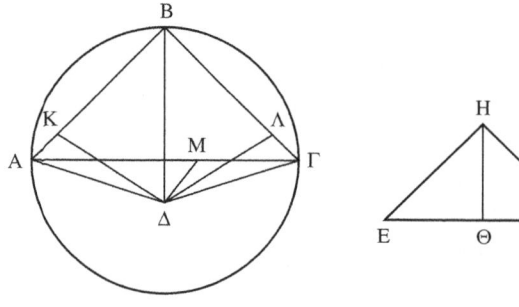

---

[58]   The interpolator simply avoids the thorny issue of identifying the *grounds* for this claim (the same as that involved in the transition from Steps 1–2 to Step 3 in the preceding argument: see general comments).

[59]   All the Steps 2, 3, 4 and 8 are based on *Elements* I.41.

TEXTUAL COMMENTS

As noted in fn. 57, Step 7 in the first part is strangely placed: the reference of "that is" is to a distant noun phrase, "the triangles," not to the antecedent "line." This adds to the probability that the Step is interpolated; but the text as a whole is not so polished: for all we know, such infelicities might be authorial.

That the alternative proof is indeed an interpolation seems likely: it is a direct copy of the proof given for Proposition 8, so it is easy to see how a scholiast could have decided to produce it. It is tempting to suggest that while the creative mathematician produces different proofs in different contexts simply to make the text less boring, the scholiast is happier to replicate. At any rate, it is strange to imagine Archimedes making the "it is clearer" claim himself (why would he bother to give a "less clear" proof, to begin with?).

GENERAL COMMENTS

Equality and identity

"A triangle having a base equal to the perimeter of the base <of the pyramid> and, <as> the height, the perpendicular drawn from the vertex on one of the sides of the base <of the pyramid>:" we shall often see expressions of this general kind, and one thing is immediately striking about many of them: the base is *equal* to a given line, the height simply *is* a given line. Here it is possible to see the reason for the distinction: the perimeter of the base does not come as a ready made line, but as a sequence of lines, whereas the perpendicular drawn from the vertex is a line to begin with. But the general rule is that identity and equality are treated as interchangeable. For us, equality is a relation in which *functions* of the objects are compared (namely, their principal measurements, length for lines, areas for two-dimensional figures, etc.; these are functions from geometrical objects to numbers), while identity is a relation between the objects themselves. For the Greeks, no function is envisaged. Objects are compared directly, not through some mediating measurement.[60]

The "grounds" for a claim

The transition from Steps 1–2 to 3 is a good example of a general phenomenon: the move is valid, but it is impossible to reconstruct Archimedes' own thought. Perhaps his argument ran like: all the triangles have their two sides equal (the sides are all sides of the equilateral cone); their base is always equal (an equilateral polygon at the base), hence by *Elements* I.8 they are congruent and obviously their heights will be equal (though this last assumption – "homologous objects in congruent objects are equal" – is not in the *Elements*). Or he imagined the axis of the cone, and perpendiculars drawn from the center of the circle of the base on the sides of the triangle of the base. So all the heights are

---

[60] Compare fn. 6 to the general introduction.

in triangles whose two other sides are (1) the axis of the cone (always equal simply because everywhere the same), (2) the perpendiculars drawn from the center of the circle on the sides of the triangle (equal through *Elements* III.14), while these two lines enclose a right angle, hence through *Elements* I.4 the triangles are congruent, the heights themselves equal.

The truth is that facts such as Step 3 are over-determined by the Euclidean material. There are many routes offered by the *Elements*, and it is difficult to say which was taken. But then again, did he "use the *Elements*" at all (whatever we mean, textually, by "the *Elements*"?). The way in which the passage 1–3 is phrased leads us to think Archimedes argued directly on the basis of symmetry, not "through Euclid" at all: the cone is symmetrical, the base is symmetrical, there is nothing to make the heights unequal.

Perhaps the following analysis should be adopted: a mathematician who is deeply acquainted with the contents of the *Elements* knows directly the truth of such facts as Step 3. He or she does not articulate to him/herself any clear arguments. Steps 1 and 2 are not "Archimedes' grounds for believing that 3." They are a way of communicating Archimedes' *intuition* concerning Step 3: the references to the *Elements* supplied by modern commentators such as myself, are no more than indications, useful for modern readers.

/ 8 /

If a pyramid is circumscribed around an isosceles cone, the surface of the pyramid without the base is equal to a triangle having a base equal to the perimeter of the base, and, <as> height, the side of the cone.

Let there be a cone, whose base is the circle ABΓ, and let a pyramid be circumscribed so that its base, that is the polygon ΔEZ, is circumscribed around the circle ABΓ; I say that the surface of the pyramid without the base is equal to the said triangle.

(1) For since [the axis of the cone is right to the base, (2) that is to the circle ABΓ, (3) and] the lines joined from the center of the circle to the touching points are perpendiculars on the tangents,[61] (4) therefore the <lines> joined from the vertex of the cone to the touching points will also be perpendiculars on ΔE, ZE, ZΔ.[62] (5) Therefore HA, HB, HΓ, the said perpendiculars, are equal to each other; (6) for they are sides of the cone.[63] (a) So let the triangle ΘKΛ be set out having: ΘK equal to the perimeter of the triangle ΔEZ, and the perpendicular ΛM equal to HA. (7) Now, since the <rectangle contained> by ΔE, AH is twice the triangle EΔH, (8) and the <rectangle contained> by

Eut. 255

---

61 *Elements* III.18.   62 See Eutocius.
63 The cone is assumed to be isosceles, hence its sides are equal.

ΔZ, HB is twice the triangle ΔZH, (9) and the <rectangle contained> by EZ, ΓH is twice the triangle EHZ,[64] (10) therefore the <rectangle contained> by ΘK, and by AH – that is by MΛ[65] – (11) is twice the triangles EΔH, ZΔH, EHZ.[66] (12) But the <rectangle contained> by ΘK, ΛM is also twice the triangle ΛKΘ; (13) through this, then, the surface of the pyramid without the base is equal to a triangle having a base equal to the perimeter of ΔEZ and, <as> height, the side of the cone.

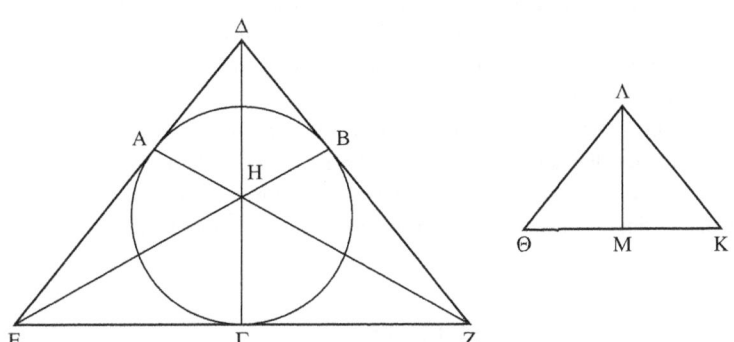

I.8
Codex G has the line
ΘK at the same height
as EZ.   Codex E
omits H.

TEXTUAL COMMENTS

Heiberg effectively brackets all of Steps 2–3. He does not give any explanation. Perhaps his grounds were the following. We have some very weak evidence[67] according to which some early Greek authors, at least in the context of producing conic sections, understood by "cone" only that cone generated by the rotation of a right-angled triangle around one of its (non-hypotenuse) sides. This is the *definition* at *Elements* XI Def. 18. The result of this rotation is (1) "right" in the sense that the axis (the side around which the rotation took place) is at right angles to the base, (2) isosceles (all the lines on the surface, drawn from vertex to base, are equal). It is clear that Apollonius already had a different conception of a cone, namely (somewhat anachronistically): the locus of the lines passing through a given point and a given circular circumference. This includes "oblique" cones where properties (1)–(2) above no longer hold. A question arises, then, who first envisaged Apollonius-style cones. The answer may conceivably be Apollonius himself and, if this is the case, and given that

---

[64] Steps 7–9 are all based on *Elements* I.41.   [65] Step a.   [66] Step a.

[67] Jones' verdict on Pappus VII.30 is "we find Pappus reconstructing his story from little more historical data than we have" (Jones [1986] II.400; but note that whatever its credentials, Pappus' testimony is in fact very vague). Eutocius, in his commentary to Apollonius, has a sustained argument, allegedly based on Geminus, for Apollonius' complete originality (II.168–74). This, however, is in a polemical context, against Heraclius, a biographer of Archimedes (on whom hardly anything is known), who, in turn, asserted that Apollonius' *Conics* was pilfered from Archimedes! All late, biased, vague sources.

Archimedes is certainly earlier than Apollonius, it might be argued that (1) Steps 2–3 ought indeed to be bracketed (there was no need for Archimedes to state the perpendicularity of the axis to the base, which was true by the *definition* of the cone), (2) *all* references to "isosceles" cones (e.g. in the enunciation of this proposition, or the first step in the proof of 7) are to be bracketed as well (for the same reason: all cones are, on this hypothesis, isosceles by definition). I suggest we keep our minds open. The evidence that early authors took only *right* cones as "cones" is weak; Archimedes' position in the historical development cannot be ascertained (modern discussions of the question[68] focus on a somewhat different topic, namely, what conception Archimedes had of conic *sections*; and it is clear that one can have a limited conception of conic *sections*, and still have a wider conception of *cones*).

Independently of this general historical background, note finally that Archimedes may point out that an axis is right to the base, even if this is true by *definition* (rather than by construction). Unpacking definitions is part of what mathematicians do. In short, then, there is no good reason to doubt our manuscripts.

GENERAL COMMENTS

The significance of different ways of naming objects

A mathematician always has a choice between several ways of naming the same object. The choice reflects specific interests.

A nice example is Archimedes' quaint gesture towards generality, where he feigns ignorance concerning the n-gonality of "the polygon ΔEZ" mentioned at the start of the setting-out: instead of calling this a "triangle," he calls it a "polygon." Such minute examples show Archimedes' worrying over generality.

Now, notice the following interesting practice: while *polygons* are usually named linearly (along a single clockwise or anti-clockwise tour of the vertices of the polygon), when a *series of lines* is given, as in Step 4, "ΔE, ZE, ZΔ," such names only rarely follow such a linear order. In this way, it is further stressed that the subject is not the lines *qua* constituents of polygons, but *qua* individual lines.

Similarly, consider the following: "(10) therefore the <rectangle contained> by ΘK, and by AH – that is MΛ." The content is the identification AH=MΛ. To clarify that the content of this step is not ("the <rectangle contained> by ΘK, AH")=(MΛ) – which otherwise could be a natural interpretation of the connector "that is" – the word "and" is inserted between the two sides of the rectangle (elsewhere, rectangles are typically given as "the <rectangle contained> by X, Y," not by "the <rectangle contained> by X and <by> Y"). Both cases are examples of the structural semantics operative in these tight texts, where every difference *signals*. Because language is used in a relatively rigid way, any difference in usage implies a special intended meaning.

---

[68] Foremost of these is Dijksterhuis (1938) 55–118, with Knorr (1987) 430.

Generality and the use of letters referring to the diagram

In the ending of the proof, at Step 13, we are given what seems, at first glance, like a general conclusion in the style of Euclid, reverting to the original enunciation and affirming that it has been proved. I quote the relevant texts. *Enunciation*: "If a pyramid is circumscribed around an isosceles cone, the surface of the pyramid without the base is equal to a triangle having a base equal to the perimeter of the base, and, <as> height, the side of the cone." *Definition of goal*: "I say that the surface of the pyramid without the base is equal to the said triangle." *End of proof*: "Through this, then, the surface of the pyramid without the base is equal to a triangle having a base equal to the perimeter of ΔEZ and, <as> height, the side of the cone."

The end of the proof is not a Euclidean-style conclusion in the strict sense, because it does not state the condition ("If a pyramid is circumscribed around an isosceles cone"). It is a return to the formulation at the (particular) definition of the goal: that it is particular and not general is shown by the presence of the letters ΔEZ at the end of the proof, referring to the particular diagram.

Here, however, comes an important complication. There are no letters at the definition of goal, although this, too, is particular – mainly because of Archimedes' impatience with detail. Throughout, Archimedes will be happy to refer to particular objects not through their lettered representation, but through some description (often as open-ended as "the said triangle"). Such references through descriptions are inherently ambiguous: are they particular or general?

In general, letters in Greek mathematics are references to the particular diagram, hence signs standing for particular objects. This is especially true in the standard case, where most objects in the diagram are referred to constantly through their lettered labels. However, when the context begins to be dominated by open-ended descriptions without letters, the ambiguity of the descriptions begins to be transferred to the letters themselves. They begin to look like a general reference of some strange sort, yet another way of picking out a quasi-general object – "the said triangle," "the triangle at the base." In short, letters begin almost to look like variables. Of course we are not yet there: but it might be argued that it is through such largely invisible processes that, much later in the history of mathematics, variables will finally emerge.

/9/

If in some isosceles cone a straight line falls through the circle which is the base of the cone, and straight lines are joined from its <=the line's> limits to the vertex of the cone, the triangle contained by both: the falling <line>; and the <lines> joined to the vertex – will be smaller than the surface of the cone between the <lines> joined to the vertex.

Let there be a base of an isosceles cone – the circle ABΓ, <let> Δ <be> its <=the cone's> vertex, and let some line AΓ be drawn through it <=the circle>, and let AΔ, ΔΓ be joined from the vertex to A, Γ; I say that the triangle AΔΓ is smaller than the conical surface between the <lines> AΔΓ.

Eut. 256

(a) Let the circumference ABΓ be bisected at B, (b) and let AB, ΓB, ΔB be joined; (1) so the triangles ABΔ, ΒΓΔ will be greater than the triangle AΔΓ;[69] (c) so let Θ be that by which the said triangles exceed the triangle AΔΓ. (2) So Θ is either smaller than the segments AB, ΒΓ, or not.

(d) First let it be not smaller. (3) Now, since there are two surfaces: the conical <surface> between the <lines> AΔB together with the segment AEB; and the <surface> of the triangle AΔB, having the same limit, <namely> the perimeter of the triangle, AΔB, (4) the container will be greater than the contained;[70] (5) therefore the conical surface between the <lines> AΔB together with the segment AEB is greater than the triangle ABΔ. (6) And similarly, also the <surface> between the <lines> BΔΓ together with the segment ΓZB is greater than the triangle BΔΓ; (7) therefore the whole conical surface[71] together with the area Θ is greater than the said triangles.[72] (8) But the said triangles are equal to the triangle AΔΓ and the area Θ. (9) Let the area Θ be taken away <as> common; (10) therefore the remaining conical surface between the <lines> AΔΓ is greater than the triangle AΔΓ.

---

[69] An extremely perplexing argument. Eutocius' commentary is very unclear, and Heiberg' footnote is mathematically false (!). Dijksterhuis (1987) 157, offers what is a rather subtle proof, which I adapt (see fig.): if we bisect AΓ at X, we may derive the result by comparing the triangles AΔB, AΔX. Now, if we draw a perpendicular from A on ΔB, to fall on the point Ψ, then we have: triangle AΔB is half the rectangle contained by ΔB, AΨ, while triangle AΔX is half the rectangle contained by ΔX, AX. Now, ΔB > ΔX (X being internal to the circle) while, AX being perpendicular to the plane ΨBX, it can be shown that AΨ > AX, as well. It thus follows that AΔB > AΔX, and Archimedes' conclusion is guaranteed. It would be amazing if Archimedes, who throughout the treatise is dealing with some very subtle relations of size between surfaces in space, would take this fundamental relation on faith. On the other hand, the rules of mathematical writing seem to be that, when such a simple step is left without argument, it is implied (in this case, misleadingly) that the argument follows in a straightforward way from elementary results. Eutocius and Heiberg were in fact misled and they assumed this was the case.

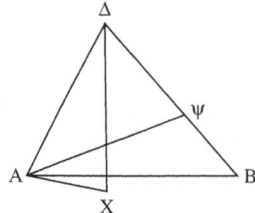

[70] *SC* I Post. 4.

[71] "Whole" in the sense that it comprises both surfaces (not in the sense that it covers all the cone).

[72] The transition from Step 6 to Step 7 substitutes Θ for the segments – based on Step d (Θ not smaller than the segments) and an *a fortiori* argument.

(e) Now let Θ be smaller than the segments AB, BΓ. (f) So, bisecting the circumferences AB, BΓ, and their halves, (11) we will leave segments which are smaller than the area Θ.[73] (g) Let there be left the <segments> on the lines AE, EB, BZ, ZΓ, (h) and let ΔE, ΔZ be joined. (12) Therefore,[74] again, according to the same <argument>,[75] (13) the surface of the cone between the <lines> AΔE together with the segment on AE is greater than the triangle AΔE, (14) and the <surface> between the <lines> EΔB together with the segment on EB is greater than the triangle EΔB; (15) therefore the surface between the <lines> AΔB together with the segments on AE, EB is greater than the triangles AΔE, EBΔ. (16) But since the triangles AEΔ, ΔEB are greater than the triangle ABΔ, (17) as has been proved,[76] (18) much more, therefore, the surface of the cone between the <lines> AΔB together with the segments on AE, EB is greater than the triangle AΔB. (19) So, through the same <argument>,[77] (20) the surface between the <lines> BΔΓ together with the segments on BZ, ZΓ is greater than the triangle BΔΓ; (21) therefore the whole surface between the <lines> AΔΓ together with the said segments is greater than the triangles ABΔ, ΔBΓ. (22) But these <=the two triangles> are equal to the triangle AΔΓ and the area Θ; (23) <while> in them <=the unit composed of the conical surface and the segments>,[78] the said segments are smaller than the area Θ; (24) Therefore the remaining surface between the <lines> AΔΓ is greater than the triangle AΔΓ.

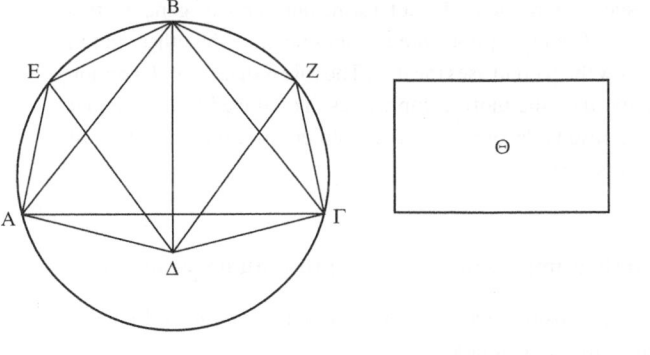

I.9
Codex D has the rectangle to the left of the triangle; it has the point Δ slightly to the right, so that line AΔ, for instance, is greater than line ΓΔ; it has the height of the rectangle a little greater than its base.   Codex G has omitted the lines AΔ, ΓΔ (inserted by a later hand).   Codex 4 has K instead of A.

---

[73] *SC* I.6 (second paragraph).

[74] Instead of *ara*, we have the particle *toinun* – a *hapax legomenon* for this work, and very rare in the Archimedean corpus as a whole.

[75] Refers, locally, to Step 4 and, beyond, to Post. 3.

[76] The reference is to Step 1, which does not seem to have been *proved* by Archimedes. See textual comments.

[77] The reference is to Step 4.

[78] That the word "them" refers to this particular composite unit, and not to the triangles mentioned in the preceding step, is a violation of natural Greek syntax. See general comments.

TEXTUAL COMMENTS

As mentioned below concerning Archimedes' language, Heiberg suspected that this text was corrupt, hence its ambiguities: if anything, ambiguities might indicate an author more than a scholiast, who is essentially interested in reducing ambiguity.

Step 17 is strange: one thing Archimedes surely did not do is to give a *proof* of the claim which is said to "have been proved." Are we to imagine a lost Archimedean proof? This is possible, but in general this kind of remark is typical of scholiasts who will then refer not to anything in Archimedes, but to Eutocius' own commentary. Heiberg does not bracket Step 17, but on the other hand he does not translate it in the Latin, either. I suspect the absence of the brackets is a mere typographic mistake in his edition.

GENERAL COMMENTS

Bifurcating structure of proof

The most important logical feature of this proposition is that it is the first to have a bifurcating structure. Instead of giving a single proof, the logical space is divided in two, and a separate proof is given for each of the two sections. Such divisions are an important technique, which is far from obvious. As is shown in Lloyd (1966), the understanding of what is involved in an *exhaustive* division of logical space is a difficult historical process. Here we see Archimedes clearly thinking in terms of "smaller"/"not smaller" (instead of "smaller"/"greater"), and in this he is indebted to a complex historical development.

Note also that such bifurcations do not form, here and in general, two hermetic *textual* units. The two options are set out one *after* the other, and in general we will see how the second uses the first. The rule is that a simpler option is dealt with first, and then the more complex case is reduced to the simpler case, or at least uses results developed in the simpler case; the advantage of this division of labor is clear.

Ambiguities and their implication for the author/audience attitude

The language of this proposition tends to be somewhat ambiguous; I believe this is authorial. Here are two examples:

1 Expressions such as "the surface of the cone between the <lines> joined to the vertex" (enunciation), or "the conical surface between the <lines> AΔΓ" (definition of goal) are doubly indeterminate. First, we are not told *where* the section of the cone should end. It has nowhere better to end than the surface of the circle, and clearly this is what Archimedes means. This could be a normal ellipsis, where the text is meant to be supplemented by the diagram. Note, however, that in the next proposition, the text will be *more* explicit – so one possibility is that our text is corrupt here: – this was Heiberg's view. Second, an indeterminacy which shows that while Archimedes may clarify his expressions occasionally, he does not *aim* at clarity. What I mean is that there are *two*

conical surfaces defined by Archimedes' expression – one in either direction of the triangle. The proof will apply to both, of course, but Archimedes uses the definite article for this surface, so he thinks of it as if it were uniquely defined. Now, the proof will be taken up, in the corollary of 12 (there, it will supply the grounds for asserting that a conical surface is greater than the surface of an inscribed pyramid). The way in which Proposition 9 will be used inside that corollary of 12 implies that Archimedes has in mind here only the *smaller* surface (the one associated with an inscribed pyramid). However, Archimedes did not set out to clarify this. The indeterminacy of the expression is thus a meaningful phenomenon, showing something about the way Archimedes aimed to use language. Language is not the ultimate object, it is merely a tool for expressing a mathematical content, and Archimedes' mind is fixated upon the level, of content. He *knows* what the references of the expressions are, and he therefore does not set out systematically to disambiguate such references.

2 Related to this is the glaring solecism towards the end of the proof (stylistically, a meaningful position!): the words "in them" (genitive plural of the relative clause) at the start of Step 23 which, syntactically, may refer most naturally to the "these" mentioned in Step 22, i.e. to the two triangles, or, possibly but less naturally, may refer to the unit composed of the triangle $A\Delta\Gamma$ and the area Θ, again mentioned in Step 22. Archimedes refers to neither: instead, he refers to the unit composed of the conical surface and the segments, mentioned in Step 21. For Archimedes, the problem must never have arisen. The words "in them" must have been charged by an internal gesturing, necessarily non-reproducible in the written mode. He pointed mentally to the relevant objects; and failed to un-notice his own mental pointing when translating his thought to the written mode – and thus, as it were, he failed to notice *us*. That is, it seems that in such cases Archimedes loses sight of any imagined audience.

/10/

If there are drawn tangents of the circle which is <the> base of the cone, being in the same plane as the circle and meeting each other, and lines are drawn, from the touching-points and from the meeting-point <of the tangents>, to the vertex of the cone, the triangles contained by: the tangents, and <by> the lines joined to the vertex of the cone – are greater than the surface of the cone which is held by them.[79]

Let there be a cone, whose base is the circle $AB\Gamma$, and its vertex the point E, and let tangents of the circle $AB\Gamma$ be drawn, being in the same plane – <namely> $A\Delta$, $\Gamma\Delta$ – and, from the point E – which is <the>

---

[79] The literal translator's nightmare. The verbs "contained" and "held" in this sentence stand for what are, in this context, near-synonymous Greek verbs (περιέχειν, ἀπολαμβάνειν, respectively). Perhaps "contained" would have been better for both.

vertex of the cone – to A, Δ, Γ, let EA, EΔ, EΓ be joined; I say that
the triangles AΔE, ΔEΓ are greater than the conical surface between:
the lines AE, ΓE, and the circumference ABΓ.

(a) For let HBZ be drawn, tangent to the circle, also being parallel
to AΓ, (b) the circumference ABΓ being bisected at B (c) and, from
H, Z, to E, let HE, ZE be joined. (1) And since HΔ, ΔZ are greater
than HZ,[80] (2) let HA, ZΓ be added <as> common; (3) therefore AΔ,
ΔΓ, as a whole, are greater than AH, HZ, ZΓ.[81] (4) And since AE,
EB, EΓ are sides of the cone, (5) they are equal, (6) through the cone's
being isosceles; (7) but similarly they are also perpendiculars[82] [(8) as
was proved in the lemma] [(9) and the <rectangles contained> by the
perpendiculars and the bases are twice the triangles];[83] (10) therefore
the triangles AEΔ, ΔEΓ are greater than the triangles AHE, HEZ,
ZEΓ [(11) for AH, HZ, ZΓ are smaller than ΓΔ, ΔA, (12) and their
heights <are> equal] [(13) for it is obvious, that the <line> drawn
from the vertex of the right cone to the tangent-point[84] of the base is
perpendicular on the tangent].[85] (d) So let the area Θ be that by which
the triangles AEΔ, ΔΓE are greater than the triangles AEH, HEZ, ZEΓ.
(14) So the area Θ is either smaller than <the remaining <segments>
AHBK, BZΓΛ or not smaller.

(e) First let it be not smaller. (15) Now since>[86] there are composite
surfaces: that of the pyramid on the trapezium HAΓZ <as> base,
having E <as> vertex; and the conical surface between the <lines>
AEΓ together with the segment ABΓ, (16) and they have <as> limit
the same perimeter of the triangle AEΓ, (17) it is clear that the surface
of the pyramid without the triangle AEΓ is greater than the conical
surface together with the ABΓ segment.[87] (18) Let the segment ABΓ
be taken away <as> common; (19) therefore the remaining triangles
AHE, HEZ, ZEΓ together with the remaining <segments> AHBK,
BZΓΛ are greater than the conical surface between the <lines> AE,

---

[80] *Elements* I.20.

[81] This argument unpacks a simple corollary from Euclid's *Elements* I.20. Both the
corollary and, indeed, I.20 itself, can be derived directly from Postulates 1–2.

[82] They are perpendiculars to the tangents. See Proposition 8, Step 5 (to which the
"similarly" refers?).

[83] *Elements* I.41.

[84] The interpolator does not use ἀφή, the word used above "touching-point," but a
variant, ἐπαφή.

[85] See textual comments on this obviously redundant Step 13 (a repetition of Steps
6–7).

[86] The "the remaining . . . since" are a lacuna in the manuscripts. In my completion
of the lacuna I essentially follow Heiberg. See also textual comments.

[87] *SC* I Post. 4.

ΕΓ. (20) But the area Θ is not smaller than the remaining <segments> AHBK, BZΓΛ. (21) Much more, therefore, the triangles AHE, HEZ, ZEΓ together with the <area> Θ, will be greater than the conical surface between the <lines> AEΓ. (22) But the triangles AHE, HEZ, ΓEZ together with the <area> Θ are the triangles AEΔ, ΔEΓ; (23) therefore the triangles AEΔ, ΔEΓ will be greater than the said conical surface.[88]

(f) So let the <area> Θ be smaller than the remaining <segments>.

Eut. 257

(g) So, circumscribing polygons ever again around the segments (the circumferences of the remaining <segments> being similarly[89] bisected, and tangents being drawn), (24) we will leave some remaining <segments>, which will be smaller than the area Θ.[90] (h) Let them be left and let them be AMK, KNB, BΞΛ, ΛOΓ, being smaller than the area Θ, (i) and let it be joined to E.[91] (25) So again it is obvious that the triangles AHE, HEZ, ZEΓ will be greater than the triangles AEM, MEN, NEΞ, ΞEO, OEΓ[92] [(26) for the bases are greater than the bases[93] (27) and the height equal].[94] (28) And moreover, similarly, again, the pyramid, having <as> base the polygon AMNΞOΓ, <and having> E <as> vertex, without the triangle AEΓ – has a greater surface than: the conical surface between the <lines> AEΓ together with the segment ABΓ.[95] (29) Let the segment ABΓ be taken away <as> common; (30) therefore the remaining triangles AEM, MEN, NEΞ, ΞEO, OEΓ together with the remaining <segments> AMK, KNB, BΞΛ, ΛOΓ will be greater than the conical surface between the <lines> AEΓ. (31) But the area Θ is greater than the said remaining <segments>, (32) and the triangles AEH, HEZ, ZEΓ were proved to be greater than the triangles AEM, MEN, NEΞ, ΞEO; (33) much more, therefore, the triangles AEH, HEZ, ZEΓ together with the area Θ, that is the triangles AΔE, ΔEΓ, (34) are greater than the conical surface between the lines AEΓ.

---

[88] Notice that the future tense of this conclusion relativizes it, reminding us that this is not a final conclusion, but an interim one – a consequent of the antecedent in Step (e), that Θ is not smaller.

[89] The "similarly" refers not to repetition in the process, but to its similarity to the earlier drawing of HBZ in Steps a–b.

[90] *SC* I.6 (second paragraph).

[91] The sense is that lines are to be drawn from the new points to the vertex. This is a drastic abbreviation, leading to a remarkable expression ("it is joined," used as an impersonal verb, rather like, say, "it rains"), which Heiberg attributes (unnecessarily, I think) to textual corruption.

[92] *Elements* I.41.   [93] *Elements* I.20.

[94] The cone is isosceles.   [95] *SC* I Post. 4.

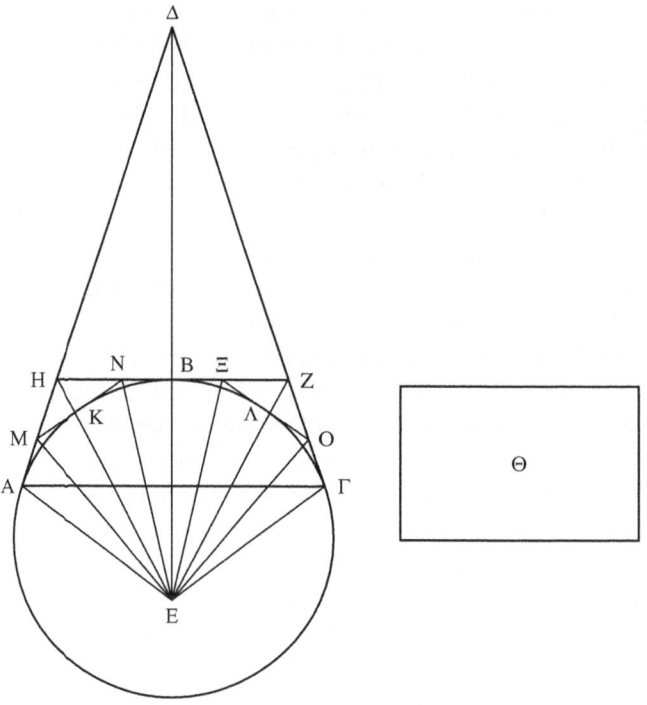

I.10
Codex A has omitted the rectangle Θ. (It is added by codex B, and by a later hand in codex G).   Codex E has K (?) instead of H, as well as (corrected by a later hand) Θ instead of O. The Ξ in codex G is a correction by a later hand, but it is difficult to say, a correction from what.

TEXTUAL COMMENTS

Step 8, asserting that Step 7 is proved in a lemma, is almost certainly an interpolation: had Archimedes himself supplied a lemma, this would be referred to in Step 4 of Proposition 8, as well. Most probably, Step 8 is by Eutocius or by one of his readers, referring to Eutocius' commentary on Step 4 of Proposition 8. Step 13 is a strangely placed, belated assertion that Step 7 is "obvious." Probably it started its life as a marginal comment, inserted into the main text in a "wrong" position. Its historical relation to Step 8 cannot be fathomed now. The one probable thing is that the interpolator responsible for Step 8 is not the same as the one responsible for Step 13. Who came first is unclear and, indeed, they could come from different traditions of the text, united in some later stage.

Heiberg has his usual doubts about 9, 11–12, 26–7, and, as usual, we can only suspend our judgment.

GENERAL COMMENTS

What is the sequence of actions inside a construction?

"(a) For let HBZ be drawn, tangent to the circle, also being parallel to AΓ, (b) the circumference ABΓ being bisected at B." This is a good example of the difficulty of parsing constructions. Should I have divided this text the way I did, into (a) and (b)? What are we to do, and in what order? Following the literal meaning of the text, we should imagine the following: first, we draw a "floating" tangent (one placed freely around the circle). We then fix it as parallel to AΓ (this leaves us with two options, either to the side of Δ or to the other side; the choice, to the side of Δ, is based on the diagram). We then call the point where the tangent touches the circle B (that it bisects the circumference is an *assertion*, perhaps, not a construction). Or again, that we bisect the circumference AΓ may be part of the construction itself: it is a specification which just happens to be equivalent to the specification that the tangent and AΓ be parallel. Or, finally, these parsings of the construction into its constituents might be misleading: perhaps we do not start with a sliding tangent at all. The whole construction is virtual: it need not be spelled out in any clear order. All we have is an unpacking of the diagram, where order is immaterial.

The role of the axiomatic discussion

As noted in a footnote to Step 3, Archimedes could in principle conceive of Euclid's I.20 (any line in a triangle being smaller than the other two) as a special case of his Postulates 1–2. We can not say of course whether he actually conceived of it in this way, but the question of principle is important: what was the role of axiomatic discussions? Were they meant to apply "retroactively," so to speak?

Step 25 might tell us something about this question. It may be seen to derive from the *Elements* (as spelled out in the possibly interpolated Steps 26–7), or directly from Archimedes' Postulate 4 (assuming that the sets of triangles are seen as two composite surfaces answering to Archimedes' postulate). Thus there is a textual problem here – whether Steps 26–7 are interpolated, or not – and a mathematical problem – what are the grounds for Step 25.

If 25 relies on the *Elements*, this would be interesting: we find that Archimedes views his postulates, at least in this particular case, as ad-hoc contrivances, designed to do a specific job, but to be dispensed with when simpler methods will do. On the other hand, if Step 25 does not rely upon the *Elements*, the role of the axiomatic discussion seems to be more profound – to supply new foundations for geometrical properties. Now, while the question cannot be settled, it seems more likely that 25 relies upon the *Elements*, simply because it is introduced by "obviously" – an adverb suiting elementary arguments better than it does Archimedes' sophisticated axiomatic apparatus. This then is a potentially important observation.

Another issue regarding the role of the axiomatic discussion is this. As noted several times above, in his commentary to the Definitions, Eutocius pointed out, correctly, that the notion of "line" used by Archimedes there (and, as an implicit consequence, the notion of "surface") covered "composite" lines and surfaces as well (although Archimedes speaks in the definitions simply of "lines" and "surfaces"). That is, in the definitions, the words "line," "surface" *meant* "composite line," "composite surface." In Step 15, however, which implicitly invokes the Definitions, the term used is "composite surface." Once again, therefore, we see that the axiomatic discussion is designed to do a specific job – to introduce a certain claim, about the relations between lines or between surfaces. That job accomplished, Archimedes lets the apparatus drop, not even relying upon the *terminology* that was implicitly sustained by the Definitions.

/11/

If in a surface of a right cylinder there are two lines, the surface of the cylinder between the lines is greater than the parallelogram contained by: the lines in the surface of the cylinder, and the <lines> joining their <=the original two lines> limits.

Let there be a right cylinder, whose base is the circle AB, and <its> opposite the <circle> ΓΔ,[96] and let AΓ, BΔ be joined; I say that the cylindrical surface cut off by the lines AΓ, BΔ is greater than the parallelogram AΓBΔ.

(a) For let each of AB, ΓΔ be bisected at the points E, Z, (b) and let AE, EB, ΓZ, ZΔ be joined. (1) And since AE, EB are greater than the [diameter] AB,[97] (2) and the parallelograms on them are of equal heights, (3) then the parallelograms, whose bases are AE, EB and whose height is the same as the cylinder, are greater than the parallelogram ABΔΓ.[98] (c) Therefore let the area H be that <by which> they are greater.[99] (4) So the area H is either smaller than the plane segments AE, EB, ΓZ, ZΔ or not smaller.

(d) First let it be not smaller. (5) And since the cylindrical surface cut off by the lines AΓ, BΔ, and the <segments> [triangles][100] AEB, ΓZΔ, have <as> a limit the plane of the parallelogram AΓBΔ, (6) but the surface composed of the parallelograms, whose bases are AE, EB and whose height is the same as the cylinder, and the <triangles> [planes][101] AEB, ΓZΔ, also have <as> a limit the plane of the parallelogram ABΔΓ, (7) and one contains the other, (8) and both are concave in the same direction, (9) so the cylindrical surface cut off by the lines AΓ, BΔ, and the plane segments AEB, ΓZΔ, are greater than the surface composed of: the parallelograms whose bases are AE, EB and whose height is the same as the cylinder; and of the triangles

---

[96] "Opposite": the "upper" base.

[97] *Elements* I.20. The square-bracketed word "diameter" is in a sense "wrong" (the line does not have to be a diameter). See textual comments.

[98] An extension of *Elements* VI.1.

[99] This translation follows an emendation suggested in the textual comments, against Heiberg's emendation.

[100] Another "wrong" interpolation. To follow the mathematical sense, read "segments," but see textual comments.

[101] Again, read "triangles" and consult the textual comments.

AEB, ΓΖΔ.[102] (10) Let the triangles AEB, ΓΖΔ be taken away <as> common; (11) so the remaining cylindrical surface cut off by the lines ΑΓ, ΒΔ, and the plane segments AE, EB, ΓΖ, ΖΔ, are greater than the surface composed of the parallelograms, whose bases are AE, EB, and whose height is the same as the cylinder. (12) But the parallelograms, whose bases are AE, EB, and whose height is the same as the cylinder, are equal to the parallelogram ΑΓΒΔ and the area H; (13) therefore the remaining cylindrical surface cut off by the lines ΑΓ, ΒΔ is greater than the parallelogram ΑΓΒΔ.

(e) But then, let the area H be smaller than the plane segments AE, EB, ΓΖ, ΖΔ. (f) And let each of the circumferences AE, EB, ΓΖ, ΖΔ be bisected at the points Θ, Κ, Λ, Μ, (g) and let ΑΘ, ΘΕ, ΕΚ, ΚΒ, ΓΛ, ΛΖ, ΖΜ, ΜΔ be joined. [(14) And therefore the triangles ΑΘΕ, ΕΚΒ, ΓΛΖ, ΖΜΔ take away no less than half the plane segments AE, EB, ΓΖ, ΖΔ].[103] (15) Now, this being repeated, certain segments will be left which will be smaller than the area H. (h) Let them remain, and let them be ΑΘ, ΘΕ, ΕΚ, ΚΒ, ΓΛ, ΛΖ, ΖΜ, ΜΔ. (16) So, similarly we will prove[104] that the parallelograms whose bases are ΑΘ, ΘΕ, ΕΚ, ΚΒ, and whose height is the same as the cylinder, will be greater than the parallelograms, whose bases are AE, EB, and whose height is the same as the cylinder. (17) And since the cylindrical surface cut off by the lines ΑΓ, ΒΔ, and the plane segments AEB, ΓΖΔ, have <as> a limit the plane of the parallelogram ΑΓΒΔ, (18) but the surface composed of: the parallelograms, whose bases are ΑΘ, ΘΕ, ΕΚ, ΚΒ, and whose height is the same as the cylinder; and . . .[105]
<(19)>
. . . the rectilinear <figures> ΑΘΕΚΒ, ΓΛΖΜΔ.[106] (20) Let the rectilinear <figures> ΑΘΕΚΒ, ΓΛΖΜΔ be taken away <as> common; (21) therefore the remaining cylindrical surface cut off by the lines ΑΓ, ΒΔ, and the plane segments ΑΘ, ΘΕ, ΕΚ, ΚΒ, ΓΛ, ΛΖ, ΖΜ, ΜΔ, are greater than the surface composed of the parallelograms, whose bases are ΑΘ, ΘΕ, ΕΚ, ΚΒ, and whose height is the same as the cylinder. (22) But the parallelograms, whose bases are ΑΘ, ΘΕ, ΕΚ, ΚΒ, and whose height is the same as the cylinder, are greater than the parallelograms,

---

[102] Post. 4.   [103] *Elements* XII.2.   [104] The "similarly" refers to Steps 1–3.

[105] An obvious lacuna, whose completion by Heiberg is practically certain. I translate this completion:

". . . of the rectilinear <figures> ΑΘΕΚΒ, ΓΛΖΜΔ; has <as> a limit the plane of the parallelogram ΑΓΒΔ, (19) so, the cylindrical surface cut off by the lines ΑΓ, ΒΔ, and the plane segments AEB, ΓΖΔ, are greater than the surface composed of: the parallelograms, whose bases <are> ΑΘ, ΘΕ, ΕΚ, ΚΒ and <their> height the same as the cylinder, and . . ." See textual comments.

[106] Post. 4.

whose bases are AE, EB, and whose height is the same as the cylinder; (23) therefore also: the cylindrical surface cut off by the lines AΓ, BΔ, and the plane segments AΘ, ΘE, EK, KB, ΓΛ, ΛZ, ZM, MΔ, are greater than the parallelograms, whose bases are AE, EB, and whose height is the same as the cylinder. (24) But the parallelograms, whose bases are AE, EB, and whose height is the same as the cylinder, are equal to the parallelogram AΔΓB[107] and the area H; (25) therefore also: the cylindrical surface cut off by the lines AΓ, BΔ, and the plane segments AΘ, ΘE, EK, KB, ΓΛ, ΛZ, ZM, MΔ, are greater than the parallelogram AΓBΔ and the area H. (26) Taking away the segments AΘ, ΘE, EK, KB, ΓΛ, ΛZ, ZM, MΔ, (27) which are smaller than the area H; (28) therefore the remaining cylindrical surface cut off by the lines AΓ, BΔ is greater than the parallelogram AΓBΔ.

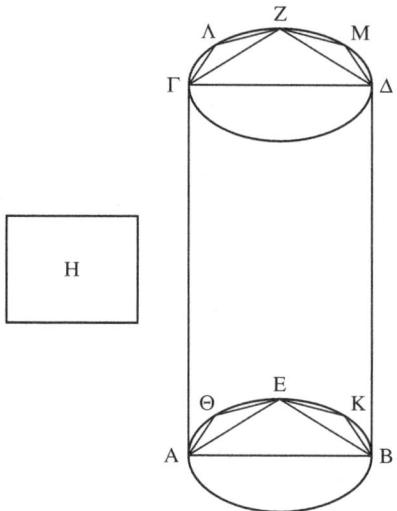

I.11
Codex E has the rectangle nearer the bottom of the cylinder, codices B4 have the rectangle nearer the top of the cylinder. Codex D has the height of the rectangle slightly greater than its base.

TEXTUAL COMMENTS

This proposition forms a special case in terms of its deviations from the Euclidean norm. It is therefore an important test-case. If we believe these deviations are authorial, we shall have one view of the process the text went through (it was gradually "standardized"). If we believe the deviations are not authorial, we may have another view (we are to some extent entitled to suppose that the text was gradually corrupted from an original Euclidean form). I will say immediately that most of the deviations, in all probability, are not authorial. This then is some (admittedly weak) argument in favor of the view that the text, originally, was at least as "Euclidean" as it is at present.

---

[107] See general comments on the weird non-linear lettering of this parallelogram.

First, there are a number of (what Heiberg saw as) one-word interpolations: "(1) . . . AE, EB are greater than the [diameter] AB,"[108] "(5) And since the cylindrical surface cut off by the lines AΓ, BΔ and the <segments> [triangles] AEB, ΓZΔ . . . ," "(6) but the surface composed of: . . . and the <triangles> [planes] AEB, ΓZΔ." In all three cases, an original elliptical phrase – typical of the Greek mathematical style – was filled up, and falsely at that, by some later hand. Archimedes can be sloppy, but the probability of his making three such mistakes is very low. On the other hand, it is easy to imagine a novice doing this. The same novice may account for the following:

1. There is a massive lacuna in Steps 18–19. The cause is obvious: the text of the lacuna repeats almost exactly the passage preceding it (a copying mistake known as homoeoteleuton). What is noteworthy is that such lacunae do not occur more often in our text (I make them all the time in my translation). So this is a tribute to the robustness of the transmission, and another indication of a lack of professionalism in this proposition.

2. Step 26 is linguistically deviant. The word ἀφαιρεθέντα, "taking away," is in the accusative (or nominative?) instead of the genitive. Heiberg ascribes this to a late Greek influence (so this cannot be Archimedes). Language history aside, the format is new: the clause is not completed to a full sentence and, most significantly, the imperative is avoided (compare, e.g., Step 20). In itself this could have been a normal authorial variation, but when coupled with the linguistic difficulty, one begins to suspect the scribe.

3. Finally there is something very weird: "(c) Therefore let the area H be that <by which> they are greater." For these words, the manuscripts have "by what then are they greater? Let it be, by the area H." Not strictly meaningless and impossible, but so radically different from normal style to merit some thought.

Some Greek is necessary. The manuscripts are τίνι ἄρα μείζονά ἐστιν; ἔστω τῷ H χωρίῳ. Heiberg suggests this was a normal ᾧ δὴ μείζονά ἐστιν, ἔστω τὸ H χωρίον. Perhaps; but then it is difficult to see what could be the source for this strange confusion. On the other hand, it is useful to note that the Greek particle ἄρα can be, with different accents, either "therefore" or an interrogative particle (Greek writing in Archimedes' time and much later was neither accented nor punctuated). Suppose the following, then: that an original "therefore" (in itself a deviation from Archimedes' common practice so far, to have δή, "so," in this context) changed into an interrogative, with a concomitant change of the relative particle at the start to an interrogative particle. It remains to explain the dative case of the area H (as against the normal nominative): this may be another corruption, or it may be authorial. So I suggest as Archimedes' "koinicized" Greek the following: ᾧ ἄρα μείζονά ἐστιν, ἔστω τὸ Hχωρίον, But it must be realized that this is a guess. The only thing which is truly probable is that the present form as it stands in the manuscripts is not Archimedes', but the scribe's.

---

[108] It is interesting – and typical of the diagrams in the manuscripts in general – that this line in fact *appears* to be a diameter. This is an important piece of evidence, then: the diagram standing in front of our mathematically hopeless scribe already had this feature.

Finally, there is Step 14, which Heiberg suspects without compelling grounds (incidentally, if this is interpolated, then the interpolator of 14 is not the main offender of the proposition – *he* would never be able to make such an apposite geometrical remark).

GENERAL COMMENTS

The significance of different ways of using letters

Letters, in this treatise as in Greek mathematics, do not usually carry meanings in a direct way. For instance, H is used here to denote the "difference" area. Θ seemed to specialize in this role until now, but now we see that such specialized roles are very localized, and that letters do not become symbols, standing for stereotypical objects. They do carry meaning, but locally.

For instance, in this proposition, letters are most fluidly used with the parallelogram ΑΓΔΒ. In two occasions it behaves strangely. In the definition of goal it is called ΑΓΒΔ, which is non-linear (i.e. you cannot trace the figure along this sequence of points). This probably reflects the fact that this mention of the parallelogram follows closely upon the mention of the parallels ΑΓ, ΒΔ (so here we see another tendency in using letters: to refer to parallels "in the same direction," in this case both parallels going *up*). Later, in Step 24, the manuscripts cannot decide quite how to call this parallelogram. A has ΑΔΓΒ, B has ΑΓΒΔ, and the Palimpsest has ΑΒΓΒ (*sic*). One of the copyists of A turned the ΑΔΓΒ he had in front of him into ΑΔΒΓ, which is hardly better. It is easy to imagine that the Palimpsest's ΑΒΓΒ is a misreading of ΑΔΓΒ, the same as A, and this is the version I translate. I cannot understand what happened here, but what I find striking is that none of the variations is the alphabetical sequence ΑΒΓΔ, the one most natural (unless this was the Palimpsest's original?). At any rate, the important principle suggested by this textual detail is that the names of objects are never mere sequences of letters: they are always oblique ways of referring to a diagrammatic reality.

Operations on phrases as a tool for argumentation

The proof starts with two inequalities, one based on *Elements* I.20 (argued in Steps 1–3, stated in 16), the other based on Post. 4 (argued in 5–9). Steps 4 and 14–15 are embedded within the construction. All the rest of the proposition (i.e. arguments 9–11, 11–13, and the entire sequence from 17 to the end) argues on the basis of operations on phrases.

A simple and instructive example is the first such argument, 9–11: "(9) so the cylindrical surface cut off by the lines ΑΓ, ΒΔ, and the plane segments AEB, ΓΖΔ, are greater than the surface composed of: the parallelograms whose bases are AE, EB and whose height is the same as the cylinder; and of the triangles AEB, ΓΖΔ. (10) Let the triangles AEB, ΓΖΔ be taken away <as> common; (11) so the remaining cylindrical surface cut off by the lines ΑΓ, ΒΔ, and the plane segments AE, EB, ΓZ, ZΔ, are greater than the surface composed of the parallelograms, whose bases are AE, EB, and whose height is the same as the cylinder."

One can ask, what is being "taken away from:" the geometric, or the linguistic object? The answer is, of course, that both are being taken away, both are being manipulated simultaneously. The interminable phrases are given meaning by reference to a geometric reality, but the argument itself is then conducted through the linguistic structure itself. What happens in such arguments is that phrases are concatenated and cut – a jigsaw puzzle of the fixed phrases of Greek mathematics.

In this case, there is a pair of objects manipulated by the argument, both introduced in Step 9:

- The cylindrical surface cut off by the lines AΓ, BΔ, and the plane segments AEB, ΓZΔ.
- The surface composed of: the parallelograms whose bases are AE, EB and their height is the same as the cylinder; and the triangles AEB, ΓZΔ.

Step 10 asks us to subtract the triangles AEB, ΓZΔ from both. The first is manipulated through the diagram to derive:

- The remaining cylindrical surface cut off by the lines AΓ, BΔ, and the plane segments AE, EB, ΓZ, ZΔ.

And the second is manipulated through the operation on phrases to derive:

- The surface composed of the parallelograms, whose bases are AE, EB, and whose height is the same as the cylinder.

All of this is contained inside the fixed expression "X is greater than Y" (in Steps 9, 11), and is mediated by the fixed expression "let . . . be taken away as common" (in Step 10). In such ways, the language serves as the basis for mathematical argument.

/ 12 /

If there are two lines in a surface of some right cylinder, and from the limits of the lines certain tangents are drawn to the circles which are bases of the cylinder, <the lines> being in their <=the circles> plane, and they meet; the parallelograms contained by the tangents and <by> the sides of the cylinder will be greater than the surface of the cylinder between the lines in the surface of the cylinder.

Let there be the circle ABΓ, base of some right cylinder, and let there be two lines in its <=the cylinder's> surface, whose limits are A, Γ, and let tangents to the circle be drawn from A, Γ, being in the same plane, and let them meet at H, and let there also be imagined, in the other base of the cylinder, lines drawn from the limits of the <lines>

in the surface, being <=the lines drawn> tangents to the circle; it is to be proved that the parallelograms contained by the tangents and <by> the sides of the cylinder are greater than the surface of the cylinder on the circumference ABΓ.

(a) For let the tangent EZ be drawn, (b) and let certain lines be drawn from the points E, Z parallel to the axis of the cylinder as far as [the surface] of the other base; (1) so the parallelograms contained by AH, HΓ and <by> the sides of the cylinder are greater than the parallel-ograms contained by AE, EZ, ZΓ and <by> the side of cylinder[109] [(2) For since EH, HZ are greater than EZ,[110] (3) let AE, ZΓ be added <as> common; (4) therefore HA, HΓ as whole are greater than AE, EZ, ZΓ]. (c) So let the area K be that by which they are greater. (5) So the half of the area K is either greater than the figures contained by the lines AE, EZ, ZΓ and <by> the circumferences AΔ, ΔB, BΘ, ΘΓ, or <it is> not.

(d) First let it be greater. (6) So the perimeter of the parallelogram at AΓ is a limit of the surface composed of: the parallelograms at AE, EZ, ZΓ, and the trapezium AEZΓ, and the <trapezium> opposite it in the other base of the cylinder. (7) The same perimeter is also a limit of the surface composed of the surface of the cylinder at the circumference ABΓ and <of> both segments: ABΓ, and the <one> opposite it; (8) so the said surfaces come to have the same limit, which is in a plane, (9) and they are both concave in the same direction, (10) and one of them contains some <parts of the other>, but some <parts> they have <as> common;[111] (11) therefore the contained is smaller.[112] (12) Now, taking away <as> common: the segment ABΓ, and the <one> opposite it, (13) the surface of the cylinder at the circumference ABΓ is smaller than the surface composed of the parallelograms at AE, EZ, ZΓ and <of> the figures AEB, BZΓ and <of> those opposite them. (14) But the surfaces of the said parallelograms together with the said figures are smaller than the surface composed of the parallelograms at AH, HΓ [(15) for together with K, which is greater than the figures <=AEB, BZΓ>, (16) they <=the surfaces of the parallelograms AE, EZ, ZΓ> were equal to them <=the parallelograms at AH, HΓ>];

---

[109] *Elements* I.20, VI.1. This is a truncated version of the argument at the start of Proposition 10 above.

[110] *Elements* I.20.

[111] By far the most complete invocation of Postulate 4 we had so far – and the first which is effectively complete. Apparently this is due to the truly complex three-dimensional configuration involved, combining curved and straight surfaces.

[112] Post. 4.

(17) now, it is clear, that the parallelograms contained by AH, ΓH and <by> the sides of the cylinder are greater than the surface of the cylinder at the circumference ABΓ.

(e) But if the half of the area K is not greater than the said figures, lines will be drawn <as> tangents to the segments, so that the remaining figures will become smaller than the half of the <area> K, (18) and the rest will be proved, the same as before.

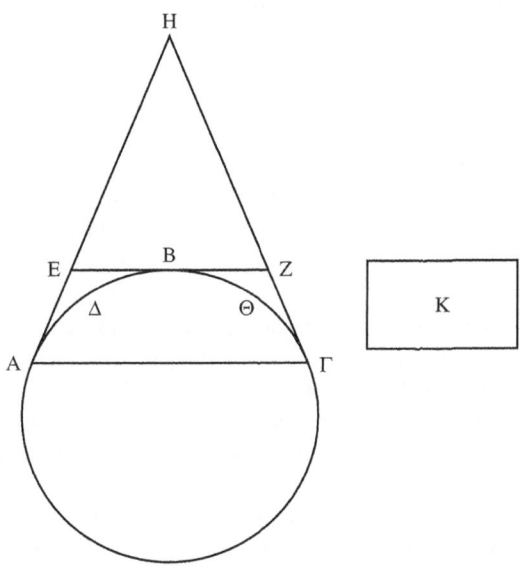

I.12
Codices BD have the small rectangle to the left of the main figure. Codex H has omitted Z.

TEXTUAL COMMENTS

Proposition 11 marked a special case for scribal intervention – repeated intervention, of repeatedly poor quality. It seems almost as if Heiberg, still on his guard, is doubly cautious now, bracketing the words "the surface" in Step b. There is little real reason, I think, for this bracketing: the base is visualized as a surface of the cylinder, and a close paraphrase of the original may be "as far as that surface *which is* the other base." Generally speaking, we seem to have moved back to normal territory, where scribal intervention is hard to detect, as it does not seem to deviate sharply from Archimedes' own spirit. There are the usual difficult cases: Steps 2–4 are a backward-looking justification, and seem redundant in light of 1. They could be a scholion. Steps 15–16, as well, are backwards-looking justifications, but they are brief and to the point; perhaps they are authorial. Heiberg argued that "them" in 16 should have been "it," referring to the composite surface; but such a mistake (if this is the right word) is very common. The interest of this "mistake" is in its letting us see how, here, Archimedes (or his interpolator) loses interest in fully spelled out phrases, and reverts more and more to abbreviating them by – more and more vague – pronouns.

GENERAL COMMENTS

The extent to which words are allowed to shift

The first shift I have to discuss here is introduced – shame to admit! – in the translation itself. It does reveal something important about the original text: that the same Greek word may correspond to a number of concepts (not necessarily easy to define). From the enunciation: "If there are two lines in a surface of *some* right cylinder, and from the limits of the lines *certain* tangents are drawn:" the two italicized words correspond to a single word in the original Greek, *tis*, usually translated "some." I vary the translation, as the word is not used according to any strict logical system (as a quantifier would be in a modern mathematical text). Consider, e.g., the construction: "(a) For let the tangent EZ be drawn, (b) and let *certain* lines be drawn from the points E, Z parallel to the axis of the cylinder as far as [the surface] of the other base." The lines mentioned in (b) are uniquely given. We would probably use just the expression "lines" ("and let lines be drawn . . ."). I do not know why Archimedes used the word *tis* here, but such examples show that the meaning of *tis* is not quite fixed in this text.

A very minor (but no less significant) case of lexical looseness is provided by the expressions for "opposite" bases. *katenantion* is used in Step 6; *apenantion* is used in Step 7 (this last was used already in the setting-out of Proposition 11). Nothing hangs on it: the interest, however, is in seeing how unstable the terminology becomes as soon as we move away from the central typical terms of Greek mathematics.

Finally, consider the expression for moving from the setting-out to the definition of goal. The expression used here is "it is to be proved that . . ." Now, this Proposition 12 is a theorem (proving the truth of a claim). Compare this to a problem (constructing a given task), say Proposition 2, where the same move is made by "I say that it is possible . . ." In Euclid's text of the *Elements* as we have it, the expression in theorems is "I say that . . .", the expression in problems is "[given . . .] to do [the task]." Archimedes has theorem-like problems ("I say that it is possible" – i.e. asserting a truth, not just doing the task), and problem-like theorems ("it is to be proved that . . ." – i.e. setting out a task to be fulfilled, not just asserting a claim). This variation is a mark of the fluidity of the second-order language – the language where geometrical discussions are discussed (as opposed to the first-order language, where geometrical objects themselves are discussed).

So we saw a logical ambiguity with the word *tis*; a variability concerning a rare term; and a fluidity of the second-order discourse. All of these contrast with a very rigid discourse as far as the geometrical, common objects of Greek mathematics are concerned.

The operation of "imagination"

The verb "to imagine" occurs for the first time in the setting-out: "and let there also be *imagined*, in the other base of the cylinder, lines drawn from the limits of the <lines> in the surface . . ." Why "imagine" instead of "draw?" To some extent this is a mark of the general tendency here, to abbreviate; more importantly, the diagram is strictly two-dimensional, concentrating on only one base of the cylinder. Without a three-dimensional diagram, the other base can be only "imagined." The diagram of Proposition 11 is highly three-dimensional, with the typical almond-shapes for representing circles seen obliquely. It is the last three-dimensional representation of the book, whose strategy from now on will involve the reduction of three-dimensionality into two-dimensionality.

Notice that the verb "to imagine" is used regularly in Greek mathematics, whenever the diagram does not represent the geometrical action. This shows how seriously the action is taken when it is *not* "imagined." By using the verb "to imagine" when the line is not drawn in the diagram, the geometer implies that the drawn line is not just "imagined" but, in some sense, is really there – as if the diagram was the actual mathematical object.

A tendency to abbreviate

The text abbreviates more and more (leading in this way to the next passage, which is a series of completely truncated arguments). For example, consider the tangent EZ: this is the same as HZ in Proposition 10, where we had an over-specification (this happened in Steps a–b in Proposition 10 where – translating into the terms of Proposition 12 – the tangent was made to be *both* parallel to AΓ *and* bisecting the circumference ABΓ). No specification at all is made in this Proposition 12. The tangent hangs in the air – and is given meaning purely by our acquaintance with earlier material.

Such abbreviations, however, pale into insignificance compared with the last paragraph, where everything is abbreviated. Construction: "lines will be drawn <as> tangents to the segments" – a virtual action, no such lines being drawn in our diagrams (the floating, "inert" letters Δ, Θ may point towards such lines: a mere visual hint, comparable to the mere textual hint). Proof: "(18) and the rest will be proved, the same as before" – a complete breakdown of exposition towards the end of this sequence. The future tense, instead of imperatives, stresses the mere virtual nature of the action required by the text (note also that the "before" refers not to the first half of this proposition, but to the second half of the preceding proposition).

/Sequel to sequence 9–12/

So, those things proved,[113] it is obvious, [first concerning the things discussed earlier],[114] that if a pyramid is inscribed inside an isosceles cone, the surface of the pyramid without the base is smaller than the conical surface [for each of the triangles containing the pyramid is

---

[113] Perhaps a reference to the entire text so far; perhaps only to Propositions 9–12.

[114] The reference is to Propositions 9–10, understood as a single unit.

smaller than the conical surface between the sides of the triangle;[115] so that the whole surface of the pyramid without the base, as well, is smaller than the surface of the cone without the base], and that, if a pyramid is circumscribed around an isosceles cone, the surface of the pyramid without the base is greater than the surface of the cone without the base [on the basis of the <proposition> following that].[116]

Second, it is obvious from the things proved already, both: that if a prism is inscribed inside a right cylinder, the surface of the prism, composed of the parallelograms, is smaller than the surface of the cylinder without the base [for each parallelogram of the prism is smaller than its respective cylindrical surface];[117] and that if a prism is circumscribed around a right cylinder, the surface of the prism, composed of the parallelograms, is greater than the surface of the cylinder without the base.[118]

TEXTUAL AND GENERAL COMMENTS

Here is an "independent" piece of text, belonging neither to Proposition 12 nor to 13. Heiberg, who believed the original had numbered propositions, was ill-equipped to deal with such pieces of text, and his textual doubts are very unconvincing (for instance, he doubts the words "first concerning the things discussed earlier," at the top of the first paragraph, because they seemed to him redundant following the very start, "those things being proved" – failing to see that the different references point to different places). Apparently he understood this passage the way he finally printed it – as an annex to Proposition 12.

I would suggest that there are no "propositions" in the text: just one Greek sentence after another. Sometimes the Greek sentences are organized in well-recognized proposition-type chunks; sometimes, they are not. Archimedes concluded one series of arguments, and he became increasingly brief as he progressed. Here abbreviation culminates in the almost complete absence of proof, which means that the text breaks away from anything resembling "proposition" structure.

The interesting feature of the text is that the most substantive claims of the entire sequence of Propositions 9–12, are left by Archimedes to this heavily truncated stage. For instance, it is indeed obvious, following Proposition 9 "that if a pyramid is inscribed inside an isosceles cone, the surface of the pyramid without the base is smaller than the conical surface." This claim is obvious – but it is also a far more interesting claim than the claim of Proposition 9 itself. Indeed, it is only now that we see the point of propositions such as 9. Such Archimedean moves are meant to take us by surprise: to make it difficult for

---

[115] *SC* I.9.

[116] I.e. while the previous argument was based on Proposition 9, this argument is based on Proposition 10. Whoever wrote this (Archimedes?) had no numbers in his manuscript.

[117] *SC* I.11.   [118] *SC* I.12.

us to realize the point of arguments when they are first offered, and then to let the significance of such arguments dawn on us, all of a sudden. This is the structure of the book in all its levels of organization, from the individual proof, through the level of a sequence of proofs, to the book as a whole.

This passage is meaningful, finally, at this level – the book taken as a whole. In its modulation of voice – moving into completely truncated writing – it makes a clear signal: we are given to understand that we have concluded yet another "chapter" of this book. The first "chapter," Propositions 2–6, dealt with the inscription and circumscription of polygons standing in certain ratios. We then moved, suddenly, into parallelograms and curved surfaces, or in fact – we now understand – into pyramids, prisms, cones and cylinders. These are the subject matter of Propositions 7–12. We now wait for chapter three to unfold – and for some illumination, concerning the relevance of all of this to the sphere.

/13/

The surface of every right cylinder without the base, is equal to a circle whose radius has a mean ratio between the side of the cylinder and the diameter of the base of the cylinder.[119]

Let there be the circle A, base of some right cylinder, and let $\Gamma\Delta$ be equal to the diameter of the circle A, and EZ <equal> to the side of the cylinder, and let H have a mean ratio between $\Delta\Gamma$, EZ, and let a circle be set out, whose radius is equal to H, <namely the circle> B; it is to be proved that the circle B is equal to the surface of the cylinder without the base.

(1) For if it is not equal, it is either greater or smaller. (a) First let it be, if possible, smaller. (2) So, there being two unequal magnitudes, the surface of the cylinder and the circle B, it is possible to inscribe an equilateral polygon inside the circle B and to circumscribe another, so that the circumscribed has to the inscribed a smaller ratio than that which the surface of the cylinder has to the circle B.[120] (b) So let a circumscribed <polygon> be imagined, (c) and an inscribed, (d) and let a rectilinear <figure> be circumscribed around the circle A, similar to the <polygon> circumscribed around B, (e) and let a prism be set up on the rectilinear <figure>; (3) so it will be circumscribed around the cylinder. (f) And also, let $K\Delta$ be equal to the perimeter of the rectilinear <figure> around the circle A, (g) and ΛZ equal to $K\Delta$, (h) and let ΓT be half $\Gamma\Delta$; (4) so the triangle $K\Delta T$ will be

Eut. 258

---

[119] A curious terminology. "X has a mean ratio between Y and Z" defines the proportion Y:X::X:Z.

[120] *SC* I.5.

equal to the rectilinear <figure> circumscribed around the circle A,[121] [(5) since it has a base equal to the perimeter <of the circumscribed figure>, (6) and a height equal to the radius of the circle A], (7) and the parallelogram EΛ <will be equal> to the surface of the prism circumscribed around the cylinder, [(8) since it is contained by the side of the cylinder and <by> the <line> equal to the perimeter of the base of the prism]. (i) So let EP be set out equal to EZ; (9) therefore the triangle ZPΛ is equal to the parallelogram EΛ,[122] (10) so that <it is>

Eut. 259

also <equal> to the surface of the prism. (11) And since the rectilinear <figures> circumscribed around the circles A, B are similar, (12) they will have the same ratio [<i.e.> the rectilinear <figures>],[123] (13) which the radii <have> in square;[124] (14) therefore the triangle KTΔ will have to the rectilinear <figure> around the circle B a ratio, which TΔ <has> to H in square[125] [(15) for TΔ, H are equal to the radii]. (16) But that ratio which TΔ has to H in square – TΔ has this ratio to PZ in length[126] [(17) for H is a mean proportional between TΔ, PZ[127] (18) through <its being a mean proportional> between ΓΔ, EZ, too; how is this? (19) For since ΔT is equal to TΓ, (20) while PE <is equal>

---

[121] That Step 4 derives from Steps 5 and 6 is apparent if we resolve the circumscribed figure into a series of triangles, each having one side of the circumscribed figure as a base, and the radius as a height (through *Elements* III.18). Their sum will be equal to a triangle whose base is the entire perimeter, and whose height is the radius.

[122] *Elements* I.41.

[123] My punctuation is explained in a textual comment; this is a textually difficult passage.

[124] A difficult moment for the translator: the first appearance of *dunamis*, a Greek noun meaning "potentiality," "power." From very early on, in Greek mathematical texts, a certain metaphorical use of this word became technical, and it came to refer to the Cheshire-cat-smile potential imaginability of *squares* upon *lines*. You see a line, and you imagine its "potential" – what it could become, if you only drew a square on it – namely, it could become a *square*. Here we have the dative of this noun (literally meaning, then, "in potential," "potentially"), which I translate by "in square." For the claim, see Eutocius (based on *Elements* XII.1).

[125] Note carefully: the "in square" qualifies the *having*. This is an adverbial construction (and not an *adjective* of any of the lines). The assertion is that the ratio-relation holds – not between the lines themselves, but between the potential squares on the lines. Anachronistically, the assertion is that: "(the triangle KTΔ):(the <figure> around the circle B)::(TΔ:H)^2."

[126] "In length" = as lines. In a context where some ratios are qualified "in square," it becomes necessary to qualify explicitly ratios that are not intended to be "in square," but are, as is usual, intended to hold between the lines *as lines* (normally, this qualification is just taken for granted). The content may be given, anachronistically, as: "(TΔ:H)^2::(TΔ:ΠZ)." The claim is argued for in the following passage.

[127] The expressions "having a mean ratio between . . ." and "being a mean proportional between . . ." are equivalent. I.e. (anachronistically): "TΔ:H::H:PZ." This indeed yields, through *Elements* VI.20, the claim of Step 16.

to EZ, (21) therefore ΓΔ is twice TΔ, (22) and PZ <is twice> PE; (23) therefore it is: as ΔΓ to ΔT, so PZ to ZE.[128] (24) Therefore the <rectangle contained> by ΓΔ, EZ is equal to the <rectangle contained> by TΔ, PZ.[129] (25) But the <rectangle contained> by ΓΔ, EZ is equal to the <square> on H;[130] (26) therefore the <rectangle contained> by TΔ, PZ, too, is equal to the <square> on H; (27) therefore it is: as TΔ to H, so H to PZ;[131] (28) therefore it is: as TΔ to PZ, the <square> on TΔ to the <square> on H; (29) for if three lines are proportional, it is: as the first to the third, the figure on the first to the figure on the second which is similar and similarly set up[132]]; (30) but that ratio which TΔ has to PZ in length, the triangle KTΔ has to the <triangle> PΛZ[133] [(31) since, indeed, KΔ, ΛZ are equal]; (32) therefore the triangle KTΔ has to the rectilinear <figure> circumscribed around the circle B the same ratio which the triangle TKΔ <has> to the triangle PZΛ.[134] (33) Therefore the triangle ZΛP is equal to the rectilinear <figure> circumscribed around the circle B;[135] (34) so that the surface of the prism circumscribed around the cylinder A, too, is equal to the rectilinear <figure> around the circle B. (35) And since the rectilinear <figure> around the circle B has to the <figure> inscribed in the circle a smaller ratio than that which the surface of the cylinder A has to the circle B, (36) the surface of the prism circumscribed around the cylinder, too, will have to the rectilinear <figure> inscribed in the circle B a smaller ratio than the surface of the cylinder to the circle B; (37) and alternately;[136] (38) which is impossible.[137] [(39) For the surface of the prism circumscribed around the cylinder has been proved to be greater than the surface of the cylinder,[138] (40) while the inscribed rectilinear <figure> in the circle B is smaller than the circle B].[139] (41) Therefore the circle B is not smaller than the surface of the cylinder.

(j) So let it be, if possible, greater. (k) So again let there be imagined a rectilinear <figure> inscribed inside the circle B and another

Eut. 259

Eut. 260

---

[128] *Elements* V.15.   [129] *Elements* VI.16.

[130] Setting-out, and *Elements* VI.17.   [131] *Elements* VI.17.

[132] A direct reference to Euclid's *Elements* VI.20 Cor. II.   [133] *Elements* VI.1.

[134] *Elements* V.11.   [135] *Elements* V.9.

[136] *Elements* V.16 (extended to inequalities). The operation "alternately" takes A:B::C:D to produce A:C::B:D. The statement can be expanded to mean "the rectilinear <figure> has to the cylinder a smaller ratio than the figure inscribed in the circle to the circle."

[137] The claim of Step 37 is: (rect. figure:cylindrical surface)<(inscribed in circle:circle). As pointed out by Step 39, rect. figure>cylindrical surface, and as pointed out by Step 40, inscribed in circle<circle. I.e. one thing is clear: (rect. figure:cylindrical surface)>(inscribed in circle:circle), and the claim of Step 37 is impossible.

[138] Sequel to *SC* I.9–12; based on *SC* I.12.   [139] Sequel to postulates.

circumscribed, so that the circumscribed has to the inscribed a smaller ratio than the circle B to the surface of the cylinder, (l) and let a polygon be inscribed inside the circle A, similar to the <figure> inscribed inside the circle B, (m) and let a prism be set up on the polygon inscribed in the circle; (n) and again let KΔ be equal to the perimeter of the rectilinear <figure> inscribed in the circle A, (o) and let ZΛ be equal to it <=KΔ>. (42) So the triangle KTΔ will be greater than the rectilinear <figure> inscribed in the circle A [(43) because it <=the triangle> has <as> a base its <=the polygon's> perimeter, (44) while <it has a> height greater than the perpendicular drawn from the center on one side of the polygon],[140] (45) and the parallelogram EΛ <will be> equal to the surface of the prism composed of the parallelograms [(46) because it <=the parallelogram> is contained by the side of the cylinder and <by> the <line> equal to the perimeter of the rectilinear <figure>, which is a base of the prism]; (47) so that the triangle PΛZ, too, is equal to the surface of the prism.[141] (48) And since the rectilinear <figures> inscribed in the circles A, B are similar, (49) they have the same ratio to each other, which their radii <have> in square.[142] (50) But the triangles KTΔ, ZPΛ also have to each other <the> ratio, which the radii of the circles <have> in square;[143] (51) therefore the rectilinear <figure> inscribed in the circle A to the rectilinear <figure> inscribed in the <circle> B, and the triangle KTΔ to the triangle ΛZP, have the same ratio. (52) But the rectilinear <figure> inscribed in the circle A is smaller than the triangle KTΔ; (53) therefore the rectilinear <figure> inscribed in the circle B, too, is smaller than the triangle ZPΛ; (54) so that <it is smaller> than the surface of the prism inscribed in the cylinder, too; (55) which is impossible [(56) for since the circumscribed rectilinear <figure> around the circle B has to the inscribed <figure> a smaller ratio than the circle B to the surface of the cylinder, (57) and

---

[140] The inscribed polygon can be resolved into a series of triangles, each having a side of the polygon as a base, and a line drawn from the center as a height. This line will be smaller than the radius (a similar claim is made in *Elements* III.15, but essentially this is just taken for granted). Other than this, the equality (ΓΔ)=(diameter of A) is at the setting-out, while the identity of T as bisection of ΓΔ is taken over from the first part of the proof (Step h).

[141] *Elements* I.41. The equality PE=EZ is taken over from the first part of the proof (Step i).

[142] *Elements* XIII.1.

[143] A recombination of results from the first part. Step 30 – KTΔ:PΛZ::TΔ:PZ. Step 16 – TΔ:PZ::(TΔ:H)². Then, through setting-out and Step h, it is known that (TΔ)=(radius of A), (H)=(radius of B), and so (KTΔ:PΛZ)::((radius of A):(radius of B))².

alternately,[144] (58) but the <figure> circumscribed around the circle B is greater than the circle B,[145] (59) therefore the <figure> inscribed in the circle B is greater than the surface of the cylinder;[146] (60) so that <it is greater> than the surface of the prism, too].[147] (61) Therefore the circle B is not greater than the surface of the cylinder. (62) But it was proved, that neither <is it> smaller; (63) therefore it is equal.

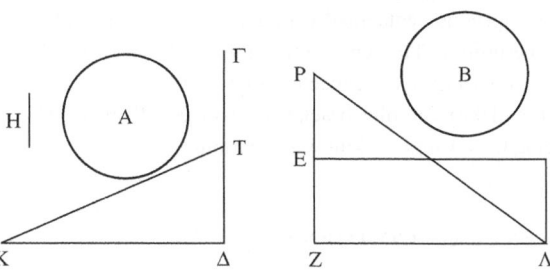

I.13
Codex D has introduced a rectangle K to the left of the line H (!).   Codices BD have the circle B greater than the circle A.   Codices GH4 have the point T higher than the point E; codices DE have them at equal height (codex B has a very different layout). Codices EH have the point P higher than the point Γ; codex G has the point Γ higher than the point P; codices D4 have them at equal height. The exact original arrangement is perhaps unascertainable.   In codex G, the line ΓT is much shorter than the line TΔ; the same, to a lesser extent, is true in codex H; while at the same codex H, the line PE is longer than the line EZ. Otherwise, the lines ΓΔ, PZ are perceptibly bisected by T, E respectively, and I suppose the same was true in codex A.

TEXTUAL COMMENTS

This is the longest proposition in the book, and it seems to have attracted considerable scholiastic attention – which has made it even more unwieldy.

Most extraordinary is the passage 19–29. It adds nothing new; instead, it *expands* an argument which was already given in Steps 16–18. Note also that Step 21 ("ΓΔ is twice TΔ") can be derived directly from Step h ("let ΓT be half ΓΔ"), and so Step 19 ("ΔT is equal to TΓ") is made otiose. Step 19 is natural for someone who did not write, himself, Step h, but just interiorized the content of Step h through the diagram. So 19–29 seem like a scholion – which would then make the direct quotation of *Elements* VI.20 Cor. 2 in Step 29 appear more natural (and would of course explain the strange question introducing Step 19).

Another special problem is the passage: "(11) And since the rectilinear <figures> circumscribed around the circles A, B are similar, (12) they will have the same ratio [<i.e.> the rectilinear <figures>], (13) which the radii <have> in square." The words "the rectilinear" in Step 12, bracketed by Heiberg, are indeed strange. They repeat Step 11 at a very small distance and (what was decisive for Heiberg), they do not occur in Eutocius' quotation of this passage. But this argument from Eutocius is weak; and even if this *is* a scholion, we must work out the purpose of repetition at such short range.

---

[144] *Elements* V.16. Expand to: "The circumscribed rectilinear <figure> around the circle B has to the circle B a smaller ratio than the inscribed <figure> to the surface of the cylinder."

[145] *SC* I.1.

[146] A central piece of the logic of proportion inequalities is the following, unproved principle, used here, that: a:b<c:d and a>b yield c>d.

[147] Through sequel to *SC* I.9–12, based on *SC* I.11.

The answer may be the following. *Elements* XII.1 states that similar polygons *inscribed* in circles are to each other as the diameters in square. *Elements* XII.2 states that circles are to each other as the diameters in square. Nowhere do the *Elements* show that similar polygons *circumscribed* around circles are to each other as the diameters in square (this is shown by Eutocius in his commentary to this passage). Thus, a keen reader of the *Elements* would feel under pressure to interpret the "they" at the start of Step 12 as referring to the *circles* – slightly incongruous syntactically, perhaps, but demanded by such a reader's geometrical erudition. The sense of the "they" must be stressed, therefore, by such a reader, to specify its unexpected reference.

The usual doubts hold for all the remaining bracketed passages (all bracketed by Heiberg merely for being backwards-looking justifications).

GENERAL COMMENTS

Equalities and inequalities

As we move on from "chapter 2," whose end was the sequence of statements following Proposition 12, we also move back from inequalities to equalities.

All the above was preliminary, in the sense that none of the results so far were at all striking: some problems were solved – whose significance we were not given (why should we bother inscribing and circumscribing polygons?); some relatively obvious inequalities were proved. But now we begin to get real results – remarkable equalities. These results are obtained in extremely indirect ways. Instead of tackling the equalities directly, we will show the impossibility of the inequalities, and the impossibility will be based on complex proportions – as well as on the inequalities proved so far. Thus, the seemingly irrelevant introductory material would become directly relevant for the real results of the treatise.

Arguments based on division of logical space

A typical feature of the complex arguments used in this treatise is the use of indirect proof. This should be contextualized, as part of the larger context of arguments based on exhaustive divisions of the logical space.

We have already come across arguments in which logical "space" was divided according to exhaustive and mutually exclusive divisions. The basic structure met so far was "A>B or not" (Proposition 9, Step 2; 10, 14; 11, 4; 12, 5). Both possibilities were real, and the two wings of the argument, for each case, were necessary to establish the universal validity of the claim, whatever the starting-point. In this proposition, three options are envisaged (Step 1): A=B, A>B, A<B. The last two are mere hypothetical options. The two wings of the argument, for each of the cases of inequality, are necessary in order to *rule out* these possibilities, so that only the first is left as possible. Thus, the *use* made of mutual exclusion and exhaustion is different from the previous propositions, while the logical principle is the same: divide logical space, survey it, and exhaust it.

Relations between parts of text

There are two types of relations between parts of text: inside propositions, and outside propositions. Since, in the original, propositions were not necessarily clearly marked, this can be seen as follows. Within propositions (as we call them), both parts refer typically to the same diagram, while between propositions (as we call them), each part must have a different diagram.

The relation of the two parts of this proposition (to Step 41 inclusive, from Step j inclusive) is based on the fact that both share the same diagram. Thus, while Steps j–o reconstruct the diagram, this reconstruction is a mere make-believe, as shown by Steps 42 and 47, where the identities of T and P, respectively, are taken over from the first part. Similarly, Step 50 is based on Steps 16 and 30, of the first part: results which were proved for the configuration of the first part are transported directly into the second, without argument. (Indeed, the manuscripts have a single diagram for both parts.) The only tangible difference between the two parts is that between "(d) and let a rectilinear <figure> be circumscribed around the circle A, similar to the <polygon> circumscribed around B," and "(l) and let a polygon be inscribed inside the circle A, similar to the <figure> inside the circle B." Significantly, both circumscribed and inscribed polygons are merely imaginary (they are not directly signaled as such, but Steps d and l follow, respectively, upon Steps b and k, which are explicitly "imaginary"). The one substantial difference between the two parts – the single reason to go into the exercise of imaginary reconstruction of the diagram – is that the basic *metrical* assumptions are different. In the first part, the circle B is assumed smaller than a given magnitude; in the second, it is assumed greater. It is clear that, in some sense, the same object cannot be both greater and smaller than a given magnitude, hence the circle B cannot be the same in both parts, and a new construction is called for. However – and here we reach a crucial result – such metrical distinctions are not even *meant* to be portrayed in diagrams, and this is the underlying reason why the reconstruction of the diagram is a mere make-believe.

So we see a close connection between the two parts, with results and assumptions directly transported from the first to the second. The relation of this proposition to Proposition 5, on the other hand, is much more indirect. Compare the text here: "(2) . . . there being two unequal magnitudes, the surface of the cylinder and the circle B, it is possible to inscribe an equilateral polygon inside the circle B and to circumscribe another, so that the circumscribed has to the inscribed a smaller ratio than that which the surface of the cylinder has to the circle B" to Proposition 5: "Given a circle and two unequal magnitudes, to circumscribe a polygon around the circle and to inscribe another, so that the circumscribed has to the inscribed a smaller ratio than the greater magnitude to the smaller." It will be seen that while the gist is the same, everything in the formulation has gone through slight mutations: most significantly, the order of the polygons is reversed, from circumscribe/inscribe (Proposition 5) to inscribe/circumscribe (Proposition 13). This happens, even though Proposition 13 does refer, obliquely, to Proposition 5, through the word "polygon" – used only in the context of the construction based on Proposition 5 (otherwise Archimedes speaks of "rectilinear <figure>"), i.e. used only in the quotation above and in Step 1. (This indeed is typical of the mechanism of reference between propositions, which is based on subtle verbal echoes, no more.)

Note in passing that Proposition 5 demands that three objects be given – two unequal magnitudes, and a circle (around which both circumscribing and inscribing take place) – the trick of Proposition 5, as it were, calls for two rabbits and a hat. Here, one of the given unequal magnitudes is the circle itself, and so the cast is reduced from three to two: one of the rabbits happens to be a hat. This sleight of hand is, I find, particularly delightful.

Some notes on operations with ratios

Ratio is perhaps the central theme in Greek mathematics. I use this opportunity to make a number of comments which apply elsewhere; all reflect the fact that ratio-talk is so pervasive in Greek mathematics, that it becomes extremely elliptical, and conveys the impression that the text involves the transfer of ratio properties from one object to another.

First, we are accustomed already to expressions such as "X has to Y the ratio which . . .", i.e. cases where a ratio seems to belong to a single object (rather than to the pair of objects which moderns may see as constituting the ratio). Now, in expressions such as the following, from the enunciation: "whose radius has a mean ratio between the side of the cylinder and the diameter of the base of the cylinder," the same principle makes an entire *proportion* belong to a single object. The object in a mean ratio has two ratios, which are the same, and this sameness is the proportion. By saying that it has the mean ratio, we assert the entire proportion.

Further, in expressions such as Step 37: "[A]nd alternately," the adverb takes over the entire sentence. This adverb means something like "and the result of the 'alternately' operation is valid too," without specifying in full what this result is. Since the result can be deduced directly from the preceding step, no information is lost.

A much more difficult verbal detail is in Step 50: "[T]he triangles KTΔ, ZPΛ . . . have to each other <the> ratio, which the radii of the circles <have> in square." Why does the English demand a definite article (before the word "ratio") which the Greek does without?

Once again, this may show something about the perception of ratio. We find it very difficult not to attach the definite article to a well-specified ratio. For us, a well-specified ratio is as definite as a ratio can get. The ratio equal to 4:3 is *the* ratio 4:3; representing it differently, say by 8:6, would not constitute, for us, a different *ratio*. It might perhaps, for us, be something different – but not a different *ratio*. Was it, perhaps, *a different ratio*, for the Greeks? Would they feel that 4:3 and 8:6 are the same ratio from some points of view, but not the same from others? This would explain the expression in 50 – and it reminds us of the primacy of the concept of ratio among the Greeks. It is not reducible to equalities and inequalities between numerical quantities – as in the modern perception – and therefore 4:3 and 8:6 may be, for the Greeks, in some sense, genuinely different.

Finally, notice the special position in this treatise of a certain kind of proportion-theory: the logic of proportional inequality. Archimedes, throughout, manipulates proportion inequalities. Such arguments are not based upon results in the *Elements* as we have them (they could, possibly, be based upon extensions of such results, in some *Elements* we do not possess but Archimedes did). Perhaps, they simply represent Archimedes' intuitive understanding of proportions.

In this treatise, at any rate, the logic of proportional inequality seems to be based on two principles:

1. Extension of manipulations on proportional equalities, generally well covered in the *Elements* (e.g. the extension of V.16, "alternately," from $(a:b::c:d) \rightarrow (a:c::b:d)$ to $(a:b>c:d) \rightarrow (a:c>b:d)$);

2. The qualitative distinction between two kinds of ratios: the ratio of the greater to the smaller (a:b when a>b) and the ratio of the smaller to the greater (a:b when a<b). This distinction leads to the simple principle, that the ratio of the greater to the smaller must be greater than the ratio of the smaller to the greater. The implications of this principle are far from trivial – they allow a translation from magnitude-inequalities to proportion-inequalities – while, on the other hand, the intuitive basis for this principle is very strong, and so, perhaps, Archimedes could simply rely on his intuition when using such a principle.

/14/

The surface of every isosceles cone without the base, is equal to a circle whose radius has a mean ratio between: the side of the cone, and the radius of the circle which is the base of the cone.

Let there be an isosceles cone, whose base <is> the circle A, and let the radius be Γ, and let Δ be equal to the side of the cone, and <let> E <be> a mean proportional between Γ, Δ, and let the circle B have the radius equal to E; I say that the circle B is equal to the surface of the cone without the base.

(1) For if it is not equal, it is either greater or smaller. (a) Let it first be smaller. (2) So there are two unequal magnitudes, the surface of the cone and the circle B, and the surface of the cone is greater; (3) therefore it is possible to inscribe inside the circle B an equilateral polygon and circumscribe another similar to the inscribed, so that the circumscribed has to the inscribed a smaller ratio than that which the surface of the cone has to the circle B.[148] (b) So let a circumscribed polygon be imagined around the circle A, too, similar to the <polygon> circumscribed

---

[148] *SC* I.5.

around the circle B, (c) and let a pyramid be set up on the polygon circumscribed around the circle A – constructed having the same vertex as the cone. (4) Now, since the polygons circumscribed around the circles A, B are similar, (5) they have the same ratio to each other, as the radii in square to each other,[149] (6) that is, <the ratio> which Γ has to E in square, (7) that is Γ to Δ in length.[150] (8) But that ratio which Γ has to Δ in length, the circumscribed polygon around the circle A has to the surface of the pyramid circumscribed around the cone[151] [(9) for Γ is equal to the perpendicular <drawn> from the center on one side of the polygon,[152] (10) while Δ <is equal> to the side of the cone;[153] (11) and the perimeter of the polygon is a common height to the halves of the surfaces];[154] (12) therefore the rectilinear <figure> around the circle A (to the rectilinear <figure> around the circle B), and the same rectilinear <figure> <=that around A> (to the surface of the pyramid circumscribed around the cone) have the same ratio; (13) so that the surface of the pyramid is equal to the rectilinear <figure> circumscribed around the circle B.[155] (14) Now, since the rectilinear <figure> circumscribed around the circle B has to the inscribed a smaller ratio than the surface of the cone to the circle B, (15) the surface of the pyramid circumscribed around the cone will have to the rectilinear figure inscribed in the circle B a smaller ratio than the surface of the cone to the circle B; (16) which is impossible[156] [(17) for the surface of the pyramid has been proved to be greater than

---

[149] Extension of *Elements* XII.1 (See Eutocius on Steps 11–13 in the preceding proposition).

[150] Setting-out and *Elements* VI.20 Cor. 2.

[151] The basic idea is *Elements* VI.1: algebraically (and thus anachronistically), that a:b::(a*c):(b*c). The "a" and "b" in question are Γ and Δ; "c" is the perimeter of the polygon, and the result is double the figures: (polygon circumscribed around the circle A), (surface of pyramid without the base), respectively, as is spelled out in Steps 9–11.

[152] Compare Steps 4–6 of the preceding proposition (and similarly, Steps 42–4 there): the polygon envisaged as a sequence of triangles, it is equal to a triangle with the perimeter of the circumscribed polygon as the base, and the radius as the height on that base; with *Elements* I.41.

[153] With *SC* I.8, the surface of the pyramid without the base is equal to a triangle whose base is the perimeter of the base, and whose height is the side of the cone circumscribed by the pyramid.

[154] The meaning is clear – if we consider areas as (base)*(height)/2, it is immaterial which line we consider as "base" and which as "height." Still, the expression is very strange; see textual comments.

[155] *Elements* V.9.

[156] The argument from Steps 17–18 and 15 to the impossibility claim of Step 16 is easier if we imagine (as in the preceding proposition) an "alternately" move, translating Step 15 into (surf. of pyramid):(surf. of cone)<(rect. ins. in circle):(circle).

the surface of the cone,[157] (18) while the rectilinear <figure> inscribed in the circle B will be smaller than the circle B].[158] (19) Therefore the circle B will not be smaller than the surface of the cone.

But I say that it will not be greater, either. (d) For if it is possible, let it be greater. (e) So again let a polygon inscribed inside the circle B be imagined and another circumscribed, so that the circumscribed has to the inscribed a smaller ratio than that which the circle B has to the surface of the cone, (f) and inside the circle A let an inscribed polygon be imagined, similar to the <polygon> inscribed inside the circle B, (g) and let a pyramid be set up on it <=the polygon inside A> having the same vertex as the cone. (20) Now, since the inscribed <polygons> in A, B are similar, (21) they will have to each other the same ratio as the radii <have> to each other in square;[159] (22) therefore the polygon

Eut. 260 to the polygon, and Γ to Δ in length, have the same ratio.[160] (23) But Γ has to Δ a greater ratio than the polygon inscribed in the circle A to the surface of the pyramid inscribed inside the cone[161] [(24) for the radius of the circle A has to the side of the cone a greater ratio than the perpendicular (drawn from the center on one side of the polygon) to the perpendicular (drawn on the side of the polygon from the vertex of the cone)]. (25) Therefore the polygon inscribed in the circle A has to the polygon inscribed in the <circle> B a greater ratio than the same polygon to the surface of the pyramid; (26) therefore the surface of the pyramid is greater than the polygon inscribed in the <circle> B.[162] (27) But the polygon circumscribed around the circle B has to the inscribed <=in B> a smaller ratio than the circle B to the surface of the cone; (28) much more, therefore, the polygon circumscribed around the circle B has to the surface of the pyramid inscribed in the cone a smaller ratio than the circle B to the surface of the cone;[163] (29) which is impossible [(30) for the circumscribed polygon <=around the circle B> is greater than the circle B,[164] (31) while the surface of the pyramid in the cone is smaller than the surface of the cone].[165] (32) Therefore neither is the circle greater than the surface of the cone. (33) But it was proved, that neither <is it> smaller; (34) therefore <it is> equal.

---

[157] Sequel to *SC* I.9–12, based on *SC* I.10.

[158] Sequel to Postulates.   [159] *Elements* XII.1.

[160] The argument of Steps 6–7 is telescoped now into a single assertion (based on the setting-out and on *Elements* VI.20 Cor.).

[161] See Eutocius.   [162] *Elements* V.8.

[163] Again imagine an "alternately" move: (circ. around B):(B)<(surf. of pyramid ins. in cone):(surf. of cone).

[164] *SC* I.1.

[165] Sequel to *SC* I.9–12, based on *SC* I.9.

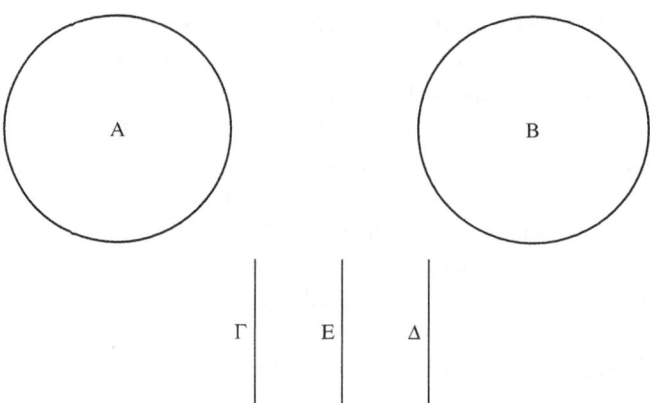

I.14
Codex B has B>A,
Δ>E>Γ.   Codex G
has the lines aligned
much nearer the circle
B.

TEXTUAL COMMENTS

There are four backwards-looking justifications in this proposition, all bracketed by Heiberg. The first (Steps 9–11) is the only one which behaves
"strangely:" "(9) for Γ is equal to the perpendicular <drawn> from the center
on one side of the polygon, (10) while Δ <is equal> to the side of the cone;
(11) and the perimeter of the polygon is a common height to the halves of
the surfaces." The difficulties are the following. First and foremost, there is a
geometrical solecism: the perimeter of the base is described in Step 11 as a
common height. In fact it functions as a common base. Further, there is what
seems to be a lapse: "[T]he perimeter of the polygon is a common height to
the halves of the surfaces." Why "halves?" If anything, it is natural to think of
a rectangle in which the hypothetical triangles are enclosed – which is *double*
the surfaces, not their *half*!

But these two arguments can be countered. One speaks of "common height,"
not of "common base," since this is the formula taken over from *Elements*
VI.1. As for double/half, this is a natural mistake for anyone to make, skilled
or unskilled. In fact I think it can be argued that this passage may be authorial.
The argument is the following: we have had the expression "the perpendicular
drawn on *one* side . . ." once already, in the enunciation of Proposition 7 – but
once only. This use of "one" is a distinctive mark of the style. It has not been
common enough to imprint itself, by this stage, on the mind of a scholiast who
annotates as he goes along. So Step 9 is possibly Archimedean and, if so, Steps
10–11 go with it. For the rest, the usual doubts remain.

Note that, immediately following Step 19, I have "But I say," taking the
manuscripts' δέ instead of Heiberg's (highly probable) conjectural δή.

GENERAL COMMENTS

Different layers of imagination

In Step b we are asked to *imagine* a polygon around A, similar to the one
around B. But what did we do to *that* original polygon, the one around B? Was
it drawn, or was it, too, merely imagined? It appears, that neither of the two.

B's polygon's only claim to existence seems to be the notice made in Step 3, of its *possibility*. One is tempted to say that to assert an object's possibility is tantamount to imagining its existence. (Needless to say, the polygons are absent from the drawn diagram: thus both polygons are in a sense "imaginary.")

There is a general make-believe about the geometrical underpinning of this proposition. For instance, Step 18: "[T]he rectilinear <figure> inscribed in the circle B will be smaller than the circle B." Why the future tense? Probably, as a signal of the hypothetical nature of the object. (Assuming that the polygon is drawn, then it *is* smaller, in the present tense; but in fact it is not drawn, it is a merely conceptual polygon.) And what I find to be a nice detail: in Step f, "inside the circle A let an inscribed polygon be imagined," the preposition "inside" qualifies the act of imagination, not any act of drawing. Our mind, imagining, should place its imagined object inside the circle A.

The diagram is in fact reduced to a minimum: three lines and two circles. All the objects are directly represented by single letters – i.e. there is nowhere any geometrical configuration of intersecting lines, lettered on the points of intersections. This is the logical culmination of the erosion of the third dimension. Since there is no attempt to represent the geometrical objects in their full, three-dimensional solidity, one might as well retain just the basic elements of the structure, juggling the objects through the machinery of inequalities established so far – almost a purely logical exercise.

/ 15 /

The surface of every isosceles cone has to the base the same ratio which the side of the cone <has> to the radius of the base of the cone.

Let there be an isosceles cone, whose base <is> the circle A, and let B be equal to the radius of the <circle> A, and Γ – to the side of the cone. It is to be proved that <they> have the same ratio: the surface of the cone to the circle A, and Γ to B.

(a) For let a mean proportional be taken between B, Γ, <namely> E, (b) and let a circle be set out, Δ, having the radius equal to E; (1) therefore the circle Δ is equal to the surface of the cone [(2) for this was proved in the previous <proposition>]. (3) But the circle Δ was proved to have to the circle A the same ratio as <that> of Γ to B in length[166] [(4) for each is the same as E to B in square, (5) through the circles being to each other as the squares on the diameters to each other,[167] (6) but similarly, also the <squares> on the radii of the circles; (7) for if the diameters, also the halves (8) that is the radii; (9) and B, E are

---

[166] Steps 4–7 in the preceding proposition, modified by *Elements* XII.2.
[167] *Elements* XII.2.

equal to the radii].[168] (10) So it is clear that the surface of the cone has to the circle A the same ratio which Γ <has> to B in length.

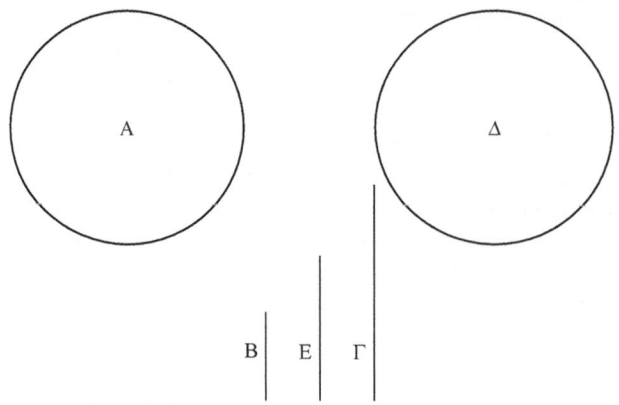

I.15
Codex B has Δ>A. Codex G aligns the lines underneath the circle Δ, and has the bottom of line B a little higher than that of lines E, Γ.   Codex H has Z instead of B.

TEXTUAL COMMENTS

Of the two bracketed passages, one (Step 2) is an explicit reference – the staple of scholia – while the other (Steps 4–9) is at a very basic mathematical level, expanding what is first introduced simply by "it was proved." Both passages are probably indeed scholia, and we should appreciate how brief the remainder is: three steps and a single argument. This proposition, then, is a corollary of 14, no more (note that both starting-points to the argument are directly taken from 14). The diagram is essentially the same, the only difference being that of lettering.

Note, incidentally, that whoever was responsible for Step 2 probably did not have numbered propositions in his text.

GENERAL COMMENTS

Reference to Proposition 14

There are two references here to Proposition 14: one, Step 2, is very explicit and is probably by a scholiast. The other, much less explicit, is closely woven into the essential material, and is probably authorial. This is Step 3, "the circle Δ was proved to have to the circle A the same <ratio> as <that> of Γ to B in length."

Nothing like this was proved. The reference is to Steps 4–7 in the preceding proposition: "[T]he polygons circumscribed around the circles A, B . . . have the same ratio to each other, as . . . Γ to Δ in length." This original statement

---

[168] A remaining part of the equation – Γ to B in length is the same as E to B in square – is apparently completely obvious to the interpolator (we need, for that, *Elements* VI.20 Cor.). Steps 4–9 are so simple, they might become confusing: all they do is to extend *Elements* XII.2 from diameters to radii.

was not about circles, but about polygons (it was based on *Elements* XII.1, not on *Elements* XII.2). The letters used were of course different – not a trivial consideration, given that this conclusion was never asserted in general terms, only in such particular lettered terms. In other words, this result was not the enunciated conclusion of Proposition 14, but an interim result, stated merely in the particular terms of the particular diagram of 14. What is the force of "it was proved," then? Perhaps this means no more than, say, that "we are familiar with this territory."

In general, references which are not to the conclusion of an earlier proposition, but to some *interim* results, are rare. Interim results are not argued in the same way as main conclusions are. Main conclusions are repeated, in enunciation, definition of goal, and proof. They are the focus of attention: they are independent of particular lettering, and so they come in identifiable units. Interim results, on the other hand, are passed through, no more. So this is a rare kind of reference – and it is probably related to the special nature of this proposition, as suggested in the textual notes above: no more than an appendix to the preceding proposition, so that the membrane separating one proposition from the next is especially thin.

/ 16 /

If an isosceles cone is cut by a plane parallel to the base, then a circle – whose radius has a mean ratio between: the side of the cone between the parallel planes, and the <line> equal to both radii of the circles in the parallel planes – is equal to the surface of the cone between the parallel planes.

Let there be a cone, whose triangle through the axis[169] is equal to $AB\Gamma$, and let it be cut by a plane parallel to the base, and let it <=the plane> make, <as> a section, the <line> ΔE, and let the cone's axis be BH, and let some circle be set out, whose radius is a mean proportional between: $A\Delta$, and ΔZ, HA taken together, and let it be the circle Θ; I say that the circle Θ is equal to the surface of the cone between ΔE, $A\Gamma$.

(a) For let there be set out circles, <namely> Λ, K, (b) and let the radius of the circle K be, in square, the <rectangle contained> by $B\Delta Z$, (c) and let the radius of Λ be in square the <rectangle contained> by BAH; (1) therefore the circle Λ is equal to the surface of the cone $AB\Gamma$,[170] (2) while the circle K is equal to the surface of the <cone>

---

[169] "The triangle through the axis" is a formulaic expression, referring to the triangle resulting from cutting the cone by a plane passing through the axis.

[170] *SC* I.14.

Eut. 262   ΔEB.[171] (3) And since the <rectangle contained> by BA, AH is equal
to both: the <rectangle contained> by BΔ, ΔZ, and the <rectangle
contained> by AΔ and <by> ΔZ, AH taken together (4) through
ΔZ's being parallel to AH,[172] (5) but the <rectangle contained> by
AB, AH is the radius of the circle Λ in square, (6) while the <rectangle
contained> by BΔ, ΔZ is the radius of the circle K in square, (7) and
the <rectangle contained> by ΔA and <by> ΔZ, AH taken together
is the radius of Θ in square, (8) therefore the <square> on the radius
of the circle Λ is equal to the <squares> on the radii of the circles K,
Θ; (9) so that the circle Λ, too, is equal to the circles K, Θ. (10) But Λ
is equal to the surface of the cone BAΓ, (11) while K <is equal> to
the surface of the cone ΔBE; (12) therefore the remaining surface of
the cone between the parallel planes ΔE, AΓ is equal to the circle Θ.

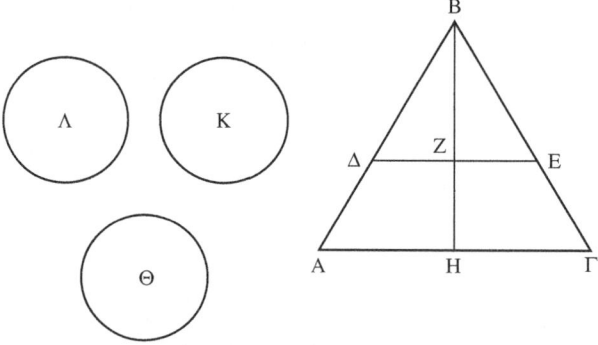

I.16
Codex B has Λ>Θ>K.
Codex G has a different
layout: the triangle
somewhat tilted, and
the circles arranged on
top of the triangle, as in
the thumbnail.

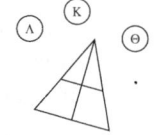

[Let the parallelogram be BAH, and let its diameter be BH.[173] Let
the side BA be cut, at random, at Δ, and let a parallel to AH be drawn
through Δ, <namely> ΔΘ, and <another parallel,> to BA, through Z,
<namely> KΛ; I say that the <rectangle contained> by BAH is equal
to: the <rectangle contained> by BΔZ, and the <rectangle contained>
by ΔA and <by> ΔZ, AH taken together.

(1) For since the <rectangle contained> by BAH is, <as a> whole,
the <area> BH,[174] (2) while the <rectangle contained> by BΔZ
is the <area> BZ,[175] (3) and the <rectangle contained> by ΔA

---

[171] *SC* I.14.   [172] See Eutocius.

[173] We have moved to a lemma attempting to prove the claim of Steps 3–4.

[174] Something went wrong. *Elements* I.47 (Pythagoras' theorem) is applied: i.e. the
author assumes that the angle BAH is right. This is although BH has been constructed
(correctly) not as a rectangle, but as a parallelogram – so that the angle is right only as
a special case. The most charitable way to read the following is as a lemma valid as far
as rectangles are concerned, misunderstood by its author to apply to parallelograms in
general.

[175] *Elements* I.47.

and <by> ΔZ, AH taken together is the gnomon MNΞ;[176] (4) for the <rectangle contained> by ΔAH is equal to KH (5) through the complement KΘ being equal to the complement ΔΛ,[177] (6) while the <rectangle contained> by ΔA, ΔZ <is equal> to ΔΛ; (7) therefore the whole BH (which is the <rectangle contained> by BAH) (8) is equal to: the <rectangle contained> by BΔZ, and the gnomon MNΞ, (9) which is <=the gnomon> equal to the <rectangle contained> by ΔA and <by> AH, ΔZ taken together.]

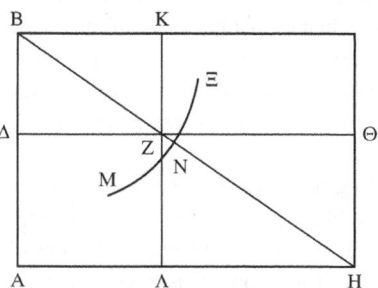

I.16 Second diagram Codex B has omitted text and diagram (see general and textual comments). Codex H has the lines ΔΘ, KΛ bisect the rectangle.

TEXTUAL COMMENTS

The lemma seems to belong to the class of late expansions of brief Archimedean arguments. Since it is mathematically wrong, there is a very strong probability indeed that it is not by Archimedes. What is most noticeable is that the author of the lemma does not use proportion-theory (the natural tool to use, as Archimedes' brief Step 4 is "through ΔZ's being parallel to AH:" parallels are usually important for the proportions they entail). It seems to come from the hand of someone who is well acquainted with Book II of the *Elements* – whose jargon is very prominent here – in other words acquainted with some of the most basic tools available to a reader of Euclid.

Eutocius has a different, correct proof. He probably did not have this lemma in front of him. Was the lemma added later? Did it reach the text from a tradition independent of Eutocius'? Both options are possible, and we cannot decide between them.

At the start of the lemma, I follow the Palimpsest (ἔστω τὸ παραλληλό-γραμμον τὸ BAH, "let the parallelogram be BAH") against A and Heiberg (ἔστω παραλληλόγραμμον το BAH, "let there be the parallelogram BAH"):

---

[176] "Gnomon" is the figure composed of any three parallelograms in this type of diagram. In this case it is signaled by the semi-circle passing through the three parallelograms that constitute it.

[177] *Elements* I.43. "Complement" is a very Euclidean word, for the "wings" of the diagonal – the rectangles (two out of four) through which the diagonal does *not* pass. The idea of the argument is this: the rectangle contained by ΔA and by ΔZ, AH taken together is equal to two rectangles, ZΔA and ΔAH (*Elements* II.1). Following Step 4, ΔAH=KΛH. Add ZΔA to KΛH, and you have the gnomon MNΞ, equal to the rectangle contained by ΔA and by ΔZ, AH taken together.

this locution is, perhaps, a way of anchoring the lemma in the antecedent proposition.

GENERAL COMMENTS

The phenomenon of arguments that are left implicit

Step 4, "through ΔZ's being parallel to AH," is meant to support Step 3. (Apparently, this is the only such support given by Archimedes himself.) The very presence of a supporting step indicates that the claim of Step 3, in itself, was not a result on a par with the propositions of the *Elements*. That is, for something like the contents of the *Elements*, it seems that Archimedes could assume their previous knowledge; they required no arguments. The claim of Step 3, however, called for an argument.

On the other hand, it is interesting that Step 4, alone, was supposed to be sufficient: for, after all, it is nothing like a proper proof. To some extent, this is a *hint*, suggesting where we should look for such a proof. "Parallels," hints Archimedes; the alert reader looks for proportions. This is what Eutocius does, successfully. The interpolator, it seemed, would require a less subtle hint.

More important than the question of which argument "Archimedes had in mind" is, I suggest, the general question of the meaning of such cryptic references. What is a phrase such as "through ΔZ's being parallel to AH" actually meant to say to its audience?

Most of all, it seems to say the following: that Archimedes himself has got a proof (which he will not divulge, probably because he does not wish to get distracted from the main line of argument). But what are we to do about this? Look for the proof for ourselves? (So did Eutocius, and the interpolator of the lemma: so does Archimedes look for this type of "interactive" reading?) Or are we just supposed to take Archimedes' word on faith? (So apparently did most other ancient readers.) Or perhaps we should suspend judgment? But then what happens to the entire deductive fabric of the treatise? What makes us *persuaded* at this stage? Partly, we are persuaded by Archimedes' authority, partly, by a reliance upon the tacit rules of the game.

The main principle seems to be those tacit rules. Greek mathematical texts were understood to have implicit arguments only where proper, fuller arguments were available to their practitioners. Such an understanding made it possible for the authors to skip details of arguments – whether or not they had the same authority as Archimedes came to have – in this way keeping the flow of the argument. Since the flow of the argument is a major component of what makes possible the phenomenon of immediate persuasion, we reach a surprising conclusion. Trust – the suspension of the desire to be actively persuaded by proof – contributes to one's ability to be persuaded by proof.

Non-pictorial diagrams

This is the first of several cases (see also Propositions 24, 32, 35 below) where circles set out by Archimedes were reproduced by Heiberg as concentric. The concentric arrangement implies a relation of size (besides, of course, saving space: which may have been Heiberg's main consideration). The manuscripts, probably following ancient authority, have no such concentric arrangement. In particular, in the diagram of this proposition, the three circles are set out as equal. I suspect this dates back to antiquity, as well. The objects of the diagram were therefore made to be, at least by some ancient authority (possibly by Archimedes himself), of a non-metrical nature. Equally sized circles in the diagram represented differently sized circles in the geometric reality depicted.

Note also a further simplification introduced by the diagram, having to do with the flattening of three-dimensional space. This may be followed through the text: "Let there be a cone, whose triangle through the axis is equal to ΑΒΓ, and let it be cut by a plane parallel to the base, and let it <=the plane> make, <as> a section, the <line> ΔΕ, and let the cone's axis be ΒΗ." The triangle ΑΒΓ is not the triangle through the axis itself, but is merely *equal* to it. The ΔΕ line, on the other hand, is already *identified* with the section of the plane with that triangle, and the same identification is held up as regards the axis. Essentially this is the usual problem of reducing the three-dimensional to the two-dimensional. We do not have the cone in our diagram, just a triangle. So is it a cross-section of the cone itself? Or is it, instead, just a copy made of a section of that cone – a triangle through which we refer to another cone, not at all represented in this diagram? The combination of text and diagram does not resolve this question, because it does not arise from the intended use of the two. The diagram is not really designed to represent any specific spatial reality. It is a logical tool, a component of the proof, but not a picture of a cone.

A related phenomenon explains the interpolator's mistake. He was misled by a combination of the following two features of the diagrams' practice:

1   The metrical properties of the diagram might be changed at will.
2   One is allowed to take the simplest representation of a geometrical situation (for instance, when no angle is specified, one is allowed to have a right angle in the diagram).

The two features of the practice form a coherent unity, when one goes on to *use* the diagram as a purely logical, and not a metrical representation. So the interpolator was "right" in manipulating the triangle ΒΑΗ, pushing Β to the left, lengthening the base ΑΗ, completing the parallelogram, and then drawing the parallelogram as if it were a rectangle. This is all within the methods of representation he would have known from the *Elements*. His one mistake was that he then trusted his own diagram: as if right angles in the diagram meant right angles in reality.

/Interlude recalling elementary results/

The cones having an equal height have the same ratio as the bases;[178] and those having equal bases have the same ratio as the heights.[179]

---

[178]   Related to (not identical with) *Elements* XII.11.
[179]   Related to *Elements* XII.14.

If a cylinder is cut by a plane parallel to the base, it is: as the cylinder to the cylinder, the axis to the axis.[180]

In the same ratio as cylinders, are the cones having the same bases as the cylinders.[181] And the bases of equal cones are in reciprocal ratio to the heights; and <the cones> whose bases are in reciprocal ratio to the heights, are equal.[182] And the cones, whose diameters of the bases have the same ratio as the axes [i.e. as the heights], are to each other in triplicate ratio of <the ratio of> the diameters in the bases.[183]

And these were all proved by past geometers.

TEXTUAL COMMENTS

The last sentence "And these were all proved by past geometers," indicates an Archimedean provenance. Any other writer would have said something like that "these are proved in the *Elements*." This sequence of text is therefore, in all probability, authentic, and is, in a sense, similar to the sequel to 9–12: a list of results in no need of elaborate proof.

Some previous modern editors gave a title to this list ("lemmas"), and numbered the individual propositions. This has no textual authority.

The words "[i.e. <as> the heights]" are indeed probably interpolated, but there can be no certainty about that. A predominant part of the argument for their being interpolated is their absence from the *Elements* but, in general, we cannot assume that Archimedes worked from a text of the *Elements* similar to ours nor, if he did, can we assume that he tried to follow it *verbatim*.

GENERAL COMMENTS

Archimedes could have assumed an acquaintance with such results as quoted here. We just saw him, in Proposition 16, merely hinting at a result more complicated than any of these. So why suddenly give such a list, in explicit detail?

Partly, perhaps, to signal another moment of transition in the treatise. A break in the text, such as this, implies that what comes before the break is qualitatively different from what comes after the break. Such passages have a stylistic significance, and help to structure the book as a whole.

It is also likely that Archimedes is deliberately putting forth results "by past geometers," to stress how far he is ahead of the tradition. The haste with which the propositions are quoted (so hastily that an error is committed – see

---

[180] Related to *Elements* XII.13.

[181] Follows from *Elements* XII.10 – provided one adds what Archimedes forgot to mention that the cylinders and cones in question must have equal (or at least proportionate) heights.

[182] Related to *Elements* XII.15.

[183] Related to *Elements* XII.12 – but the term "similar," used there, is avoided here. "Triplicate ratio" can be understood as (although of course this is anachronistic) the cube of a fraction.

n. 181) – could even suggest, perhaps, an air of superiority: "This is what past geometers have proved. You just saw what *I* proved – and there is more to come immediately!" In effect, "chapter 3," on conical surfaces, has now ended; we move on: from the surfaces, to the solid cones themselves: a step closer to our subject (almost lost sight of!) – the volumes of curvilinear solids.

/ 17 /

If there are two isosceles cones, and the surface of one of the cones is equal to the base of the other, and the perpendicular, drawn from the center of the base on the side of the cone, is equal to the height, the cones will be equal.

Let there be two isosceles cones ABΓ, ΔEZ, and let the base of the <cone> ABΓ be equal to the surface of the <cone> ΔEZ, and let the height AH be equal to the perpendicular, KΘ, drawn from the center of the base, Θ, on one side of the cone (such as ΔE); I say that the cones are equal.

(1) For since the base of the <cone> ABΓ is equal to the surface of the <cone> ΔEZ [(2) but equals have the same ratio to the same thing],[184] (3) therefore as the base of the <cone> BAΓ to the base of the <cone> ΔEZ, so the surface of the <cone> ΔEZ to the base of the <cone> ΔEZ. (4) But as the surface to its own base, so EΘ to ΘK [(5) for this was proved: that the surface of every isosceles cone has to the base the same ratio, which the side of the cone <has> to the radius of the base,[185] (6) that is ΔE to ΔΘ. (7) But as EΔ to ΘE, so ΔΘ to ΘK; (8) for the triangles are equiangular].[186] (9) But ΘK is equal to AH; (10) therefore as the base of the <cone> BAΓ to the base of the <cone> ΔEZ, so the height of the <cone> ΔEZ to the height of the <cone> ABΓ. (11) Therefore the bases of the <cones> ABΓ, ΔEZ are in reciprocal ratio to the heights; (12) therefore the <cone> BAΓ is equal to the cone ΔEZ.[187]

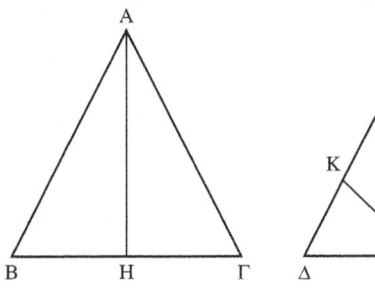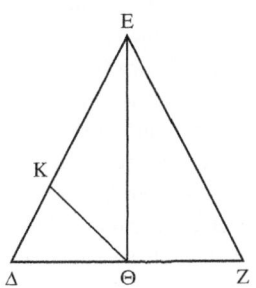

I.17
The angle ΔKΘ is acute in codices EH4, right (or nearly so), in codices BG, and obtuse in codex D. I suspect it was acute in codex A. Codex B has line AH smaller than line EΘ, line BΓ greater than line ΔZ. Codex G has line EΘ slightly smaller than line AH. Codex A, followed by all extant codices and modern editions, has permuted Δ, E. See textual comments.

---

[184] *Elements* V.7. [185] *SC* I.15.
[186] *Elements* VI.8 (and then *Elements* VI.4).
[187] *Elements* XII.15 (recalled in Interlude).

TEXTUAL COMMENTS

I translate the text of the original manuscripts, which is different from Heiberg's in Steps 4, 6, 7, always for the same reason: the text as it stands takes E as the vertex of the second cone, and Δ as its base, *against* the diagram as it stands. Bear in mind however that only steps 4, 6 and 7 make any reference, in the text, to the precise identity of Δ/E. (The explicit construction – typically for such constructions – leaves their identity underspecified: they are on a certain triangle, but we are not told which position each letter is taken to have on that triangle.) So this is a case of the diagram against the text. It is much more economic to assume that the *diagram* at some stage has been altered out of synchronization with the text, than the other way around. A change in the diagram is a single event; a change in the text involves four separate events. I therefore keep the text and correct the diagram.

Notice that three out of the four textually difficult events occur in what Heiberg sees as an interpolation. Heiberg may well be right. On the other hand, nothing in Steps 5–8 *must* be interpolated: 8 is the prime suspect (making a very trivial claim), 6–7 are pretty suspect, but 5 is something we can imagine Archimedes bringing in, to stress the applicability of his earlier results.

GENERAL COMMENTS

Underspecification of diagram by text

I have already said something on this subject in the textual notes, and so this is a good opportunity to look at the diagrammatic practices of the text. The proposition has no construction (no "Latin lettered" steps), so the diagram is set up by the text in the setting-out only. This is: "Let there be two isosceles cones ABΓ, ΔEZ . . ." (So we now know how to divide the letters ABΓ ΔEZ between the two cones. But the text gives us no further information as to the *internal* distribution of letters inside the cones – as mentioned already in the textual notes.) "[A]nd let the base of ABΓ be equal to the surface of ΔEZ" (fine: but what *is* the base of ABΓ? which two of the three points?) "and let the height AH . . ." (now we learn – by accident as it were – that A is the vertex of ABΓ, and a new letter H is introduced. Notice that this is not meant to *define* the objects: on the contrary, the height is introduced with a definite article, i.e. it is assumed to have been part of the already established discourse.) ". . . be equal to the perpendicular, KΘ, drawn from the center of the base Θ on one side of the cone (such as ΔE)." (It is only the afterthought, "such as ΔE," which gives a textual basis for judging that the perpendicular is in the cone ΔEZ [and not in ABΓ]. Notice, incidentally, that the text of the enunciation itself is just as underspecified: it is nowhere said explicitly that the height and the perpendicular should each come from a different cone. Further, we do not know which of ΔE is which, though we can say now that *one* of them must be the vertex.)

So we see in general, that there is not the slightest attempt to determine the diagram by the text, and that the author is happy to let the diagram supply the references of the letters. In this case, this has resulted in a textual corruption that we cannot easily solve – so that we learn, so to speak the hard way, that the text did not set out to give us any clues for the original form of the diagram.

/ 18 /

Every rhombus[188] composed of isosceles cones is equal to a cone having a base equal to the surface of one of the cones containing the rhombus,[189] and a height equal to the perpendicular drawn from the vertex of the other cone to a side of the other cone.[190]

Let there be a rhombus composed of isosceles cones, ABΓΔ, whose base <is> the circle around the diameter BΓ, and <its> height <is> AΔ, and let some other <cone>, HΘK, be set out, having the base equal to the surface of the cone ABΓ, and a height equal to the perpendicular drawn from the point Δ on AB or on the <line produced> in a straight line to it <=to AB>, and let it <=the perpendicular> be ΔZ, and let the height of the cone ΘHK be ΘΛ; (1) so the <height> ΘΛ is equal to ΔZ; I say that the cone is equal to the rhombus.

(a) For let there be set out another cone, MNΞ, having the base equal to the base of the cone ABΓ, the height equal to AΔ, and let its height be NO. (2) Now since NO is equal to AΔ, (3) therefore it is: as NO to ΔE, so AΔ to ΔE.[191] (4) But as AΔ to ΔE, so the rhombus ABΓΔ to the cone BΓΔ,[192] (5) while as NO to ΔE, so the cone MNΞ to the cone BΓΔ[193] [(6) through their bases being equal]; (7) therefore as the cone MNΞ to the cone BΓΔ, so the rhombus ABΓΔ to the cone BΓΔ; (8) therefore the <cone> MNΞ is equal to the rhombus ABΓΔ.[194] (9) And since the surface of the <cone> ABΓ is equal to the base of the <cone> HΘK, (10) therefore as the surface of the <cone> ABΓ to its own base, so the base of the <cone> HΘK to the base of the <cone> MNΞ, [(11) for the base of the <cone> ABΓ is equal to the base of the <cone> MNΞ]. (12) But as the surface of the <cone> ABΓ to its own base, so AB to BE,[195] (13) that is AΔ to ΔZ[196] [(14) for the triangles

[188] *SC* I Def. 6. The adjective "solid," interestingly, is dropped here.

[189] The rhombus is understood to be *contained* (not *constituted*) by its two cones.

[190] The structure of "other" cones is confusing but essentially simple. With cones 1 and 2, we take the surface of cone 1, and then the perpendicular drawn from the vertex of cone 2 on the side of cone 1.

[191] *Elements* V.7.   [192] Interlude, recalling *Elements* XII.14; also *Elements* V.18.

[193] Interlude, recalling *Elements* XII.14.   [194] *Elements* V.9.

[195] *SC* I.15.   [196] *Elements* VI.4.

are similar];[197] (15) therefore as the base of the <cone> HΘK to the base of the <cone> MNΞ, so AΔ to ΔZ.[198] (16) But AΔ is equal to NO [(17) for <so> it was laid down], (18) while ΔZ <is equal> to ΘΛ; (19) therefore as the base of the <cone> HΘK to the base of the <cone> MNΞ, so the height NO to the <height> ΘΛ. (20) Therefore the bases of the cones HΘK, MNΞ are in reciprocal ratio to the heights; (21) therefore the cones are equal.[199] (22) But MNΞ was proved equal to the rhombus ABΓΔ; (23) therefore the cone HΘK, too, is equal to the rhombus ABΓΔ.

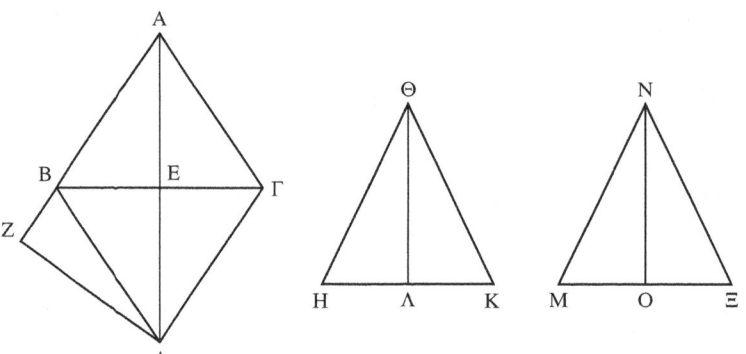

TEXTUAL COMMENTS

The proposition has four brief and pertinent backwards-looking justifications (Steps 6, 11, 14, 17). Heiberg brackets all, and of course it is difficult to form a judgment. What must be said is that the nature of three of these justifications is special: Steps 6, 11 and 17 all refer to the construction and do no more than recall it. Partly this is due to the nature of the proposition, based on unpacking a construction. That construction stipulated many equalities, so that those equalities yield, very directly, the result. So it is natural that we will have many justifications of the form "for this was the construction."

Heiberg's objection to Step 6 – that it should have appeared earlier, to justify Step 4 – is unconvincing. Step 6 can naturally be construed to cover both Steps

I.18
Codex B has the line NO greater than the line ΘΛ, the line HK greater than the line MΞ.   Codex G has the angle AZΔ obtuse. Codex H has the line BΓ aligned with the bases of the triangles. Codices EH4 have Π instead of H. Codex A had the same mistake, or else it had the upper part of the H "clipped" so that it accidentally resembled a Π.

---

[197] That the triangles AEB, AΔZ are similar can be seen like this: the angles E, Z are both right (setting-out), and the angle A is common to both triangles; then apply *Elements* I.32.

[198] The argument is: (10) The surface of ABΓ:its own base::the base of HΘK:the base of MNΞ. (12–13) The surface of ABΓ:its own base::AΔ:ΔZ. Conclusion. (15) The base of HΘK:the base of MNΞ::AΔ:ΔZ. (We move through an implicit (15*) AΔ:ΔZ::the base of HΘK:the base of MNΞ, and then an implicit move a:b::c:d→c:d::a:b, probably considered a mere notational equivalence.) See general comments.

[199] Interlude, recalling *Elements* XII.15.

4 and 5 (and probably this is why Step 6 speaks about "their" bases, leaving the identity of "their" open, while Step 11 refers to specific cones).

In the setting-out, the words "so the <height> ΘΛ is equal to ΔZ" are curious. They occur between the setting-out and the definition of goal, but they are formally an assertion (and thus belong to the proof). Such episodes help to show that the proposition is not tightly compartmentalized between such categories as "construction" and "proof." But perhaps not too much should be based on such episodes: it must be said that these words could come from a scholiast, finding the setting-out difficult to follow.

GENERAL COMMENTS

A diagram that is less general than the text

The text has: "Let there be . . . a height equal to the perpendicular drawn from the point Δ on AB or on the <line produced> in a straight line to it <=to AB>." In this expression, Archimedes refers to an inherent degree of freedom in the situation: whether the angle ABΔ is obtuse (and then the perpendicular from Δ to AB falls outside the cone) or not (and then the perpendicular falls inside the cone).

That ABΔ is obtuse is, however, assumed by the diagram as it stands. Did Archimedes produce another diagram, as well, representing the "or" in the setting-out? More probably, he supplied only one diagram, the one we have. But what happens to generality, then?

The answer is that the problem of generality does not really emerge in this case. In this sequence of demonstrations, the relations studied are essentially proportions, not geometrical configurations (hence the schematic nature of the diagrams). Where ΔZ falls is immaterial to the proof itself. All ΔZ does it provide a tag for proportion-manipulations. In this case, the diagram can support generality, because the relevant features are specified not by the visual representation, but by the verbal specifications of proportion. Let us therefore look at those proportions, and their transformations.

Overall structure of transformations of proportions

The proposition has three main arguments: 3–7, 10–15, 15–19. All start from a proportion, and transform it.

Steps 3–7 start from (NO:ΔE::AΔ:ΔE) to yield (MNΞ:BΓΔ::ABΓΔ: BΓΔ).

Steps 10–15 start from (surface of ABΓ:base of ABΓ::base of HΘK:base of MNΞ) to yield (base of HΘK:base of MNΞ::AΔ:ΔZ).

Steps 15–19 start from the conclusion of Steps 10–15 (base of HΘK:base of MNΞ::AΔ:ΔZ), to yield (base of HΘK:base of MNΞ::NO:ΘΛ).

In all three arguments, the transformation is achieved by the addition of two extra premises, each allowing some substitution in the original proportion (4, 5 in the argument 3–7; 12, 13 in the argument 10–15; 16, 18 in the argument 15–19). In Steps 3–7, both ratios are substituted. In 10–15 and 15–19, only one ratio is substituted, but it is substituted twice. Further, each argument has one, brief backwards-looking justification (6 in the argument 3–7; 14 in the argument 10–15;[200] 17 in the argument 15–19). (There seems to be an attempt to squeeze in as much information as possible into a single line of argument.)

A noticeable feature is the freedom as regards order of elements in the proportion. Step 4 introduces a substitution in the second ratio of Step 3, Step 5 then introduces a substitution in the first ratio; the sequence of ratios is inverted without a comment, between Steps 10 and 15; argument 10–15 transforms the first ratio of Step 10, while argument 15–19 transforms the second ratio of Step 15. Clearly the sequence of ratios in a proportion (not to be confused with the sequence of objects in a ratio) is considered immaterial.

/ 19 /

If an isosceles cone is cut by a plane parallel to the base, and from the resulting circle[201] a cone is set up having <as> a vertex the center of the base, and the created rhombus is taken away from the whole cone, the remainder is equal to a cone having a base equal to the surface of the cone between the parallel planes, and a height equal to the perpendicular drawn from the center of the base on one side of the cone.

Let there be an isosceles cone $AB\Gamma$, and let it be cut by a plane parallel to the base, and let it make <as> a section the <line> ΔE, and let Z be the center of the base, and on the circle around the diameter ΔE let a cone be set up having Z <as> a vertex; (1) so there will be a rhombus $B\Delta ZE$ composed of isosceles cones. Let some cone be set out, $K\Theta\Lambda$, <and> let its base be equal to the surface between ΔE, $A\Gamma$, and (after the perpendicular ZH is drawn from the point Z on AB) let <its> height be equal to ZH; I say that if the rhombus $B\Delta ZE$ is imagined taken away from the cone $AB\Gamma$, the cone $\Theta K\Lambda$ will be equal to the remainder.

(a) For let two cones be set out, $MN\Xi$, $O\Pi P$, so that the base of the <cone> $MN\Xi$ is equal to the surface of the cone $AB\Gamma$, while the height is equal to ZH [((2) so through this the cone $MN\Xi$ is equal to the cone $AB\Gamma$; (3) for if there are two isosceles cones, (4) and the surface of one of the cones is equal to the base of the other, (5) and, further, the perpendicular drawn from the center of the base on the side of the cone is equal to the height <of the other> (6) the cones will be equal)],[202] (b) <and so that> the base of the cone $O\Pi P$ is equal to the

---

[200] Step 11 supports the argument 9–10, not the argument 10–15.
[201] I.e. the circle resulting from the intersection of the plane and the cone.
[202] *SC* I.17.

surface of the cone ΔBE, while <the> height <is equal> to ZH [(7) so through this, also: the cone OΠP is equal to the rhombus BΔZE; (8) for this was proved above].²⁰³ (9) And since the surface of the cone ABΓ is composed of both: the surface of the <cone> ΔBE and the <surface> between ΔE, AΓ, (10) but the surface of the cone ABΓ is equal to the base of the cone MNΞ, (11) while the surface of the <cone> ΔBE is equal to the base of the <cone> OΠP, (12) and the <surface> between ΔE, AΓ is equal to the base of the <cone> ΘKΛ, (13) therefore the base of the <cone> MNΞ is equal to the bases of the <cones> ΘKΛ, OΠP. (14) And the cones are under the same height; (15) therefore the cone MNΞ is equal to the cones ΘKΛ, OΠP.²⁰⁴ (16) But the cone MNΞ is equal to the cone ABΓ, (17) while the <cone> ΠOP <is equal> to the rhombus BΔEZ;²⁰⁵ (18) therefore the remaining cone ΘKΛ is equal to the remainder.

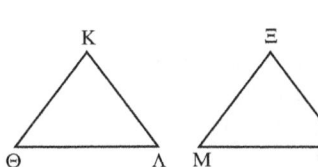 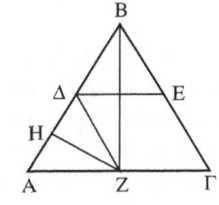

TEXTUAL COMMENTS

The proposition contains two claims followed by backwards-looking justifications, both referring to recent Archimedean results. The first of these, Steps 2–6, creates an impossible gap in the Greek: Step a begins with "for let two cones be set out, MNΞ, OΠP, so that the base of the <cone> MNΞ . . . ," and then the same syntactic construction governs Step b "<and so that> the base of the cone OΠP . . ." It is almost certain therefore that Steps a and b were not intended to be separated by such a long passage. Indeed, Steps 3–6 are extremely pedantic and use strange tenses ("(3) for if there *are* . . . (6) the cones *will* be equal"). They look like a scholion, and for once this impression can nearly be proved from the syntax. Step 2 may be a scholion as well, but need not be (by Greek standards, it does not create an impossible gap between Steps a and b). By analogy, Step 8 is probably a scholion, while Step 7 may, but need not, be one as well.

If those passages are indeed scholia, so that the proof proper begins at Step 9, the proposition becomes very difficult: very brief, assuming close acquaintance

I.19
Codex G has a mirrored order of objects, as in the thumbnail. It also has the base ΘΛ somewhat greater than the remaining two small bases. Codex B has the base MN rather greater than the remaining three bases. All codices, except B, introduce Φ at the intersection BZ/ΔE and (except H) have N instead of H. Codices DGH4 have Z instead of Ξ, and codices DEH have IZ instead of Z (!). All of this must derive from codex A, but the IZ, in particular, suggests an accidental stroke of the stylus rather than a genuine mistake.

²⁰³ *SC* I.18.

²⁰⁴ *Elements* XII.14 (in sequence following Proposition 16).

²⁰⁵ This either shows the redundancy of 2–8 (see textual comments), or is a neat recapitulation of the main results just prior to the conclusion.

with the immediately preceding propositions. This is a possible picture of Archimedes' practice here – which would help to explain, in turn, the later intervention by scholiasts.

GENERAL COMMENTS

Relations between parts of the proposition

Even with the scholiastic material, the proposition is dominated by the enunciation and by its translation into the diagram in the setting-out and definition of goal. I will therefore use this opportunity to note a few features of those parts of the proposition.

First, note that once again we have an argument inside the setting-out: "so there will be a rhombus BΔZE composed of isosceles cones." This is an assertion, not an imperative (compare the end of the setting-out in the preceding proposition). I point this out, again, to show that the division into parts is not absolutely rigid.

Another fuzzy border is in the definition of goal: "I say that if the rhombus BΔZE is imagined taken away from the cone ABΓ, the cone ΘKΛ will be equal to the remainder." The *claim* is "ΘKΛ=remainder." The first part of the sentence, however, is still effectively part of the setting-out – still in the business of preparing the geometrical object. (There are two other related, interesting features to this sentence. One is the use of the conditional form itself – in general surprisingly rare in Greek mathematics, almost limited to the formula of the *reductio* argument "for if possible . . ." The other is the role of "imagination." Both stress the fact that the taking away is only a virtual action, and therefore it is not part of the setting-out proper. The taking away does not "feel" to the reader as a veritable geometrical action, it is but the virtual shadow of an action: perhaps because it has no effect on the visible diagram.)

The sequence enunciation/setting-out/definition of goal is thus not made of rigid borders. But is it at all meant to be read in sequence? I cannot answer this question, but it should be said that most modern readers "cheat," and read the enunciation (if at all) only after they have read the setting-out and the definition of goal. The latter parts are easier to read than the first, because they are depicted in the diagram. The problem with this sort of reading is that it blurs the fact that the setting-out is not the general case, but is just a particular case of the enunciation (for instance, two qualitatively different diagrams are possible in this proposition, with H falling either below or above Δ; we saw a similar situation in the preceding proposition, and the next proposition has an interesting complication along the same theme). But it may be that the ancient readers "cheated" in the same way: for instance, we know that the diagrams were located at the *end* of the proposition, so that readers were used to looking ahead in this way. It is tempting to imagine this as a sequence: the enunciation is read through the (later) setting-out, itself read through the (later) diagram.

This however is a speculation, and in fact, the ancients might have had somewhat less difficulty with reading the enunciation than moderns have. Perhaps most importantly, they would approach it as a formulaic set of expressions,

having an expected syntactic form. This enunciation, for instance, belongs to
the class of conditionals with verbs in the subjunctive in the protasis (condition), and verbs in the indicative in the apodosis (result). The reader expects
such a structure, and therefore looks for the mood of the verb and, by picking it
up, immediately sees the condition (a cone is *cut*, a cone is *erected*, a rhombus
is *taken away*), and the result (a certain object *is equal* to a certain cone). Unfortunately, this sequence of moods can not be represented in English without
undue archaism.

/20/

If one of the cones of a rhombus composed of isosceles cones, is cut by
a plane parallel to the base, and a cone is set up on the resulting circle,
having the same vertex as the other cone, and the resulting rhombus
is taken away from the whole rhombus, the remainder will be equal
to the cone, having a base equal to the surface of the cone between
the parallel planes, and <its> height equal to the perpendicular drawn
from the vertex of the other cone on the side of the other cone.

Let there be a rhombus composed of isosceles cones, ABΓΔ, and
let one of the cones be cut by a plane parallel to the base, and let it
make <as> section the <line> EZ, and let a cone be set up on the
circle around the diameter EZ, having the point Δ <as> the vertex;
(1) so there will be a resulting rhombus, <namely> EBΔZ. And let it
be imagined taken away from the whole rhombus, and let some cone,
ΘKΛ, be set out, having the base equal to the surface between AΓ, EZ,
and <its> height equal to the perpendicular drawn from the point Δ
on BA or on a <line produced> in a straight line to it <=BA>; I say
that the cone ΘKΛ is equal to the said remainder.

(a) For let there be set out two cones, MNΞ, OΠP, (b) and let the
base of MNΞ be equal to the surface of the <cone> ABΓ, and <its>
height equal to ΔH [(2) so through the <things> proved above (3)
the cone MNΞ is equal to the rhombus ABΓΔ],[206] (c) and let the
base of the cone OΠP be equal to the surface of the cone EBZ, and
<its> height equal to ΔH [(4) so similarly the cone OΠP is equal to
the rhombus EBΔZ].[207] (5) But since, similarly,[208] the surface of the
cone ABΓ is composed of both: the <surface> of the <cone> EBZ,
and the <surface> between EZ, AΓ, (6) but the surface of the cone

---

[206] *SC* I.18.   [207] *SC* I.18.

[208] Apparently a reference to Step 9 in Proposition 19 above, or perhaps to the entire
argument of the preceding proposition.

ABΓ is equal to the base of the <cone> MNΞ, (7) while the surface of the cone EBZ is equal to the base of the cone OΠP, (8) and the <surface> between EZ, AΓ is equal to the base of the <cone> ΘΚΛ, (9) therefore the base of the <cone> MNΞ is equal to the bases of the <cones> OΠP, ΘΚΛ. (10) And the cones are under the same height. (11) Therefore the cone MNΞ, too, is equal to the cones ΘΚΛ, OΠP.[209] (12) But the cone MNΞ is equal to the rhombus ABΓΔ,[210] (13) while the cone OΠP <is equal> to the rhombus EBΔZ;[211] (14) therefore the remaining cone ΘΚΛ is equal to the remaining remainder.

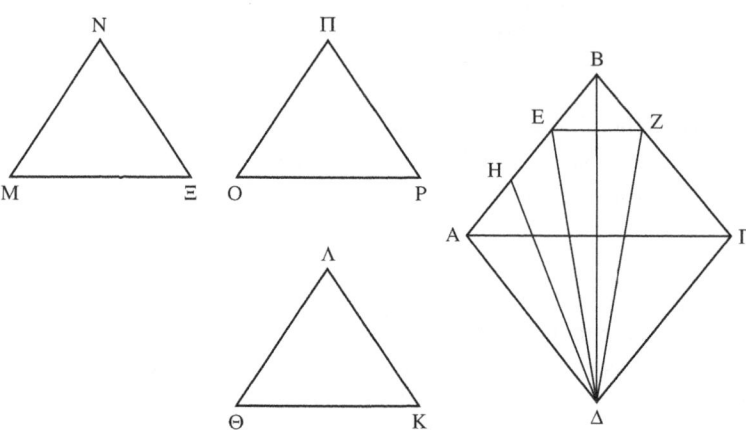

I.20
Both codices DG have a different arrangement, with the triangles in a single row aligned with the middle of the rhombus, as in the thumbnail. However, their internal arrangement is different: D has the triangles, from left to right, OΠP, MNΞ, ΘΛK, whereas G has, in the same order, ΘΛK, MNΞ, OΠP. D further permutes N/Ξ (followed, curiously, by Heiberg). Thus codices D and G must have independently ordered in a single row the arrangement printed, preserved identically in codices EH4. (Codex B, as usual, has a different layout altogether.)
Codices DEH4 have Σ instead of O. Codex E has Z instead of Ξ.

TEXTUAL COMMENTS

Heiberg printed a diagram where the sequence of the letters A and H is reversed, H projecting below A to allow a "true" right angle. His apparatus on this reads: "In fig. litteras A et H permutant AB." This is somewhat misleading, since the natural reading would be that C – the Palimpsest – had Heiberg's diagram. In fact, we now know that this is among the propositions where the Palimpsest did not have the diagrams filled in. I print therefore the manuscripts' diagram, which, typically, does not aim to represent "correctly" right angles.

GENERAL COMMENTS

Formulaic expression and definitions

"Rhombus composed of isosceles cones." Has this expression become a formulaic expression already? Notice that the word "solid" is consistently dropped, so that this expression can no longer be seen as a natural way of referring to this object. Further, contributing to the formulaic "feeling" of the expression, it is relatively long and cumbersome; most importantly, it is repeated *verbatim*.

---

[209] Interlude, recalling *Elements* XII.14.   [210] *SC* I.18.   [211] *SC* I.18.

Note that the expression does not follow naturally from the definitions. If anything, the definitions provide a special meaning for the adjective "solid" (when applied to "rhombus," it refers to a special kind of object). This adjective, however, is precisely what gets dropped in the actual use of the expression. Further, the definitions introduce a general object – any co-based and co-axial pair of cones – whereas the development of the treatise demands a special kind of object (the isosceles – where, incidentally, co-axiality follows from co-basedness). Thus the definitions introduce an expression that is not used later on, and refer to an object that is not used later on. In both sides of the semiotic equation – both sign and signified – the definitions do not fit easily with the treatise as a whole. This is indeed similar to the way in which the axiomatic material (especially concerning the notion of "concave in the same direction") never gets directly applied. In general, then, the introductory material does not really govern the main text, which is instead governed by internal, "natural" processes, such as the evolution of formulaic expressions.

/21/

If an even-sided and equilateral polygon is inscribed inside a circle, and lines are drawn through, joining the sides of the polygon[212] (so that they are parallel to one – whichever – of the lines subtended by two sides of the polygon),[213] all the joined <lines> have to the diameter of the circle that ratio, which the <line> (subtending the <sides, whose number is> smaller by one, than half <the sides>) <has> to the side of the polygon.

Let there be a circle, ABΓΔ, and let a polygon be inscribed in it, AEZBHΘΓMNΔΛK, and let EK, ZΛ, BΔ, HN, ΘM be joined; (1) so it is clear that they are parallel to the <line> subtended by two sides of the polygon;[214] Now I say that all the said <lines>[215] have to the diameter of the circle, AΓ, the same ratio as ΓE to EA.

(a) For let ZK, ΛB, HΔ, ΘN be joined; (2) therefore ZK is parallel to EA (3) while BΛ <is parallel> to ZK (4) and yet again ΔH to BΛ, (5) and ΘN to ΔH (6) while ΓM <is parallel> to ΘN [(7) and since

---

[212] In this formula, "sides" mean "vertices" (or, as Heiberg suggests, "angles").

[213] This means, in diagram terms, that the choice of EK is arbitrary: any other line "subtended by two sides of the polygon," e.g. AZ, could do as a starting-point for the parallel lines.

[214] Why is this clear? Perhaps through the equality of angles on equal chords (*Elements* III.27, 28), also the equality of alternate angles in parallels (*Elements* I.27). However obtained, an equality such as "(angle EKZ) = (angle KZΛ)" establishes "EK parallel to ZΛ."

[215] The reference is to all the *parallel* lines.

EA, KZ are two parallels, (8) and EK, AO are two lines drawn through];
(9) therefore it is: as EΞ to ΞA, KΞ to ΞO.[216] (10) But as KΞ to ΞO,
ZΠ to ΠO, (11) and as ZΠ to ΠO, ΛΠ to ΠP, (12) and as ΛΠ to ΠP,
so BΣ to ΣP, (13) and yet again, as BΣ to ΣP, ΔΣ to ΣT, (14) while
as ΔΣ to ΣT, HY to YT, (15) and yet again, as HY to YT, NY to YΦ,
(16) while as NY to YΦ, ΘX to XΦ, (17) and yet again, as ΘX to XΦ,
MX to XΓ[217] [(18) and therefore all are to all, as one of the ratios to
one];[218] (19) and therefore as EΞ to ΞA, so EK, ZΛ, BΔ, HN, ΘM to
the diameter AΓ. (20) But as EΞ to ΞA, so ΓE to EA;[219] (21) therefore
it will be also: as ΓE to EA, so all the <lines> EK, ZΛ, BΔ, HN, ΘM
to the diameter AΓ.

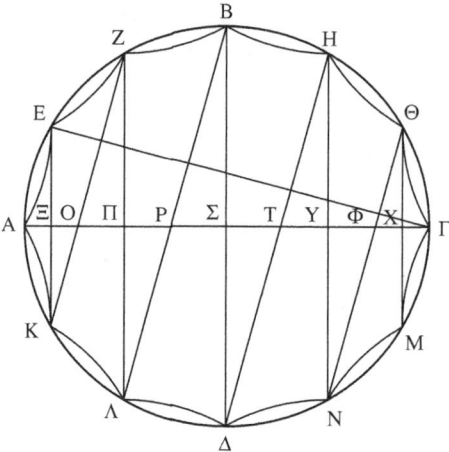

I.21
Codices BG, followed
by Heiberg, have
straight lines instead of
arcs in the polygon.
Codices EH4 have Σ
instead of E. Codex E
has K instead of X.

TEXTUAL COMMENTS

Heiberg brackets Steps 7 and 8. If this is indeed an interpolation then it is
noteworthy: an interpolated argument *preceding* its result – which is also a
very *cryptic* interpolation. It seems to state something like "if two lines are
drawn through two parallel lines, so that the two lines drawn through cut each
other between the parallels, the two resulting triangles are similar" – which
may be proved through *Elements* I.29, 32, VI.4. At any rate, the result implicit

---

[216] *Elements* I.29, 32, VI.4.

[217] In Steps 9–17, the odd steps are based on *Elements* I.29, 32, VI.4 (besides, of
course, a conceptualization of the configuration similar to that suggested at Steps 7–8).
The even steps are based on Step 1, as well.

[218] *Elements* V.12. The formulation is literally meaningless: instead of "one of the
ratios," it should have been "one of the terms" or "one of the lines."

[219] *Elements* VI.4 (but more than this is required to see that the triangles are similar,
e.g. we may notice that, since AB is one-quarter the circle, so following *Elements* III.31
the angle at Σ must be right, and since the lines are parallel it follows that the angle at
Ξ, too, is right, and then we apply *Elements* VI.8).

here is *not* in the *Elements*. Are we to imagine a late interpolator assuming extra-Euclidean knowledge? Perhaps, then, this could be Archimedean (and similar to Proposition 16, Step 4). Once again, then, this would be a result, not from some lost, extra-Euclidean *Elements*, but simply one that Archimedes had no patience to go through (even though it was nowhere proved as such), opting instead for a vague indication of its possible grounds.

Step 18 is strange as well. "All" appears in the neuter – signifying what? (Antecedents to consequents? Anyway, not "ratios," which are masculine.) Further, "a ratio to a ratio" is meaningless: we should have had "as one of the lines to one" or "as one of the terms to one." Such solecism may be the mark of the clumsy interpolator – unless, of course, it is a mark of the careless author.

GENERAL COMMENTS

The strange nature of the proposition

Now to the main feature of this proposition, already mentioned in the textual comments: it is strange.

Take the diagram. In this proposition a new, striking diagrammatic practice appears for the first time: the sides of the polygon are represented by curved, concave lines. This is probably done to aid the resolution of the arcs and the chords (a considerable problem with dodecagons), but at any rate, this novel practice marks a radical departure from simple, "pictorial" representation. It is, I think, somewhat improbable that a scribe would invent such a practice, in defiance of his sources. If so, we may have in this practice a hint of Archimedes' own diagrammatic practices. At any rate, the strange diagram marks a strange proposition.

The diagram leads immediately to another strange feature of this piece of text, this time in the reference to the diagram in the name of the dodecagon with its twelve letters. The longest we had so far was five letters, in Proposition 11, Steps 19–20; in general, names are usually within the range of one to three letters, with occasionally four letters: longer names are very rare. (Thus, for instance, the polygon of the first proposition is nowhere named, partly perhaps to avoid an extremely long name.) I believe this long name is intentionally playful – a long, strange name, in a strange proposition.

More important is the logical structure of the proposition. It enumerates facts, in long lists (similar to the strange, long name of the dodecagon); beyond listing the facts, the proposition practically does not *argue*. Its arguments are implicit (and difficult to reconstruct): that the lines are parallel (see n. 214 above), that this entails a certain set of proportions (see textual comments), and that those proportions are reducible to the ratio ΓE:EA (see n. 219 above). So a strange logical flow, as well.

Most importantly, the subject matter itself is strange: it has nothing to do with anything we know, from this book or from, say, the *Elements*. Instead, this proposition is a complete break. Specifically, it has nothing to do with sphere and cylinder. We had thought we had got closer, with conic solids treated by "chapter 4" (Propositions 17–20), but we now move suddenly to an unexpected, unconnected interlude, once again signaling a major transition in the work.

/22/

If a polygon is inscribed inside a segment of a circle, having the sides
(without the base) equal and even-numbered, and lines are drawn parallel to the base of the segment, joining the sides of the polygon, all
the drawn <lines> and the half of the base have to the height of the
segment the same ratio, which the <line> joined from the diameter
of the circle[220] to the side of the polygon[221] <has> to the side of the
polygon.

For let some line, AΓ, be drawn through the circle ABΓΔ, and let a
polygon be inscribed on AΓ, inside the segment ABΓ, even-sided and
having the sides equal (without the base AΓ), and let ZH, EΘ be joined
(which are parallel to the base of the segment); I say that it is: as ZH,
EΘ, AΞ to BΞ, so ΔZ to ZB.

(a) For similarly again let HE, AΘ be joined; (1) therefore they are
parallel to BZ;[222] (2) so through the same,[223] (3) it is: as KZ to KB,
HK to KΛ and EM to MΛ and MΘ to MN and ΞA to ΞN [(4) and
therefore as all to all, one of the ratios to one];[224] (5) therefore as ZH,
EΘ, AΞ to BΞ, so ZK to KB. (6) But as ZK to KB, so ΔZ to ZB;[225]
(7) therefore as ΔZ to ZB, so ZH, EΘ, AΞ to ΞB.

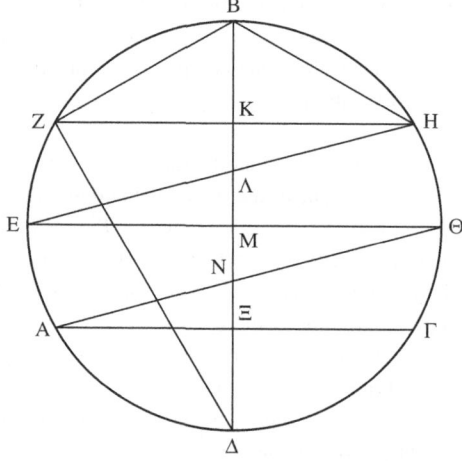

I.22
Codex B, followed by
Heiberg, adds straight
lines AE, EZ, HΘ, ΘΓ.
At the intersection of
AΘ, ZΔ, codex E has a
letter K, while codex H
has a letter M.

---

[220] "From the diameter of the circle" means here: "from that extreme point of the
diameter perpendicular to the base of the segment, which is outside the segment." Or
simply: "from Δ."

[221] "The side of the polygon" means here: "to an angle of the polygon just next to the
other end of the same diameter." Or simply: "to Z or H."

[222] *Elements* I.27, III.27, 28.

[223] *Elements* I.29, VI.4. This is not a reference to Step 1, but to the preceding proposition.

[224] *Elements* V.12.   [225] *Elements* VI.4 (see comment on Proposition 21).

TEXTUAL COMMENTS

The only textual question is Step 4, which obviously stands or falls together with Step 18 of the preceding proposition. I would suggest that the variation in the formulation between the two makes it slightly more probable that both are authorial (scholiasts are somewhat more conservative than authors). This of course is a very weak argument.

GENERAL COMMENTS

Underspecification of objects by the text

In the enunciation we see the text trying to keep up with the speed by which Archimedes introduces new geometrical objects: the result is that many objects are loosely described. Some examples: "If a *polygon* is inscribed inside a segment of a circle, having the sides (without the *base*) equal and even-numbered, and lines are drawn parallel to the *base* of the *segment* . . ." The first mention of "base" seems to be a reference to the base of the *polygon*. This clashes with the immediately following base of the *segment*: the same object, but two different descriptions. (Incidentally, it is not accidental that AΓ is in the diagrams at the *bottom* of the polygon – adopting an orientation different from that of the preceding proposition. This is typical of the way in which these "bases" are understood, literally, as supporting the objects of which they are bases.) Further: ". . . joining the sides of the polygon." This is a strange expression, already seen in the preceding proposition ("angles" or "vertices" would be more correct). Finally (as noted in nn. 220–1) line ZΔ is defined in an extremely compressed way.

Another kind of underspecification in this proposition has to do with the diagram. In the setting-out, notice the expression: "[A]nd let a polygon be inscribed on AΓ, inside the segment ABΓ, even-sided and having the sides equal (without the base AΓ)." This polygon is not set out as a sequence of letters. As we should expect, therefore, *there is no such polygon drawn in the diagram.* This is a virtual polygon, a scaffold for the parallel lines of the proof. In general, no letter in this proposition is explicitly attached to a point by the text. To a great extent, the diagram is organized spatially and not textually (clockwise with ABΓΔ, then again with EZHΘ, and then top-down with the rest).

All of this is related, of course, to the generally compressed nature of this proposition. As in the other "chapters" of the book, the end of the chapter is marked by an extremely abbreviated form: here, the chapter being very brief, abbreviation takes place immediately as the objects are being introduced.

Other aspects of the brevity of the proof

EΘ is taken by the manuscripts to be the diameter of the circle; i.e. if a polygon were constructed and completed, it would have been the minimal case of 4n-gon, namely octagon (always assuming that a square is not considered as a legitimate case at all of "a polygon with 4n sides"). This should be compared with the dodecagon of the preceding proposition and once again, this is easy to understand in light of the position of Proposition 22: no more than an extension of Proposition 21. At this stage, as the generality of Proposition 21 is secured, one may well take the minimum case of n=2 (in the 4n sided polygon). A similar unconcerned air is to be noticed, as well, in the quick surveying attitude of Step 3 (as against the laborious sequence of Steps 9–17 in the preceding proposition).

Finally and most importantly, that AΔ is at right angles to the parallel lines is not asserted, although this is required by Step 6. This is a rare case of an essential geometrical property, left completely implicit by the text.

Is Archimedes perhaps impatient with this trivial extension? Alternatively, in a "simple" extension of the preceding proposition, he may intentionally try to mystify his audience.

/ 23 /[226]

Eut. 262

Let there be a great circle in a sphere, <namely> ABΓΔ, and let an equilateral polygon be inscribed inside it <=the circle>, and let the number of its sides be measured by four, and let AΓ, ΔB be diameters. So if the circle ABΓΔ is carried in a circular motion, holding the polygon (the diameter AΓ remaining fixed), it is clear that its circumference will be carried along the surface of the sphere, while the angles of the polygon[227] (except those next to the points A, Γ) will be carried along circumferences of circles in the surface of the sphere, drawn <in such a way that they are> right to the circle ABΓΔ; and their <=the circles created by the movement of the angles> diameters will be the <lines> joining the angles of the polygon (which are parallel to BΔ). And the sides of the polygon will be carried along certain cones:[228] AZ, AN along a surface of a cone, whose base <is> the circle around the diameter ZN, and <whose> vertex <is> the point A; while ZH, MN will be carried along a certain conical surface, whose base <is> the circle around the diameter MH, and <whose> vertex <is> the point at which ZH, MN, produced, meet both each other and the <line> AΓ; and the sides BH, MΔ will be carried along a conical surface, whose base is the circle, around the diameter BΔ, <which is> right to the

---

[226] This proposition forms the exact middle of the first book; its Archimedean point.

[227] "Angles" here finally means "vertices" (the same objects, confusingly, were called "sides" in the preceding propositions, in such locally established formulaic expressions as "lines joining the sides").

[228] "Cones" in this instance means "surfaces of cones."

circle ΑΒΓΔ, while <its> vertex <is> the point at which ΒΗ, ΔΜ, produced, meet both each other and the <line> ΓΑ; and similarly, the sides in the other semicircle, too, will be carried along conical surfaces <which>, again, <will be> similar to these. So there will be a certain figure inscribed in the sphere, contained by conical surfaces, <namely> those mentioned above, whose surface will be smaller than the surface of the sphere.

(a) For, the sphere being divided by the plane <which>, at the <line> ΒΔ, <is> right to the circle ΑΒΓΔ, (1) the surface of one hemisphere and the surface of the figure inscribed in it, have the same limits in a single plane; (2) for the circumference of the circle, <which is> around the diameter ΒΔ <and is> right to the circle ΑΒΓΔ, is a limit of both surfaces; (3) and they are both concave in the same <direction>, (4) and one of them is contained by the other surface and by the plane having the same limits as itself.[229] (5) And similarly, the surface of the figure in the other hemisphere, too, is smaller than the surface of the hemisphere;[230] (6) so the whole surface of the figure in the sphere, too, is smaller than the surface of the sphere.

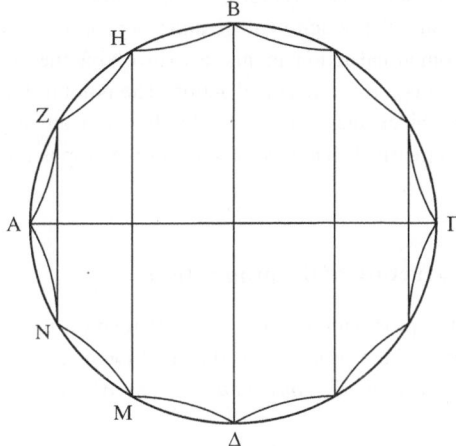

I.23
Codices BG, followed by Heiberg, have straight lines instead of arcs in the polygon. (Codex 4, which has the lines drawn freely without a compass, has a few of the lines roughly straight.) All codices, except B, introduce a letter Π at the intersection ΑΓ/ΗΜ.   Codices EH4 have N instead of H.   Codex H has P instead of Z.

TEXTUAL AND GENERAL COMMENTS

Heiberg did not bracket anything in this proposition. It is a direct, almost minimal presentation of a striking idea; nothing seems superfluous. Mathematical genius speaks, as it were, and the textual critic has to remain silent.

---

[229] And then Steps 1–4, through Post. 4, may yield the result that the figure in the hemisphere is smaller than the hemisphere. This is not asserted.

[230] Post. 4.

Producing objects by rotation

This proposition introduces objects constructed by rotation. This is an interesting kind of construction, since it is irreducible to ruler-and-compass construction. However, it should be understood that generating three-dimensional
objects by revolutions of two-dimensional objects is a Greek mathematical
commonplace. It is in this way that Euclid introduces the sphere, the cone,
and the cylinder (*Elements* XI. Defs. 14, 18, 21 respectively), and Archimedes
devotes a whole treatise, *On Conoids and Spheroids*, to solids produced in this
way from conic sections: Greek mathematical objects are regularly perceived
as the result of action and movement.

I see one small puzzle, however, with this particular construction. If we
have a circular motion that generates a sphere, why do we start from "a great
circle in a sphere?" Why not simply start from a circle, and generate a sphere
from that circle? Perhaps this may be because, starting with just a circle, one
would need to imagine in more vivid detail the actual sphere – one would need
its construction much more. Starting with a hypothetical ghost of a sphere
(implied by "great circle in a sphere"), the scaffolds for the construction are
already there, and one does not need to imagine the sphere in detail. (As usual,
this proposition has a strictly two-dimensional diagram, without any indication
of depth, representing a three-dimensional structure; but here, the verb "to
imagine" is not used.) At any rate, the action of generating objects in three
dimensions is not so much a geometrical action, in this case (involving the true
construction of geometrical objects), as a conceptual action. The point of this
proposition is not to show how objects may come to be, but how objects may
be conceived. It is this, meta-geometrical character, that makes the proposition
so strikingly original.

The internal structure of the proposition

The key stylistic feature of this proposition is its (deviant) relation to the
Euclidean norm of the structure of the proposition. (General *enunciation* –
particular *setting-out* and *definition of goal* – *construction* and *proof* – general
conclusion.)

Archimedes has had many variations on the Euclidean structure. General
conclusions are avoided, and construction, setting-out, and proof are often
intermingled. This Proposition 23, however, is radically original: a proposition without a general enunciation. Is it a "proposition" at all? The Greek
manuscripts do not number it (i.e. the Greek scribes who added numbers did
not think this was a "proposition"). It is perhaps comparable to the sequences
of text following Propositions 12 and 16, lying outside the normal sequence
of propositions. Whereas the sequences of text following Propositions 12 and
16 were completely *general* (they had no particular, lettered material), this
"Proposition 23" is completely *particular* (it has no general, unlettered material). The normal, unmarked proposition is a *combination* of particular and
general. For a text to be marked in such a context, and set apart, it must be either
wholly general or wholly particular. This material of "Proposition 23," unlike

the sequences of text following Propositions 12 and 16, cannot be expressed in a wholly general format (one must refer to a diagram, to explain the contents of "Proposition 23"), hence the complete particularity, which is then extremely marked in context.

This text would come as a shock to a reader used to Greek mathematical texts, and we must make an effort to feel this shock ourselves. The text has its meaning inside this structure of related mathematical texts – to which it is so very *different*.

(We can say then that the style of this particular text is a negative reflection of the general Greek mathematical style – no more. The style is not intended to carry any special significance, besides its *difference*. Note, however, that the potential implications of this new stylistic departure are great, and it is perhaps possible to argue that the modern variable emerged through such structural transformations of the use of letters in particular/general contexts.)

So what is left of the Euclidean structure? There being no general enunciation, the proposition effectively starts as a setting-out. Almost immediately at the start, a *conditional* is introduced ("if the circle ABΓΔ is carried . . .") and, at first, the impression is that the force of the conditional will be of the form "if an action is done, a certain object results." Only at the end of the setting-out, it is seen that the force of the conditional is that "if an action is done, *the resulting object is smaller than a sphere*." This is a second, minor shock – the conditional changes its meaning, with the conclusion: "so there will be a certain figure inscribed in the sphere, contained by conical surfaces, <namely> those mentioned above, whose surface will be smaller than the surface of the sphere." This sentence functions in this proposition similarly to the *definition of goal* in a "normal" proposition: this is what we shall seek a proof *of* (note however the final twist to this structure: for whatever reason this may have happened, this last sentence is unlettered, and therefore can be read not only as a particular definition of goal, but also as the end of a general *enunciation*). This leads on to a very brief (and, as it were, "virtual") added *construction*, that of Step a; and then a direct, elementary application of a postulate – no more – in the *proof*. (Incidentally, the added construction and proof are based on a bisection of the sphere into two hemispheres, when, in fact, any other division of the sphere along any of the planes could do. The hemisphere is chosen as the most natural case – and probably also the one where the concavity is most obvious. Also, as ever, the concavity is taken for granted rather than proved. In such ways, the proof is at the level of a direct appeal to intuition, similar to the earlier applications of the postulate.)

Only a virtual enunciation, and not much of a proof. So what *has* happened in this proposition?

Something very important has happened – as Archimedes stresses, by radically marking this proposition. The sphere, the real hero of the book, has been introduced for the first time; and it has been re-identified with the revolution of a circle inscribed with a polygon. The re-identification immediately suggests how the results we already possess, on cones and cylinders on the one hand, and on circles and polygons on the other hand, may finally be integrated and brought to bear on the sphere. Following the brief "interlude" of Propositions

21–2, Proposition 23 begins "chapter 5" – the first substantial chapter of the book: as it were, the first chapter of *On the Sphere and the Cylinder* proper.

/ 24 /

The surface of the figure inscribed inside the sphere[231] is equal to a circle, whose radius is, in square, the <rectangle> contained by the side of the figure[232] and <by> the <line> equal to all the <lines> joining the sides of the polygon (<the lines> being parallel to the line subtended by two sides of the polygon).

Let there be ABΓΔ, greatest circle in a sphere, and let an equilateral polygon whose sides are measured by four be inscribed in it, and, on the inscribed polygon, let some figure inscribed inside the sphere be imagined, and let EZ, HΘ, ΓΔ, KΛ, MN be joined, being parallel to the line subtended by two sides; so let some circle be set out, Ξ, whose radius is, in square, the <rectangle> contained by AE and <by> the <line> equal to EZ, HΘ, ΓΔ, KΛ, MN; I say that this circle is equal to the surface of the figure inscribed inside the sphere.

(a) For let there be set out circles, O, Π, P, Σ, T, Y, (b) and let the radius of O be, in square, the <rectangle> contained by EA and <by> the half of EZ, (c) and let the radius of Π be, in square, the <rectangle> contained by EA and <by> the half of EZ, HΘ, (d) and let the radius of P be, in square, the <rectangle> contained by EA and <by> the half of HΘ, ΓΔ, (e) and let the radius of Σ be, in square, the <rectangle> contained by EA and <by> the half of ΓΔ, KΛ, (f) and let the radius of T be, in square, the <rectangle> contained by AE and <by> the half of KΛ, MN, (g) and let the radius of Y be in square the <rectangle> contained by AE and <by> the half of MN. (1) So through these, the circle O is equal to the surface of the cone AEZ,[233] (2) and Π <is equal> to the surface of the cone between EZ, HΘ,[234] (3) and P <is

---

[231] Already established as a formula, the reference is to the figure described in the preceding proposition.

[232] This "figure" refers to the two-dimensional polygon inscribed in the circle.

[233] Step b and *SC* I.14. This requires some unpacking through *Elements* VI.17, since *SC* I.14 is phrased in terms of "having a mean ratio" (algebraically: X is such to a, b that a:X::X:b), while here the data are phrased in terms of "being in square a rectangle" (algebraically: X is such to a, b that X^2=ab).

[234] Step c and *SC* I.16. Another implicit argument is that, the polygon being equilateral, AE is equal to the sides of each of the truncated cones: in particular, here, it is equal to EH.

equal> to the <surface> between HΘ, ΓΔ,[235] (4) and Σ <is equal> to the <surface> between ΔΓ, ΚΛ,[236] (5) and yet again, T is equal to the surface of the cone between ΚΛ, MN,[237] (6) and Y is equal to the surface of the cone MBN;[238] (7) therefore all the circles are equal to the surface of the inscribed figure. (8) And it is obvious that the radii of the circles O, Π, P, Σ, T, Y are, in square, the <rectangle> contained by AE and <by> twice the halves of EZ, HΘ, ΓΔ, ΚΛ, MN,[239] (9) which are (<as> wholes) EZ, HΘ, ΓΔ, ΚΛ, MN; (10) therefore the radii of the circles O, Π, P, Σ, T, Y are, in square, the <rectangle> contained by AE and <by> all the <lines> EZ, HΘ, ΓΔ, ΚΛ, MN. (11) But the radius of the circle Ξ, too, is, in square, the <rectangle contained> by AE and <by> the <line> composed of all the <lines> EZ, HΘ, ΓΔ, ΚΛ, MN; (12) therefore the radius of the circle Ξ is, in square, the radii of the circles O, Π, P, Σ, T, Y <in square>; (13) and therefore the circle Ξ is equal to the circles O, Π, P, Σ, T, Y.[240] (14) But the circles O, Π, P, Σ, T, Y were proved equal to the surface of the said figure;[241] (15) therefore the circle Ξ, too, will be equal to the surface of the figure.

I.24
Codices BG, followed by Heiberg, have straight lines instead of arcs in the polygon. Codex B has a different layout altogether, and codex D has the somewhat different lay-out shown in the thumbnail. All, however, keep the circles from overlapping, against Heiberg.   Codex B has the order of sizes Ξ>T=P>Σ>Π>O=Y; codex D has the order of sizes Ξ>T>Y=Σ=Π=O=P; codex G has the order of sizes Ξ>T=Y=Σ=Π=O=P; while codices EH4 have the order of sizes printed here, Ξ>T=Y>Σ=Π=O=P.

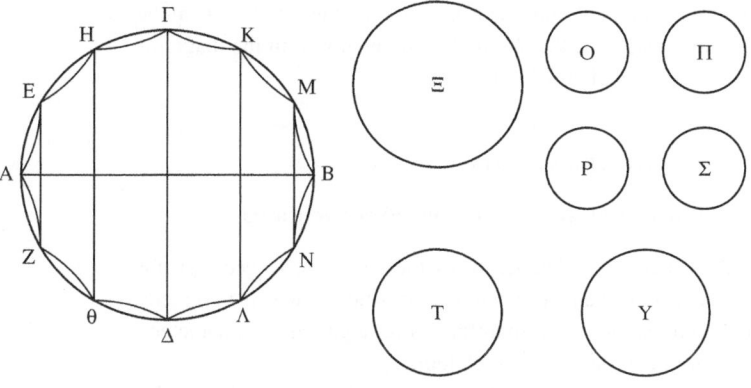

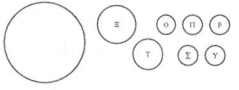

---

[235] Step d and *SC* I.16.   [236] Step e and *SC* I.16.
[237] Step f and *SC* I.16.   [238] Step g, *SC* I.14.
[239] "Obvious" is probably an overstatement, but really all you need to do is to sum Steps b–g: each half is mentioned exactly twice. It is interesting that the Greek does not distinguish here at all between summing the radii and then taking their square, and summing the separate squares of the radii (i.e. between $(a+b+c)^2$ and $a^2+b^2+c^2$): the second interpretation is the one intended here. (In other words, not only is the text non-algebraic: the language is not even equipped to deal with the simplest algebraic distinctions.)
[240] *Elements* XII.2.   [241] Greek: "the said surface of the figure."

TEXTUAL COMMENTS

In two cases, I prefer the manuscripts' reading to Heiberg's emendation. The first case from the setting-out, is relatively trivial: ". . . two sides; so let some circle be set out," which in Heiberg's emendation becomes ". . . two sides of the polygon, and let some circle be set out." Heiberg's version adds a greater uniformity to the text as a whole; I suggest the original need not have been uniform.

The second is Step 12: ". . . the radius of the circle Ξ is, in square, the radii of the circles O, Π, P, Σ, T, Y <in square>," which Heiberg corrected to ". . . the radius of the circle Ξ is in square the <squares> on the radii of the circles O, Π, P, Σ, T, Y." This involved a massive invasion of the text by Heiberg (taking a τας and changing it into τὰ ἀπὸ τῶν). The substantial question is whether Archimedes' style could allow an omission of the second appearance of the crucial adverbial phrase "in square" (which is what appears in the manuscripts). Was he that careless, that informal? I believe so, but Heiberg did not.

A word on the diagram. Heiberg draws the circles in two concentric clusters, partly perhaps to save space, partly to signal relations of size. In the original, all the circles are independent (no intersections either). The one clear metrical relation in the diagrams is that Ξ is greater than the rest of the circles (all appearing practically equal). But this is not so much metrical as qualitative: Ξ is indeed different in nature, not only in size, from the other circles. It can be said to be more important, to be the *theme*. Relations of size in the diagrams signify, I suggest, mostly relations of importance.

GENERAL COMMENTS

The absence of algebraic and logical conventions

As noted in n. 239 above, Archimedes lacks the basic tools necessary for algebraic manipulations. In general, such a proposition – which is directly based upon simple summations and is therefore more "algebraic" in character – may serve as an example of the lack of such tools.

Consider the setting-out: ". . . and let an equilateral polygon whose sides are measured by four be inscribed . . ." Compare this to the setting-out in the previous proposition: "and let the number of its sides be measured by four."

In this proposition, the sides, directly, and not their number, are measured by four. This is a typical lack of care for detail but, more significantly, this is an example of Greek confusion on number. In Greek mathematics, remarkably, it is not clear whether "four" is a kind of number, or a kind of collection (in other words, there is no distinction between the concepts "four" and "quartet"). Thus both a set of sides, and a number, may be understood to be the kind of object that may be "measured by four."

There are several other ways in which we can see the text falling short of modern understandings of algebraic and logical relations. For instance, I have noted in n. 239 the lack of distinction between $a^2+b^2+c^2$ and $(a+b+c)^2$. This is partly the limitation of the notation, partly a case of a more general Greek "confusion:" what is a sum? Is it the plurality of the objects, or is it a certain new object created by the plurality? Sometimes the phrase "A and B" is taken to mean a single object, the sum of A and B; sometimes it is taken to mean two objects, considered together. This is seen elsewhere, e.g. in Step 8: "... the <rectangle> contained by AE and <by> twice the halves of EZ, HΘ, ΓΔ, KΛ, MN ...", as against Step 11: "... the <rectangle contained> by AE and <by> the <line> composed of all the <lines> EZ, HΘ, ΓΔ, KΛ, MN ..." So do we look at the plurality of different objects, each keeping its own identity (as in Step 8) or at that one-over-many which is composed of the plurality (as in Step 11)? This particular ambivalence is repeated throughout this book.

Even more problematic is the use of logical terms. Most difficult is "some," whose use in Greek mathematics is very hard to disentangle. Why, for instance, "let *some* figure inscribed inside the sphere be imagined" in the setting-out? Why "some?" Perhaps because the figure is unlettered (and therefore, in some sense, not specific)? But this is just a guess. What is clear is that "some" and similar words are not used as strict quantifiers in the modern sense.

All of the above, of course, is not intended as criticism of Archimedes: the importance of all those examples is precisely that, while we stop to worry about them, Archimedes did not. The clear setting-out of notational conventions was less important than the geometrical understanding itself.

/ 25 /

The surface of the figure inscribed inside the sphere, contained by the conical surfaces,[242] is smaller than four times the greatest circle of the <circles> in the sphere.

Let there be in a sphere a great circle, ABΓΔ, and let an [even-sided] equilateral polygon be inscribed in it, whose sides are measured by four, and let a surface be imagined upon it, contained by the conical surfaces; I say that the surface of the inscribed <figure> is smaller than four times the greatest circle of the <circles> in the sphere.

(a) For let the <lines> subtended by two sides of the polygon – EI,[243] ΘM – and the parallels to them – ZK, ΔB, HΛ – be joined, (b) and let some circle be set out, P, whose radius is, in square, the <rectangle contained> by EA and <by> the <line> equal to all the <lines> EI, ZK, BΔ, HΛ, ΘM; (1) so through the <result> proved earlier,[244] (2) the circle <=P> is equal to the surface of the said figure

---

[242] As we have already seen e.g. in Proposition 18 above, "is contained by" in this context may mean something like "consists of." (It is clearly the surface, not the figure, that "is contained by" the conical surfaces.)

[243] The first time that the letter I – the Cinderella of mathematical letters – appears in this book.

[244] *SC* I.24.

<=the surface "contained" by the conical surfaces>. (3) And since it was proved that it is: as the <line> equal to all the <lines> EI, ZK, BΔ, HΛ, ΘM to the diameter of the circle, AΓ, so ΓE to EA,[245] (4) therefore the <rectangle contained> by the <line> equal to all the said <lines>, and by EA, that is the <square> on the radius of the circle P, (5) is equal to the <rectangle contained> by AΓ, ΓE.[246] (6) But the <rectangle contained> by AΓ, ΓE is also smaller than the <square> on AΓ;[247] (7) therefore the <square> on the radius of the <circle> P is smaller than the <square> on AΓ. [(8) Therefore the radius of the <circle> P is smaller than AΓ; (9) so that the diameter of the circle P is smaller than twice the diameter of the circle ABΓΔ, (10) and therefore two diameters of the circles ABΓΔ are greater than the diameter of the circle P, (11) and four times the <square> on the diameter of the circle ABΓΔ, that is AΓ (12) is greater than the <square> on the diameter of the circle P. (13) But as four times the <square> on AΓ to the <square> on the diameter of the circle P, so four circles ABΓΔ to the circle P;[248] (14) therefore four circles ABΓΔ are greater than the circle P;] (15) therefore the circle P is smaller than four times the great circle. (16) But the circle P was proved equal to the said surface of the figure; (17) therefore the surface of the figure is smaller than four times the greatest circle of the <circles> in the sphere.

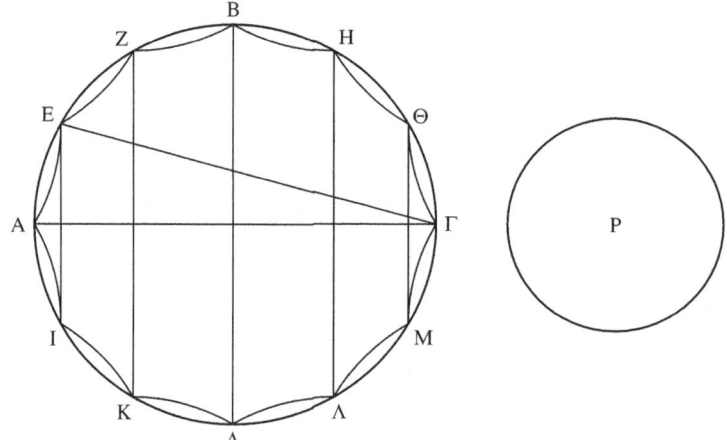

I.25
Codices BG, followed by Heiberg, have straight lines instead of arcs in the polygon. Codex G (as well as B) has the smaller circle P to the left of the main circle.

---

[245] *SC* I.21.   [246] *Elements* VI.16.

[247] *Elements* III.15: the diameter is the greatest line in the circle. (*Elements* VI.1. is in the background, too.)

[248] *Elements* XII.2.

TEXTUAL COMMENTS

First, consider the strange adjective "even-sided" in the setting-out. Since in the very same sentence the polygon is said to have its sides measured by four, to introduce it as "even-sided" is redundant. But then why would an interpolator add this? This is best understood as an authorial slip: Archimedes is so used to speaking of "even-sided polygons," that he adds the adjective even when it is superfluous.

Steps 8–14 belong to the very small class of passages that seem so incongruous that they could not possibly be by Archimedes. The passage is neither false nor irrelevant, but it is extraordinarily elementary, with a fascinating passage from P<Q to Q>P (Steps 9–10). We imagine a reader with minimal confidence, in need of verifying every bit of the argument. Note, however, that this passage deals with arithmetical relations, never easy to convey in the discursive style of Greek mathematics. Given that, the passage 8–14 is seen to have some value, and even the P>Q→Q>P argument makes some sense in a non-algebraic setting. Still, the probability is, indeed, very strongly against this passage being by Archimedes.

GENERAL COMMENTS

On "that is," and the nature of the proposition

This proposition may serve as an opportunity to analyze the operator "that is." The meaning of this operator is "identity." The effect of the operator is to squeeze an argument into a single assertion, thus reducing a couple of arguments into a single argument.

Let us look at Steps 3–5:

(3) (EI+ZK+BΔ+HΛ+ΘM):AΓ::ΓE:EA (taken from *SC* I.21).

(4) "That is" operator: (the <rectangle contained> by (EI+ZK+BΔ+HΛ+ΘM) and by EA) is identical with (the <square> on the radius of the circle P) (taken from Step b).

(5) Result as asserted: (the <rectangle contained> by (EI+ZK+BΔ+HΛ+ΘM) and by EA) is equal to (the <rectangle contained> by AΓ, ΓE. (Step 3 + *Elements* VI.16: if a:b::c:d then (a*d)=(b*c)).

However – and this is the force of the "that is" operator – the effective result which we take away from this passage may be represented as (5*) (the <square> on the radius of the circle P) is equal to (the <rectangle contained> by AΓ, ΓE).

In other words, instead of having two separate arguments (from 3 to 5 via *Elements* VI.16, and then from 5 to 5* via 4), both arguments are combined in a single argument from two premises: the argument from 5 to 5* is squeezed into the single assertion 4.

It seems that Archimedes is frequently interested in "smoothing" a passage, by simplifying its logical structure, reducing the number of arguments and getting more directly to the desired result. The "that is" operator is a tool used for this purpose.

Note that, assuming Steps 8–14 are to be bracketed, the result is an extremely brief proof (with, indeed, a minimal figure: no intersection is named by a letter, because no geometrical configuration is required). It is interesting, therefore, that even in such a brief argument, Archimedes wishes to abbreviate it even further, with the "that is" operator: essentially, this is somewhat less than a proposition, more a corollary to the preceding proposition.

/26/

The figure inscribed in the sphere, contained by the conical surfaces, is equal to a cone having, <as> a base, the circle equal to the surface of the figure inscribed in the sphere, and <its> height equal to the perpendicular drawn from the center of the sphere on one side of the polygon.[249]

Let there be the sphere, and a great circle in it, AΒΓΔ, and the rest the same as before, and let there be a right cone, P, having, <as> base, the surface of the figure inscribed in the sphere, and <its> height equal to the perpendicular drawn from the center of the sphere on one side of the polygon; it is to be proved that the cone P is equal to the figure inscribed in the sphere.

(a) For let cones be set up on the circles whose diameters are ZN, HM, ΘI, ΛK,[250] having <as> a vertex the center of the sphere; (1) so there will be a solid rhombus <composed> of the cone whose base is the circle around ZN, and <its> vertex the point A, and of the cone whose base <is> the same circle <=the circle around ZN>, and <its> vertex the point X; (2) it <=the rhombus> is equal to the cone having, <as> a base, the surface of the <cone> NAZ, and <its> height equal to the perpendicular drawn from X.[251] (3) And again, the remainder of the rhombus, too, contained by: the surface of the cone between the parallel planes at ZN, HM, and <by> the surfaces of both cones: ZNX, and HMX – is equal to the cone having a base equal to the surface of the cone between the parallel planes at MH, ZN, and <its> height equal to the perpendicular drawn from X on ZH;[252] (4) for these were proved.

---

[249] "Polygon" – referring to the inscribed equilateral polygon, known from previous propositions.

[250] Heiberg (and the manuscripts) have ΘΛ, IK. See textual comments.

[251] *SC* I.18. That the perpendicular is meant to be on AN/AZ is left implicit.

[252] *SC* I.20. The object referred to is best visualized once the overall plan to exhaust the entire figure is understood. Then the solid rhombus NAZX is seen as the ice-cream scoop inside the ice-cream cone MNXHZ (in turn – to extend the metaphor – inside your hands, ΔMXHB).

(5) Yet again, the remainder of the cone, too, contained by: the surface of the cone between the parallel planes at HM, BΔ, and <by> the surface of the cone MHX and <by> the circle around the diameter BΔ, is equal to the cone having a base equal to the surface of the cone between the planes at HM, BΔ and <its> height equal to the perpendicular drawn from X on BH.[253] (6) And similarly also in the other hemisphere: the rhombus XKΓΛ[254] and the remainders of the cones are equal to cones of the same number and of the same kind just as they were described before; (7) now it is clear, as well, that the whole figure inscribed in the sphere is equal to all the said cones. (8) But the cones are equal to the cone P, (9) since the cone P has a height equal to each <height> of the said cones, while <its> base <is> equal to all their bases;[255] (10) so it is clear that the <figure> inscribed in the sphere is equal to the cone set out.

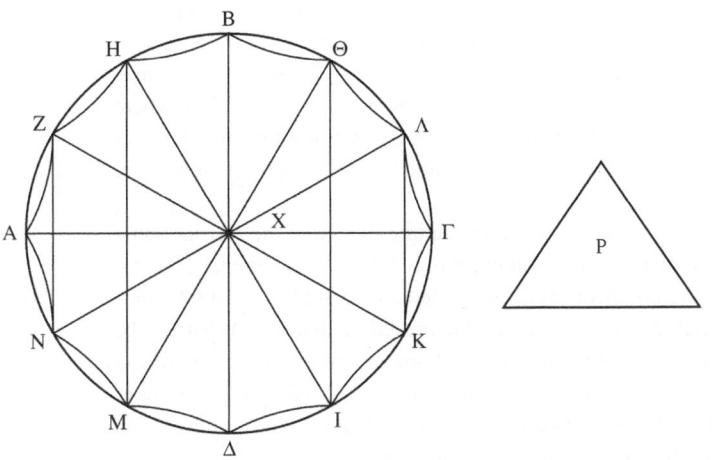

I.26
Codices BG, followed by Heiberg, have straight lines instead of arcs in the polygon. Codex G (as well as B) has the triangle to the left of the main circle. Codices DH4 have K instead of X, while EG have omitted it altogether (introduced, as X, by a later hand at G).   Heiberg, following a later hand at codex G, permutes I/Λ (see textual comments).

TEXTUAL COMMENTS

The only real textual difficulty here is the couple of letters IΛ. Heiberg has them the other way round from the manuscripts (whose version I follow). Since they are on "the dark side" of this sphere (the hemisphere where results are simply transferred) there are only two references to them in the text: in Step a, where the manuscripts have "circles whose diameters are . . . ΘΛ, IK," And Step 6, where the manuscripts have "the rhombus XKΓΛ." Step a agrees with Heiberg's diagram, Step 6 agrees with the manuscripts' diagram. One of the steps must reflect a corruption: either purely textual (perhaps in Step a, where the corruption could be a permutation of the order of letters in

---

[253] *SC* I.19.   [254] Heiberg has XKΓI. See textual comments.
[255] Interlude, recalling *Elements* XII.11, 14.

a list – an easy mistake), or a corruption based on an earlier corruption of the diagram. Heiberg's diagram produces a smooth clockwise sequence of letters. The *lectio difficilior*, therefore, is the manuscripts' diagram. The argument is not conclusive, and the issue in itself is not intrinsically significant: but it is worth noticing, at least for the kinds of textual corruption that may afflict text and diagram.

GENERAL COMMENTS

A move to a more general space

Just as Proposition 25 was very clearly no more than a corollary to Proposition 24, so Proposition 26 is clearly no more than an extension into three dimensions of the kind of claim made in Proposition 24. Thus, following the break of Proposition 23, we move into a sequence of propositions that are felt as a sequence, leading on to the main claim of the treatise.

These propositions thus lose the "independence" of individual propositions. In particular, Proposition 26 clearly harks back to preceding propositions, which constitute a shared background of assumptions: thus, for instance, the truly remarkable phrase "the polygon" in the general enunciation, where a reference is made to a definite object which has not at all been defined in the context of this proposition (it is only definite, in relation to the preceding proposition).

A similar case is the expression in the setting-out, "and the rest the same as before." Notice that strictly speaking this is false: the diagrams, even for the sphere alone, are not identical (different letters are used). So the reference of "the same as before" is not given by the preceding diagram. What is "the same?" It represents a general mathematical understanding of the sphere and the figure inscribed. The diagram at hand is the *kind* of diagram we are used to. Still, a certain continuity between the diagrams begins to form. Indeed, that the space of the proposition is not fully autonomous can also be seen from the interesting fact that, in this proposition, the letter X starts a brief career as the center – a role it will play in a number of later diagrams. (The letter is possibly chosen on the basis of iconicity: the letter X may represent an intersection, which is where the center stands.)

In such ways, then, the proposition acts not so much within its individual space, defined by its letters, but also in the generalized space of the "mathematical situation" seen in the last few propositions. This allows Archimedes to be even more sparing with his words: we see, for instance, that he uses letters in the text only where this allows him to refer to objects with more clarity and brevity (e.g. in Step 3). Elsewhere (e.g. Steps 7–10) letters are almost wholly avoided, as they will complicate the expressions. Starting at Proposition 23, then, a certain blurring of the particular and the general begins to be felt. The propositions are simultaneously about specific objects, created for their particular propositions, *and* about a more general set of objects which we may call "the sphere and cylinder objects:" the typical polygon, the strange ice-cream cones and scoops, etc.

/27/

The figure inscribed in the sphere, contained by the conical surfaces, is smaller than four times the cone having a base equal to the greatest circle of the <circles> in the sphere, and a height equal to the radius of the sphere.

For let the cone P be made equal to the figure inscribed in the sphere, having the base equal to the surface of the inscribed figure, and the height equal to the perpendicular drawn from the center of the circle on one side of the inscribed polygon; and let the cone Ξ be: having a base equal to the circle ABΓΔ, and, <as> height, the radius of the circle ABΓΔ.

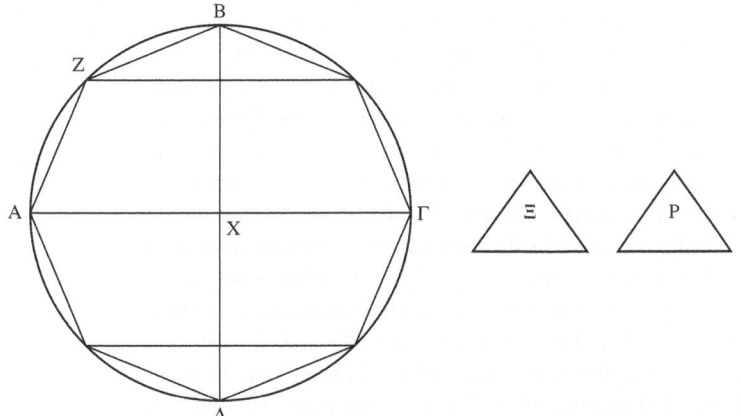

(1) Now since the cone P has a base equal to the surface of the figure inscribed in the sphere, and a height equal to the perpendicular drawn from X on AZ, (2) and the surface of the inscribed figure was proved to be smaller than four times the greatest circle <of the circles> in the sphere,[256] (3) therefore the base of the cone P will be smaller than four times the base of the <cone> Ξ.[257] (4) But the height of the <cone> P, as well, is smaller than the height of the cone Ξ.[258] (5) Now, since the cone P has the base smaller than four times the base of the <cone>

---

[256] *SC* I.25.

[257] Setting-out. A re-identification (great circle <=> base of Ξ) is left implicit.

[258] Through the setting-out, this is equivalent to the claim that, in the case of lines drawn from the center: a radius is always greater than a perpendicular on a chord. This can be proved, e.g., by "greater angle in triangle subtended by greater side" (*Elements* I.19), plus the right angle where the perpendicular meets the side of the polygon, plus "two angles at a triangle are less than two right angles" (*Elements* I.17). Clearly, however, no reader goes through this explicitly: the claim is visually compelling.

I.27
There is a fundamental error on codex A, producing in effect the diagram of I.35 instead of that of I.27 (see the first thumbnail for the outline of the figure as it appears in most manuscripts; I shall not go into details of differences between codices). The authors of codices BG realized the error. B produced a minimal figure: a triangle with the (Latin equivalent of the) letter Ξ; G produced the figure of the second thumbnail. Surely some earlier version of the diagram of I.27 was lost, perhaps because it had some confusing similarity with the diagram of I.35. Indeed it is likely, from the text of the proposition, that the diagram of I.27 should have had an octagon inscribed within a circle, with two extra triangles. I.35 has an octagon inscribed within a circle, with four extra circles. In arranging the figure, I assume that the extra triangles of the original I.27 diagram were to the right of the main circle, and that this is what influences the position of the four extra circles in the codices' diagrams. (In the diagram of I.35

Ξ, while <its> height is smaller than the height, (6) it is clear that the cone P itself, too, is smaller than four times the cone Ξ. (7) But the cone P is equal to the inscribed figure; (8) therefore the inscribed figure is smaller than four times the cone Ξ.

GENERAL COMMENTS

The particular and the general, and the "toy universe"

This brief proposition is yet another example of an important theme of this book – another variation on the theme of the particular and the general.

The proposition has no particular definition of goal. There is a general enunciation: an object of a certain kind is always equal to another. Then there is a particular setting-out, where particular objects are being constructed. However, there is no particular definition of goal, asserting for the particular objects that the equality holds between them. The equality carries over implicitly, from the general to the particular. Instead of the general being read through the particular, then – as is more often the case – the particular is, here, read through the general.

But is "the general" quite *general*? Perhaps an altogether different perception should be adopted (see also the preceding general comments). The letter-less, "general" words describe, in a sense, a particular, because they do not describe objects of general, wider significance. You just do not come across "the polygon" or "the figure inscribed in the sphere" anywhere else. The text refers throughout to a toy universe, a specific, strange space, where one meets those specific monsters, "the polygon" (meaning an equilateral, 4n-sided polygon inscribed in a circle), "the figure inscribed in a sphere," etc. Instead of individuating through letters, much of the text individuates by referring to peculiar things through peculiar formulae. So not quite "general" – and yet, not quite particular either.

This is related, once again, to the special relation between various propositions in this part of the book: very often we read, as we do here, truncated propositions, with brief arguments, essentially no more than an unpacking of earlier results. In other words, the borders demarcating individual propositions tend to dissolve and, related to that, the borders between the "particular" (the level of the single proposition) and the "general" (the cross-proposition level) get blurred, as well.

The creativity of formulaic language

Look at step 5: ". . . the cone P has the base smaller than four times the base of the <cone> Ξ, while <its> height is smaller than the height."

That the phrase ends with "the height" instead of "the height of Ξ" has a formulaic ring: this is the sort of omission you get in formulaic expressions, and the symmetry of "height equal to height," once again, appears formulaic. In particular, one is reminded of some very common expressions concerning pairs of triangles such as "and base is equal to base" (e.g. *Elements* I.4, Heiberg 16.21–2). In all probability, Archimedes is not following here an established

I.27 (*cont.*) itself, the circles are arranged symmetrically around the main figure.) I go on to distribute the letters in the figure according to the practice of the preceding diagrams. The entire operation cannot claim any real credibility.

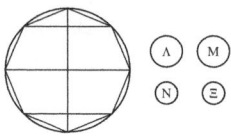

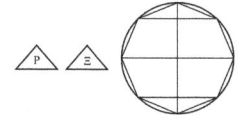

formula for cones. It seems, then, that he is extending formulae, creatively, by an analogy from triangles to cones.

/28/

Let there be, in a sphere, a great circle, ΑΒΓΔ, and let an equilateral and equiangular polygon be circumscribed around the circle ΑΒΓΔ, and let the number of its sides be measured by four, and let a circumscribed circle (created around the same center as ΑΒΓΔ) contain the polygon circumscribed around the circle. So let the plane ΕΖΗΘ, in which <are> the polygon and the circle, be carried in a circular motion (ΕΗ remaining fixed); so it is clear that the circumference of the circle ΑΒΓΔ will be carried along the surface of the sphere, while the circumference of the <circle> ΕΖΗΘ will be carried along another surface of a sphere having the same center as the smaller <sphere>, and the touching points, at which the sides are tangents, draw, in the smaller sphere, circles <which are> right to the circle ΑΒΓΔ, and the angles of the polygon, except the <angles> at the points Ε, Η, will be carried along circumferences of circles in the surface of the greater sphere, drawn right to the circle ΕΖΗΘ, and the sides of the polygon will be carried along conical surfaces, just as in the first case. Now, the figure contained by the conical surfaces will be circumscribed around the smaller sphere, and inscribed in the greater. And that the surface of the circumscribed figure is greater than the surface of the sphere, will be proved like this: (a) for let there be ΚΔ, a diameter of some circle among the <circles> in the smaller sphere ((b) there being the points Κ, Δ, at which the sides of the circumscribed polygon touch the circle ΑΒΓΔ). (1) So, the sphere being divided by the plane at ΚΔ <which is> right to the circle ΑΒΓΔ, the surface of the figure circumscribed around the sphere will be divided, too, by the plane; (2) and it is obvious that they[259] have the same limits, in a plane; (3) for the limit of both planes[260] is the circumference of the circle around the diameter ΚΔ

---

[259] The plural "they" stands not for the two segments of the sphere, or of the figure contained by conical surfaces. Rather, it stands for a pair of segments, one of them a segment of a sphere, the other a segment of the figure contained by conical surfaces – so that both "are in the same direction" (there are two such possible pairs, to the "left" or "right" of ΚΔ). This is understood through the later application of Post. 4.

[260] As Heiberg notes, this should properly have been "surfaces." The thought is "the limit of both surfaces is a plane . . . ," and this is then written down as "the limit of both planes . . .": an Archimedean lapse? (Compare the similar solecism, to which we have become accustomed, "the figure *contained* by conical surfaces.")

<which is> right to the circle ABΓΔ; (4) and they are both concave in
the same direction, and one of them is contained by the other surface
and by the plane having the same limits; (5) so the contained surface
of the segment of the sphere is smaller than the surface of the figure
circumscribed around it. (6) And similarly, the surface of the remaining segment of the sphere, too, is smaller than the surface of the figure
circumscribed around it; (7) so it is also clear that the whole surface
of the sphere is smaller than the surface of the figure circumscribed
around it.

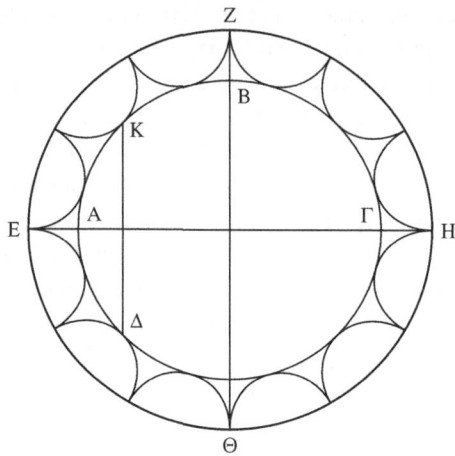

I.28
Codices BG, followed
by Heiberg, have
straight lines instead of
arcs in the polygon.
(Starting here, Heiberg,
following BG, will
occasionally simplify
the original dodecagon
to an octagon; I shall
not point this out in the
following.)   Codex
G, curiously, has the
entire figure rotated 90
degrees clockwise, so
that not only the figure,
but the characters
appear rotated.

TEXTUAL COMMENTS

The words "just as in the first case" clearly appear in the manuscripts. Editors
wanted to change them, why, I find hard to say. (Heiberg: "just as in the preceding cases," adding a reference to Propositions 23–7.) Perhaps the problem was
the erroneous belief that the original text had numbered propositions (hence
"first" may seem misleading, as if referring to I.1?). At any rate the reference
is clear: to the first application of this technique of demonstration, at *SC* I.23.

GENERAL COMMENTS

Proposition I.23 gave rise to a locally established form

We have again the deviant structure of Proposition 23: no general enunciation.
However, by now the shock-value has been lost. Instead of being a violation
of a form, this is already an application of a local, Archimedean form. It is
Archimedes' method of introducing a new visualization (in this case: a sphere,
as comprehended by a figure composed of conical surfaces). In both cases,
as well, the motivating mathematical principle is Post. 4, and the fact that
both propositions are only quasi-propositions is related to the fact that this
postulate is only partially applied (i.e. that no checking is made of the validity
of its application). Post. 4 is used merely as a tool for translating a certain

visualization into an inequality: propositions based on it are therefore merely visualizations, followed by a statement of the inequality.

/29/

The surface of the figure circumscribed around the sphere is equal to a circle, whose radius is, in square, <an area> equal to the <rectangle> contained by: one side of the polygon, and <by> the <line> equal to all the <lines> joining the angles of the polygon (<the lines> being parallel to some <line> among the <lines> subtended by two sides of the polygon).

(1) For the <figure> circumscribed around the smaller sphere is inscribed inside the greater sphere;[261] (2) and it has been proved that the surface of the <figure> inscribed in the sphere, contained by the conical surfaces, is equal to the circle, whose radius equals, in square, the <rectangle> contained by: one side of the polygon, and <by> the <line> equal to all the <lines> joining the angles of the polygon (<the lines> being parallel to some <line> among the <lines> subtended by two sides <of the polygon>);[262] (3) so what has been said above is clear.

GENERAL COMMENTS[263]

Quotations and other textual procedures

The wording of Step 2 follows exactly that of the enunciation, but it is different from that of I.24 (on which the claim relies). This may tell us something about the making of the text, especially since this proposition is really nothing but an application of I.24. What we see is that Archimedes looked up the earlier *enunciation* when writing Step 2. But he did not look up the earlier *proposition*, I.24. This may be relevant for the wider role of the written text in Greek mathematics. The possibilities of the written text are opened up (so some *verbatim* quotations are made) but they are not yet systematically used (so *verbatim* quotations at great distances are still very rare). At any rate, the sense in which the claim "follows from" I.24 is not based on any verbal identity between the formulations of the propositions, but is based instead on a geometrical understanding of the relevant situations. In other words, in this kind of mathematics, the relations between pieces of text are not constitutive to the relations between geometrical objects.

---

[261] *SC* I.28.   [262] *SC* I.24.

[263] Notice that this is not a fully fledged proposition and, for this reason, it carries no diagram.

/30/

The surface of the figure circumscribed around the sphere is greater than four times the greatest circle of the <circles> in the sphere.

For let there be the sphere and the circle and the rest the same as set out before, and let the circle Λ be equal to the surface of the <figure> circumscribed around the smaller sphere <which was> set out before.

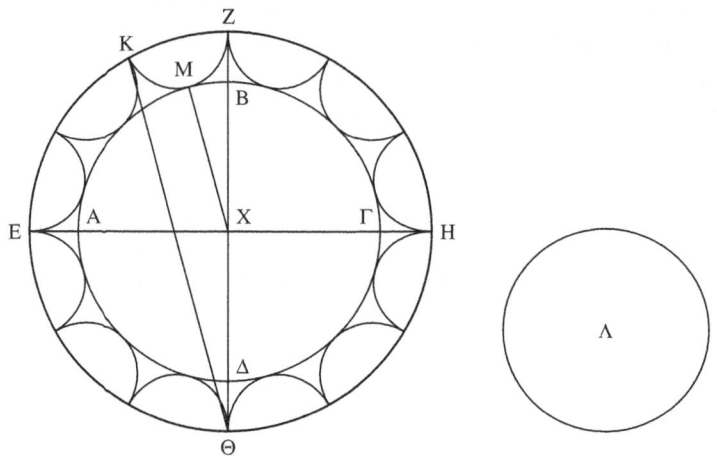

I.30
Codex B, followed by Heiberg, has straight lines instead of arcs in the polygon. (Codex G, however, succumbs to the arcs, probably because of the inherent difficulty of drawing a polygon both inscribed by, and circumscribing, a circle.)   Codices G4 (as well as B) have the circle Λ higher relative to the main circle.   All codices have Σ instead of M. Codices DEH omit Δ (if not missing from A, it was at least faint), and codex E further omits H.

(1) Now, since an equilateral and even-angled polygon has been inscribed in the circle EZHΘ, (2) the <lines> joining the sides of the polygon (being parallel to ZΘ) have to ZΘ the same ratio, which ΘK <has> to KZ;[264] (3) therefore the <rectangular> figure contained by: one side of the polygon, and <by> the <line> equal to all the <lines> joining the angles of the polygon, is equal to the <rectangle> contained by ZΘK;[265] (4) so that the radius of the circle Λ is equal, in square, to the <rectangle contained> by ZΘK;[266] (5) therefore the radius of the circle Λ is greater than ΘK.[267] (6) But ΘK is equal to the diameter of the circle ABΓΔ [(7) for <it is> twice XM which is a radius of the circle ABΓΔ];[268] (8) so it is clear that the circle Λ, that is the surface of

Eut. 263

---

[264] *SC* I.21.

[265] *Elements* VI.16. (KZ from Step 2 is now re-identified as the more general object "one side of the polygon.")

[266] *SC* I.29.

[267] ZΘ>ΘK (*Elements* III.15), and the square on the radius of Λ equals the rectangle contained by ZΘ, ΘK. So through *Elements* VI.17 we have: (ZΘ:radius of Λ::radius of Λ:ΘK), hence the radius of Λ is smaller than ZΘ, but greater than ΘK.

[268] Step 7 undoubtedly refers to Eutocius' argument. See Eutocius, therefore, for the textual problem. Mathematically, the tools required are apparently (in this order) *Elements* III.18, I.47, I.8, VI.2, VI.4.

the figure circumscribed around the smaller sphere (9) is greater than four times the greatest circle of the <circles> in the sphere.[269]

TEXTUAL COMMENTS

I discuss Step 7 – a special case – in a footnote to Eutocius' commentary.

A strange phenomenon occurs in Step 3: "the <rectangular> *figure* contained by: one side of the polygon, and <by> the <line>." The usual expression is "the contained by . . . ," and we expand this (based on *Elements* II. Def. 1) into "the <rectangle> contained by . . ." (to be even more complete, we could have "the <rectangular parallelogram> contained by"). Here, unexpectedly, someone fills in the gap, and fills it with "figure." In fact, the expression "figure contained by . . ." is common in this treatise – referring to the figure contained by conical surfaces. A scribe could get carried away, then, and add it in where it is redundant – but so could Archimedes himself, in a moment of absent-mindedness.

GENERAL COMMENTS

Reduced setting-out

There is a tendency, here as in some preceding propositions (e.g. 26), to condense the act of setting-out to a minimum. Such propositions are thus the complement to propositions such as 23 and 28, where there is nothing but setting-out and general enunciation. Here the setting-out is reduced to "the same as before." This is not to say that the diagram is identical to the preceding one: in Greek mathematics, you cannot step into the same diagram twice.

So the text does not "draw" the diagram, it does not recreate it step-by-step. Notice one result: since the parallel lines are not mentioned in the text, they are absent, as well, from the diagram (which contains only the objects *named by letters*). It is perhaps because of this that a curious mistake occurred here. Instead of Step 2 as I translate it, "the <lines> joining the sides of the polygon (being parallel to ZΘ) have to ZΘ the same ratio, which ΘK <has> to KZ," the manuscripts have "the <lines> joining the sides of the polygon (being parallel to ZE) have to ZE the same ratio, which ΘK <has> to KZ" – substituting E for Θ (this cannot represent a textual problem with the diagram, since Θ is often referred to, correctly, elsewhere). The parallel lines being "invisible," as it were, mistakes are liable to happen.

There is another, subtle twist here. *SC* I.21, on which this Step 2 is based, speaks of the ratio of the lines "parallel to one – whichever – of the lines subtended by two sides of the polygon," to the diameter. Now a line subtended by two sides of the polygon may well be parallel itself to the diameter, e.g. the diameter ZΘ in this proposition, and so it is possible to speak of "the <lines>

---

[269] The passage from $a > 2b$ to $a^2 > 4b^2$ is implicit, as is the background assumption that the areas in question are proportional to squares on radii (*Elements* XII.2).

joining the sides of the polygon (being parallel to ZΘ)," as happens here. But then the result is that now, unlike what one would expect from *SC* I.21, ZΘ figures twice: as the line to which the lines are parallel, and as the line to which the lines are said to have the ratio. So now the claim *looks* substantially different from that of *SC* I.21, and it is possible to see how even a mathematically competent reader may have become confused.

We are of course already used to the narrative significance of growing abbreviation: the chapter draws towards its conclusion, which obviously – now that spheres are so explicitly mentioned – has to be the *dénouement* for the treatise as a whole.

/31/

The figure circumscribed around the smaller sphere is equal to a cone having, <as> base, the circle equal to the surface of the figure, and a height equal to the radius of the sphere.

(1) For the figure circumscribed around the smaller sphere is inscribed in the greater sphere; (2) and the inscribed figure, contained by the conical surfaces, has been proved to be equal to a cone having, <as> base, the circle equal to the surface of the figure, and a height equal to the perpendicular drawn from the center of the sphere on one side of the polygon;[270] (3) but this <=the perpendicular> is equal to the radius of the smaller sphere; (4) so the claim is obvious.

/Corollary/

And from this it is obvious that the figure circumscribed around the smaller sphere is greater than four times a cone having, <as> base, the greatest circle of the <circles> in the sphere, and, <as> height, the radius of the sphere. (1) For since the figure is equal to a cone having a base equal to its <=the figure's> surface, and a height equal [to the perpendicular drawn from the center of the sphere on one side of the polygon, (2) that is] to the radius of the smaller sphere,[271] (3) and the surface of the figure circumscribed around the sphere is greater than four times the greatest circle of the <circles> in the sphere,[272] (4) therefore the figure circumscribed around the sphere will be greater than four times the cone having, <as> base, the great circle and, <as> height, the radius of the sphere, (5) since the cone equal to it

---

[270] *SC* I.26.

[271] *SC* I.31, Steps 2–3, are repeated here as Steps 1–2.   [272] *SC* I.30.

<=to the figure> will then also be greater than four times the said cone [(6) for it has a base greater than quadruple, and an equal height].[273]

TEXTUAL AND GENERAL COMMENTS[274]

The title "Corollary" has no basis in the manuscripts and was added by Heiberg. It may be that no gap should be associated with the words "and from this it is obvious, that . . ." Before and after that gap, the proposition has the same character: letter-less and devoid of real geometrical activity. Hardly a proposition at all, this text was not numbered by any manuscript, as if it were no more than an interlude between Propositions 30 and 32. A moment of rest, as it were – the final rest before the final climb. The next sequence of proofs will get us to the top.

This said, a real textual problem remains, in the corollary. It is repetitive, repeating both itself (Step 4 reverts to the enunciation, while Steps 5–6 recap the argument), and the proposition proper (Steps 1/2 of the corollary repeat Steps 2/3 of the proposition). Unless Archimedes is positively trying to numb us, to put us off our guard (which cannot be ruled out), this is just too plain unintelligent, too much like a scholiast. Probably the best explanation is that offered by Heiberg: the original structure of the corollary had only Steps 1, 3, 4 and 5. Archimedes' text erred, perhaps, in pushing brevity too far for some readers' comfort – readers who then reacted with several marginal comments. Those marginal comments were at a later stage incorporated into the text, which as a result errs to the side of explicitness. Let this serve as an example of a general principle. A manuscript tradition works as a feedback mechanism, adjusting between over-implicitness and over-explicitness, often erring to one side or the other.

/32/

If there is a figure inscribed in a sphere and another circumscribed, constructed of similar polygons, in the same manner as the above, the surface of the circumscribed figure has to the surface of the inscribed figure a ratio duplicate[275] of the side of the polygon circumscribed around the great circle to the side of the polygon inscribed in the same

---

[273] Interlude, recalling *Elements* XII.11.

[274] Notice, once again, that this is not a fully fledged proposition and, for this reason, it carries no diagram.

[275] A ratio "duplicate:" what we would call "the square of the ratio." The duplicate ratio is the ratio taken twice: the ratio composed from the original ratio repeated with itself, e.g., 4:3 repeated by 4:3, which, composed, yields 16:9 (we may say it yields (4*4):(3*3)). So the duplicate ratio of 4:3 is 16:9. The extension to "triplicate" ratio, to follow below, is obvious.

circle, and the [circumscribed] figure itself[276] has to the figure a ratio triplicate of the same ratio.

Let there be a circle[277] in a sphere, ABΓΔ, and let an equilateral figure be inscribed inside it, and let the number of its sides be measured by four, and let another <figure>, similar to the inscribed, be circumscribed around the circle, and yet again, let the sides of the circumscribed polygon touch the circle at the middles of the circumferences cut by the sides of the inscribed polygon, and let EH, ZΘ, at right angles to each other, be diameters of the circle containing the circumscribed polygon, and, similarly placed as the diameters AΓ, BΔ, and let <lines> be imagined joined to the opposite angles of the polygon (which will then be parallel to each other and to ZBΔΘ). So, as the diameter EH remains fixed, and the perimeters of the polygons (around the circumference of the circle) are carried in a circular motion, one <perimeter> will be a figure inscribed in the sphere, the other <will be> a circumscribed <figure>; now, it is to be proved that the surface of the circumscribed figure has to the surface of the inscribed a ratio duplicate of EΛ to AK, while the circumscribed figure has to the inscribed a ratio triplicate of the same ratio.

(a) For let there be the circle M, equal to the surface of the <figure> circumscribed around the sphere, (b) and the <circle> N, equal to the surface of the inscribed <figure>; (1) therefore the radius of the <circle> M is, in square, the <rectangle> contained by EΛ and <by> the <line> equal to all the <lines> joining the angles of the circumscribed polygon,[278] (2) while the radius of the <circle> N <is, in square,> the <rectangle contained> by AK and <by> the <line> equal to all the <lines> joining the angles.[279] (3) And since the polygons are similar, (4) the areas contained by the said lines[280] will also be similar,[281] [(that is, <the areas contained by> the <lines joined> to

---

[276] The figure "itself:" i.e. the figure *qua* solid (rather than *qua* its surface).

[277] Note the omission of "great." This, however, may be mere textual corruption, rather than formulaic brevity.

[278] Step a, *SC* I.29.

[279] Step b, *SC* I.24. Note that the "angles" intended must now be those of the *inscribed* polygon.

[280] The expression "said lines" is interesting, given that this is the first occurrence of the word "line." At any rate, the reference is to the two rectangles mentioned in Steps 1 and 2, and the "said lines" are the lines containing those rectangles (or, indeed, the lines "constituting" those lines, i.e. the lines drawn inside the polygons, as seems to be the view taken in the bracketed passage).

[281] According to *Elements* VI. Def. 1, similar rectilinear figures are such that (a) they are respectively equiangular, (b) they have the sides corresponding to the angles proportional. As both areas are understood to be rectangles, condition (a) applies automatically, and therefore the substantive claim made here is that (EΛ:AK::(summed lines in greater

the angles or the sides of the polygons) (5) so that they <=the areas>
have the same ratio to each other which the sides of the polygons have
in square.[282] (6) But also, the <areas> contained by the said lines
have that ratio which the radii of the circles M, N have to each other
in square; (7) so that the diameters of the <circles> M, N, too, have
the same ratio as the sides of the polygons. (8) But the circles have to
each other a ratio duplicate of the diameters[283] – (9) <circles> which,
moreover, are equal to the surfaces of the circumscribed <figure> and
the inscribed <figure>]; (10) so it is clear that the surface of the figure circumscribed around the sphere has to the surface of the figure
inscribed inside the sphere a ratio duplicate of EΛ to AK.

   (c) So let two cones be taken, O, Ξ, (d) and let the cone Ξ be <a
cone> having, <as> base, the circle Ξ (equal to the <circle> M) (e)
and <let> O <be a cone> having, <as> base, the circle O (equal
to the <circle> N), (f) and <let> the cone Ξ <have as> height the
radius of the sphere, (g) and <let> the <cone> O <have as height>
the perpendicular drawn from the center on AK; (11) therefore the cone
Ξ is equal to the figure circumscribed around the sphere,[284] (12) and the
<cone> O <is equal> to the inscribed[285] [(13) indeed, these have been
proved]. (14) And since the polygons are similar, (15) EΛ has to AK
the same ratio which the radius of the sphere has to the perpendicular
drawn from the center of the sphere on AK;[286] (16) therefore the height
of the cone Ξ has to the height of the cone O the same ratio which
Eut. 263 EΛ has to AK. (17) But the diameter of the circle M, also, has to the
diameter of the circle N a ratio which EΛ has to AK;[287] (18) therefore
the diameters of the bases of the cones Ξ, O have the same ratio as
the heights [(19) therefore they are similar],[288] (20) and, through this,
the cone Ξ will have to the cone O a ratio triplicate of the diameter of the
circle M to the diameter of the circle N.[289] (21) So it is clear that the

---

circle):(summed lines in smaller circle)). I now follow Heiberg: through *Elements* VI.20,
similar polygons are constituted of similar triangles. Take for example ZΛΘ, BKΔ. They
are similar: so ZΛ:BK::ZΘ:BΔ. But ZΛ:BK is the same as EΛ:AK (EΛ, AK are both
sides of equilateral polygons – setting-out – i.e. they can be interchanged with any other
side of their respective polygons); and so for each of the lines joining the angles. Through
Elements V.12, all are to all as one is to one, and the claim is seen to be true.

   [282] *Elements* VI.20.   [283] *Elements* XII.2.   [284] Steps d, f, *SC* I.31.

   [285] Steps e, g, *SC* I.26.   [286] *Elements* VI. 4.

   [287] This is effectively guaranteed by the first part of the proof (Step 10, as interpreted
by Steps a, b, asserts that the circles M, N are to each other as EΛ:AK in square, i.e.
through *Elements* XII.2, the diameters of M, N, in square, are to each other as EΛ:AK in
square, and Step 17 is seen directly. Eutocius' comment to this is much longer, but this
is because he argues, simultaneously, for both claims of Steps 4 and 17.

   [288] *Elements* XII. Def. 24.

   [289] *Elements* XII.12 (in the sequence following Proposition 16).

circumscribed figure, too, will have to the inscribed a ratio triplicate of EΛ to AK.

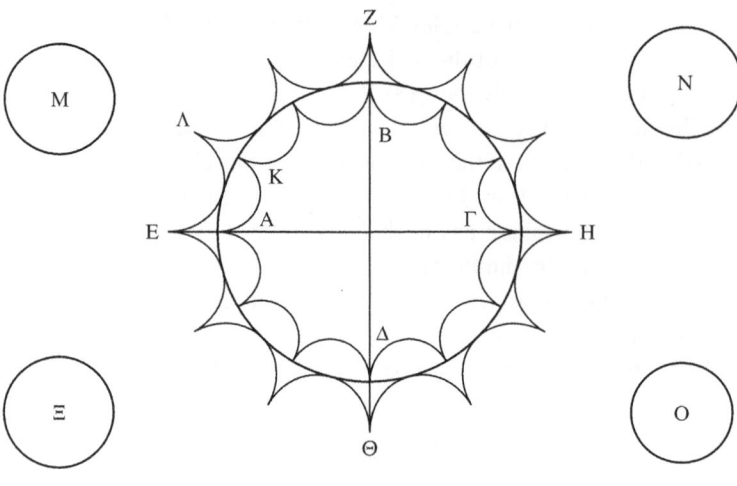

I.32
Here codex C begins to have the diagrams for *SC* (mere gaps in the column of writing are all there is until now). Note that its readings are still incomplete and are likely to remain this way.   Codex B, followed by Heiberg, has straight lines instead of arcs in the polygon.   Codices BD, followed by Heiberg, introduce an extra circle, circumscribing the external polygon. Strangely, Heiberg has introduced two lines parallel to ZΘ (one going down from K, the other, symmetrically on the other side).   In codices DEH4 a Π is introduced on the arc between K and A, and a Σ is introduced on the arc between Λ and E. This surely says something about codex A itself; codex C may not have anything similar at the same area, but this is far from certain. Codices D4 have N instead of H. Codex D further has Z instead of Ξ, while codex 4 further has ΘE (!) instead of E. (I suspect that this ΘE may be original to Codex A, reported only in codex 4.)   The basic arrangement is indeed identical between the two main traditions, A and C. This should be understood in what follows, unless otherwise stated.

TEXTUAL COMMENTS

The main problem here are Steps (end of 4)–9. The words "the angles or the sides of the polygons" (end of 4) prove that some interpolation at least occurred here: the interpolator, facing Archimedes' irritating interchangeable use of "angle" and "side," in this formula, decided not to take sides (so to speak), and to mention "angles *or* sides." Most probably, the entire passage, then, is interpolated (it is indeed very basic: laboring hard to move from ratio of radii to ratio of diameters, just because "diameters" are the objects explicitly mentioned by Euclid!).

Steps 13, 19 may, but need not, be interpolated.

GENERAL COMMENTS

A fully fledged proposition, finally

This is the closest to a fully spelled-out proposition we have had for quite some time (at least since Propositions 24–5). There are a few deviations from the "Euclidean norm:" the enunciation refers backwards to earlier constructions (but this is typical of this treatise, where the same toy objects are referred to again and again), and the construction is resumed inside the proof (that is, inside Steps c–g). (But this is common elsewhere in Greek mathematics, and indeed follows naturally from the fact that the proposition consists of two separate claims.) Still, the general impression is of a spelled-out argument, with both enunciation and careful setting-out. For the first time, the objects resulting from the revolution of the circles-and-polygons are described in a "proper" proposition, not in a proposition with no enunciation such as 23, 28

above. Furthermore, there is some non-trivial mathematical argument. Finally –
to strengthen the sense that we are back into the hard mathematical business –
the theme of proportions between surfaces and between volumes is revived,
after a long pause (it was last seen in Proposition 14 – which took some hard
mathematical work, as well).

Thinking in ratios and in geometrical relations

The way in which the cones are constructed (Steps d–g) is very peculiar:
instead of doing them one cone at a time, they are made through a certain
division of labor, the factory making, first, both bases, and then both heights.
Is Archimedes' thought here based more on the theory of proportions than
on geometrical intuition – so that "bases" and "heights" are mere ciphers for
ratios? This would be an example of a case where thinking with ratios supplants
geometrical thinking. In other ways, however, geometrical thinking seems to
be fundamental to this proposition.

Consider a general problem, arising especially from Steps 4, 15 and 17.
Essentially, the problem is that while the argument is non-trivial, it is also
non-existent. Even with Steps 5–9 assumed authorial (which is very unlikely),
still the central assertions of the proportions of this proposition are not argued
sufficiently. Now, my notes offer (generally following Heiberg) ways to fill in
the gaps, as Eutocius did. But these need not represent Archimedes' own mind.
The fact that the argument can be filled in in various ways does not change the
more basic fact that Archimedes did *not* fill in the argument, himself. The hints
that the text does provide are the two identical Steps 3, 14: "since the polygons are similar." Could Archimedes have relied on some general assumption
such as, e.g., "homologous lines in similar polygons are proportional" (from
which 4, 15 and 17 derive easily)? Possibly but once again, more likely, he
relies here not on some result, proved elsewhere, but on the basic intuition that
similar objects preserve their ratios: that ratios are what is kept intact, when
sizes alone are manipulated. We see here the role of ratio, as the essential *geometrical* feature of objects: it is the one feature that is independent of mere
size.

Letters used as indices

It is interesting to note the equivocating nature of the letters Ξ, Ο: they refer to
both cones and circles, in both Steps d and e. This is an important indication
of the way in which letters are used in Greek mathematics. Such letters are
not algebraic-like symbols, possessing definite references. They are more like
indices: signposts present in the diagram, which may refer to whatever is put,
spatially, next to them, be it "cone" or "circle" (this equivocation is further
aided by the reduction, typical of this treatise, of the three dimensional to the
two dimensional, which tends to blur the distinction between cones and circles
even further).

/ 33 /

The surface of every sphere is four times the greatest circle of the
<circles> in it.

For let there be some sphere, and let the <circle> A be four times
the great circle; I say that A is equal to the surface of the sphere.

(1) For if not, it is either greater or smaller. (a) First let the surface of the sphere be greater than the circle <=Λ>. (2) So there are
two unequal magnitudes: the surface of the sphere, and the circle A;
(3) therefore it is possible to take two unequal lines, so that the greater
<line> has to the smaller a ratio smaller than that which the surface
of the sphere has to the circle.[290] (b) Let the lines B, Γ be taken,
(c) and let Δ be a mean proportional between B, Γ, (d) and also, let
the sphere be imagined cut by a plane <passing> through the center,
at the circle EZHΘ, (e) and also, let a polygon be imagined inscribed
inside the circle, and circumscribed, so that the circumscribed is similar to the inscribed polygon, and the side of the circumscribed has
<to the inscribed> a smaller ratio than that which B has to Δ[291]
[(4) therefore the duplicate ratio, too, is smaller than the duplicate ratio.[292] (5) And the duplicate <ratio> of B to Δ is the <ratio> of B to
Γ,[293] (6) while, duplicate <the ratio> of the side of the circumscribed
polygon to the side of the inscribed, <is> the <ratio> of the surface of
the circumscribed solid to the surface of the inscribed];[294] (7) therefore
the surface of the figure circumscribed around the sphere has to the
surface of the inscribed figure a smaller ratio than the surface of the
sphere to the circle A;[295] (8) which is absurd; (9) for the surface of
the circumscribed is greater than the surface of the sphere,[296] (10) while
the surface of the inscribed <figure> is smaller than the circle A [(11)
for the surface of the inscribed has been proved to be smaller than four
times the greatest circle of the <circles> in the sphere,[297] (12) and the

---

[290] *SC* I.2.   [291] *SC* I.3.

[292] In modern terms, the step argues from (circumscribed side:inscribed side<B:Δ)
to ((circumscribed side:inscribed side)^2 <(B:Δ)^2). (This intuitive derivation is not proved
in the *Elements*.)

[293] Step c, *Elements* VI. 20 Cor. 2.   [294] *SC* I.32.

[295] By applying the two substitutions of Steps 5, 6 on the claim of Step 4, we get the
(implicit) claim that ((surface of circumscribed solid:surface of inscribed solid)<(B:Γ)),
but by Step b (interpreted through Step 3) ((B:Γ)<(surface of sphere:circle A)), and the
claim of Step 7 is seen to hold. As we were used to do in Propositions 13–14, the absurdity
argued in the following few steps is better seen if we apply an implied "alternately" operation (*Elements* V.16) on Step 7, to yield (7*) ((surface of circumscribed solid:surface
of sphere)<(surface of inscribed solid:circle A)).

[296] *SC* I.28.   [297] *SC* I.25.

circle A is four times the great circle]. (13) Therefore the surface of the sphere is not greater than the circle A.

So I say that neither is it smaller. (f) For if possible, let it be <smaller>. (g) And similarly let the lines B, Γ be found, so that B has to Γ a smaller ratio than that which the circle A has to the surface of the sphere, (h) and <let> Δ <be> a mean proportional between B, Γ, (i) and again let it be inscribed and circumscribed,[298] so that the <side> of the circumscribed has <to the inscribed> a smaller ratio than the <ratio> of B to Δ [(14) therefore the duplicates, too];[299] (15) therefore the surface of the circumscribed has to the surface of the inscribed[300] a smaller ratio than [B to Γ.[301] (16) But B has to Γ a smaller ratio than] the circle A to the surface of the sphere; (17) which is absurd; (18) for the surface of the circumscribed is greater than the circle A,[302] (19) while the <surface> of the inscribed is smaller than the surface of the sphere.[303]

(20) Therefore neither is the surface of the sphere smaller than the circle A. (21) And it was proved, that neither is it greater; (22) therefore

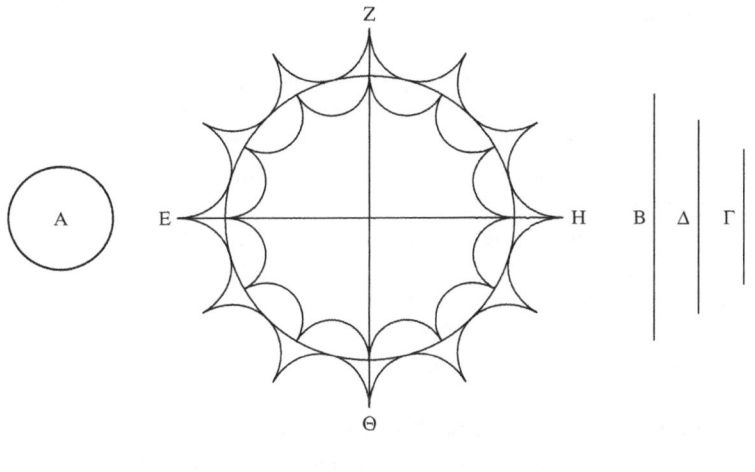

I.33
Codex B, followed by Heiberg, has straight lines instead of arcs in the polygon. Codex D inserts an extra, circumscribing circle. Codex G has the three lines (similarly arranged internally) between the two circles (and not to their right). Codex D has the circle A lower relative to the main figure, while Codex B has a different layout altogether. Codices AC have omitted Θ.

---

[298] No subject for the verb is specified in the original. The text as it stands is very confusing, failing to distinguish between sides, areas, or volumes, or between circum-scribed/inscribed polygons and circumscribed/inscribed solids. The reference would be to the sides of the polygons, if we follow the line of argument from Step e above but, alternatively, the text may assume that all the inequalities can be achieved simultaneously, so that no need is felt to distinguish between the various circumscribed and inscribed pairs.

[299] The same as Step 4 above.

[300] This time the reference is clearly to circumscribed and inscribed *solids* (since only solids have *surfaces*).

[301] This is the same as the *implicit* result of Steps 4–6 above, as explained in note to Step 7.

[302] *SC* I.30.   [303] *SC* I.23.

the surface of the sphere is equal to the circle A, (23) that is to four times the great circle.

TEXTUAL COMMENTS

There are two problems here. The first is Steps 11–12, which are a very elementary unpacking of known results and constructions. They could be genuine, but are likely to be scholiastic.

The second, very complex problem arises with the main argument in both parts of the proof. Heiberg's minimal reading gets rid of Steps 4–6 and 14, and reduces Steps 15–16 to a single step. So in both cases there is, according to Heiberg, no argument at all: the construction gives rise, directly, to the absurd result (in Steps 7, 15+16).

The strongest evidence against Heiberg's hypothesis is that, even with all the bracketed steps considered genuine, the argument is still very sketchy. The passage from 4–6 to 7 leaves out an important implicit claim (see note to Step 7); Step 14 is extremely condensed (to the point of being misleading and ambiguous Greek: "the duplicates," neuter, does not refer to anything clear); Steps 15+16, as I shall explain, are elliptic as well. I therefore tend to believe all those steps are genuine.

Steps 15–16 form an especially intriguing structure. The Palimpsest has: "(15) therefore the surface of the circumscribed has to the surface of the inscribed a smaller ratio than B to Γ. (16) But B has to Γ a smaller ratio than the circle A to the surface of the sphere," while A had: "(15) therefore the surface of the circumscribed has to the surface of the inscribed a smaller ratio than the circle A to the surface of the sphere."

The claim made by manuscript A is left implicit in the Palimpsest. A, on the other hand, omitted the argument for this claim (namely, Step 16 of the Palimpsest). The Palimpsest leaves a *result* unsaid; A left an *argument* unsaid. Mathematically speaking, A seems the more likely reading: Steps 18–19 refer to the claim asserted at A. Purely textually, however, the case of the Palimpsest seems stronger – and textual considerations must take precedence.

The textual argument is this. There are no parallels of a scholiastic expansion of this kind in the text of C, while, on the other hand, the omission of A can be easily understood as an homoeoteleuton ("a smaller ratio than:" in the Greek, a sequence of twenty-one letters, repeated exactly!).

It then seems that Archimedes left a result unsaid at the second part of the proof, while, at the first part of the proof, he (or the interpolator?) has left an argument unsaid (at Step 7). This might be intentional: one thing we have seen throughout this work is a tendency to have parallel passages display as much variation between them as possible.

GENERAL COMMENTS

A sense of accomplishment

Note that the enunciation is much briefer than many preceding enunciations. Mathematical significance tends to be in inverse proportion to length: the key results are important just because they state some direct, simple relation, whereas interim results, of no inherent significance, may often involve cumbersome, ungainly complications. And indeed we have now finally reached a result whose interest is self-evident.

The sense of accomplishment – of the book having reached a goal – is conveyed in several ways. First, the texture of the language is particularly crisp, the brevity of the enunciation being carried over to the argument itself: I return to this in a note below.

Secondly, the narrative placement of the proposition is remarkable. To begin with, we return to use Propositions 2–3 (not directly used since Proposition 5!). The return, here, to such an early proposition, is a majestic display of design. Further, Propositions 2–3 gave rise to a specific procedure, of proof through a double absurdity, manipulating proportions (last used in Proposition 14). We thought this too was a secondary tool, necessary to produce no more than some interim results. Now we discover how crucial the tool is for the main result itself. The entire mechanism is recalled: assuming both "greater" and "smaller;" obtaining absurdities through proportion inequalities; relying on the assumption that the ratio of the greater to the smaller is greater than the ratio of the smaller to the greater. Nothing superfluous: all the threads of the argument are gathered together – a pulling together that contrasts with the seemingly haphazard, centrifugal structure of the first half of the book. For, besides the use of Propositions 2–3, and the recalling of the procedure of Propositions 13–14, one notes the use of many recent results on surfaces and solids, such as *SC* I.23, 25, 28, 30, 32, which in turn of course depended on previous results for polygons and cones. This is the culmination of "chapter 5," the main chapter of the book, and the sphere has been measured: the surface here and, in the following proposition, the even more important volume.

Elegance of expression

Consider the following two details, both typical of the way in which this proposition, while involving so many threads, is still presented as a single, unified argument. Mathematical elegance is obtained through some specific verbal tools.

First, the setting-out: "let the <circle> A be four times the great circle." Notice that the only purpose of this construction is to allow Archimedes to avoid repeating again and again the cumbersome expression "four times the great circle" (or, worse, "four times the greatest circle of the <circles> in the sphere"). It is thus more a verbal than a geometrical construction.

Second, consider Step e: ". . . the side of the circumscribed has <to the inscribed> a smaller ratio than that which B has to Δ" (similarly, Step i). That the words "to the inscribed" can be omitted is a mark of how formulaic this expression has become by now. (Besides revealing, once again, that a ratio is felt to be a property belonging to the antecedent. Incidentally, the repetition of the omission in both Steps e and i proves that this is not a textual corruption: see also Step 8 in the following proposition.) Once again, a certain brevity is allowed. Because we can sometimes predict what the ratio will be *to*, such predictable ratios – the main ratios negotiated in this proposition – may be directly assigned

to individual objects, simplifying the usual complication arising from the fact
that ratio and proportion are many term relations.

In many previous propositions, Archimedes, it seemed, practically reveled
in cumbersome objects standing to each other in complex relations. (Recall the
line equal to all lines parallel to one line – whichever – of the lines subtended
by two sides of the polygon! Remember the proportions in which it partici-
pated! Consider, e.g. the enunciation of Proposition 21.) The elegance of this
proposition is designed to contrast with such cumbersome propositions – just
as its pulling together of earlier results is designed to contrast with an earlier,
seemingly haphazard structure.

/34/

Every sphere is four times a cone having a base equal to the greatest
circle of the <circles> in the sphere, and, <as> height, the radius of
the sphere.

For let there be some sphere and in it a great circle ABΓΔ. (1) Now,
if the sphere is not four times the said cone, (a) let it be, if possible,
greater than four times; (b) and let there be the cone Ξ, having a base
four times the circle ABΓΔ, and a height equal to the radius of the
sphere. (2) Now, the sphere is greater than the cone Ξ.[304] (3) So there
will be two unequal magnitudes: the sphere and the cone. (4) Now, it is
possible to take two unequal lines, so that the greater has to the smaller
a smaller ratio than that which the sphere has to the cone Ξ.[305] (c) Now,

Eut. 265   let them <=the two unequal lines> be K, H, (d) and I, Θ taken so that
they exceed each other, K <exceeding> I, and I <exceeding> Θ, and
Θ <exceeding> H, by an equal <difference>,[306] (e) and let also a
polygon be imagined inscribed inside the circle ABΓΔ (let the number of its sides be measured by four), (f) and another, circumscribed,
similar to the inscribed, as in the earlier <constructions>, (g) and let
the side of the circumscribed polygon have to the <side> of the inscribed a smaller ratio than that which K has to I,[307] (h) and let AΓ,
BΔ be diameters in right <angles> to each other. (i) Now, if the plane,
in which are the polygons, is carried in a circular motion (the diameter AΓ remaining fixed), (5) there will be figures, the one inscribed
in the sphere, the other circumscribed, (6) and the circumscribed will
have to the inscribed a ratio triplicate of the side of the circumscribed

---

[304] Steps a, b; Interlude, recalling *Elements* XII.11.   [305] *SC* I.2.
[306] The resulting sequence of lines is an *arithmetical* progression, with equal *differences*. See Eutocius' commentary for further discussion.
[307] *SC* I.3.

<polygon> to the <side> of the <polygon> inscribed inside the circle ΑΒΓΔ.[308] (7) But the side has to the side a smaller ratio than K to I; (8) so that the circumscribed figure[309] has a smaller ratio than triplicate <of the ratio> of K to I. (9) But, also, K has to H a greater ratio than triplicate that which K has to I[310] [(10) for this is obvious through the lemmas];[311] (11) much more, therefore, that which was circumscribed has to the inscribed a smaller ratio than that which K has to H. (12) But K has to H a smaller ratio than the sphere to the cone Ξ;[312] (13) and alternately;[313] (14) which is impossible; (15) for the circumscribed figure is greater than the sphere,[314] (16) while the inscribed is smaller than the cone Ξ [(17) through the fact that the cone Ξ is four times the cone having a base equal to the circle ΑΒΓΔ, and a height equal to the radius of the sphere,[315] (18) while the inscribed figure is smaller than four times the said cone].[316] (19) Therefore the sphere is not greater than four times the said <cone>.

(j) Let it be, if possible, smaller than four times; (20) so that the sphere is smaller than the cone Ξ.[317] (k) So let the lines K, H be taken, so that K is greater than H and has to it a smaller ratio than that which the cone Ξ has to the sphere,[318] (l) and let Θ, I be set out, as before, (m) and let a polygon be imagined inscribed inside the circle ΑΒΓΔ, and another circumscribed, so that the side of the circumscribed has to the side of the inscribed a smaller ratio than K to I, (n) and the rest constructed in the same way as before; (21) therefore, the circumscribed solid figure will also have to the inscribed a ratio triplicate the side of the <polygon> circumscribed around the circle ΑΒΓΔ to the <side> of the inscribed.[319] (22) But the side has to the side a smaller ratio than K to I; (23) so the circumscribed figure will have to the inscribed a smaller ratio than triplicate that which K has to I. (24) And K has to H a greater ratio than triplicate that which K has to I;[320] (25) so that the circumscribed figure has to the inscribed a smaller ratio than

---

[308] *SC* I.32.

[309] Again, the words "to the inscribed" are omitted. (Compare Steps e, i in the preceding proposition.)

[310] See Eutocius' commentary on Step d above.

[311] Most probably, a reference to Eutocius.

[312] Step c. The implicit result of Steps 11–12 taken together is (circumscribed: inscribed<sphere:cone). It is to this implicit result that Step 13 refers.

[313] I.e: (circumscribed:sphere<inscribed:cone) (*Elements* V.16).   [314] *SC* I.28.

[315] Step b; Interlude recalling *Elements* XII.11.   [316] *SC* I.27.

[317] Steps b, j; Interlude, recalling *Elements* XII.11.   [318] *SC* I.2.

[319] *SC* I.32. As for the "also," it refers to the earlier use of the same principle in the first part of this proposition.

[320] See Eutocius' commentary.

K to H. (26) But K has to H a smaller ratio than the cone Ξ to the sphere[321]; (27) which is impossible; (28) for the inscribed is smaller than the sphere,[322] (29) while the circumscribed is greater than the cone Ξ.[323] (30) Therefore neither is the sphere smaller than four times the cone having the base equal to the circle ΑΒΓΔ and, <as> height, the <line> equal to the radius of the sphere. (31) And it was proved that neither is it greater. (32) Therefore <it is> four times.

/Corollary/

And, these being proved, it is obvious that every cylinder having, <as> base, the greatest circle of the <circles> in the sphere, and a height equal to the diameter of the sphere, is half as large again as the sphere, and its surface with the bases is half as large again as the surface of the sphere.

(1) For the cylinder mentioned above is six times the cone having the same base, and a height equal to the radius <=of the sphere>,[324] (2) and the sphere has been proved to be four times the same cone;[325] (3) so it is clear that the cylinder is half as large again as the sphere. (4) Again, since the surface of the cylinder (without the bases) has been proved equal to a circle whose radius is a mean proportional between: the side of the cylinder, and the diameter of the base,[326] (5) and the side of the said cylinder (which is around the sphere) is equal to the diameter of the base [(6) it is clear that their mean proportional will then be equal to the diameter of the base],[327] (7) and the circle having the radius equal to the diameter of the base, is four times the base,[328] (8) that is <four times> the greatest circle of the <circles> in the sphere, (9) therefore the surface of the cylinder without the bases, too, will be four times the great circle; (10) therefore the whole surface of

---

[321] Step k. The implicit result of Steps 25–6 is: ((the circumscribed figure:the inscribed)<(cone Ξ:sphere)). It is to this implicit result that Step 27 refers.

[322] *SC* I.28.   [323] *SC* I.31.

[324] *Elements* XII.10: the cone with the same base and height as the cylinder is one third the cylinder. Interlude, recalling *Elements* XII.14: cones with the same base are to each other as their height (the cylinder mentioned here has the *diameter* as its height, the cone has the *radius* as its height).

[325] *SC* I.34.   [326] *SC* I.13.

[327] The assumption is that if A:B::B:C, and A=C, then A=B=C. (See general comments.) The implicit result of Steps 4–6 is that "the surface of the cylinder (without the bases) is equal to a circle, whose radius is the diameter of the base." It is this result which, together with Step 8, yields Step 9.

[328] *Elements* XII.2.

the cylinder, with the bases, will be six times the great circle. (11) And also, the surface of the sphere is four times the great circle.[329] (12) Therefore the whole surface of the cylinder is half as large again as the surface of the sphere.

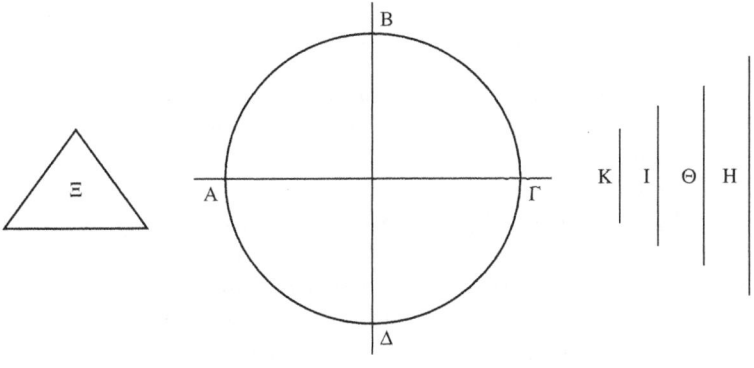

Codex C is not preserved for this diagram.   Codex B has switched, in a sense correctly, the order of sizes among the four lines. Heiberg naturally does the same. A mistake may have been made on codex A; or perhaps all the diagram sets out to do is to represent "four lines in a sequence of sizes," the actual metrical relations being ignored. Codex B represents the cone Ξ, ingeniously, as in the thumbnail. Heiberg introduces circumscribing and inscribed polygons.

TEXTUAL COMMENTS

As usual, the title "Corollary" is a modern editorial intervention, and the relation between the main proposition and its obvious conclusion is an issue we need to investigate based on the text itself.

Step 10 is clearly an interpolation (probably later than Eutocius). Steps 17–18 are "trivial" and worrying in another way too: they are a backwards-looking justification to a backwards-looking justification. Such a double break of the generally onwards-pressing argument is remarkable, and would be much easier to understand if they were scholiastic. Still, they *could* be genuine.

Heiberg argues that Step 6 in the corollary is unclear and leaves the result implicit, and so non-Archimedean: but would omitting it make things any more clear or explicit? If an image of Archimedes emerges from this treatise, it is of a mathematician who enjoyed the enigmatic: the unexpected result emerging from the shadows. (The Roman soldier was asked to extinguish his lamp, not to interfere with the darkness – and then lost his temper.)

GENERAL COMMENTS

Why state the result in a corollary?

The corollary of this proposition is the climax of the book. The words at the start of the corollary, "these being proved, it is obvious . . . ," should therefore be perceived in their full sense of drama: as referring back to *everything proved so far* – to the entire book. Having gone this way, the result is not merely convincing: it is obvious. The amazing has become obvious, and we have

---

[329] *SC* I.33.

achieved the essential goal of Greek science. Thus the decision to place the main result in a corollary is in a sense natural. The book is a tool, designed to make us see the inevitability of the main result: the entire demonstrative apparatus is a tool for overcoming the need for demonstration.

Strict proposition structure versus ease of geometrical manipulation

This proposition is on the whole "complete" – with a clear enunciation and setting-out, and a real proof. One element is missing, though: the definition of goal. In other words, Archimedes does not stop before the proof to say "it is to be proved, that . . ." This can be explained as follows. The enunciation requires a sphere, a great circle, and a cone (with base=*great circle*, height=radius of the sphere). The brief setting-out specifies only the sphere and the great circle. The cone of the enunciation, however, is not constructed in this proof. What is constructed is another cone (base=*4*great circle*, height=radius of the sphere). Archimedes needs this other cone, because (1) the proposition deals with a relation 1:4, (2) his basic tool, Proposition 2, deals with inequalities of the form a>b, not a>4*b. He therefore prefers to utilize in the demonstration proper a cone different from that of the general enunciation (there, the cone (base=great circle) was naturally preferred, as being meaningful in terms of the geometrical configuration). The bottom line is that Archimedes needs one cone for the purposes of the geometrical configuration, and another cone for the purposes of the tools of proportion-theory. He is economic – in both enunciation and demonstration he uses only the cones he really requires. And he prefers this economy to a strict adherence to the ideal structure of the proposition.

Various ways of "taking as obvious"

As in the preceding proposition, there are several cases where important interim results remain implicit. Take for instance the structure of Steps 11–13: (11) circumscribed:inscribed<K:H. (12) K:H<sphere:cone Ξ. (13) "Alternately." We are asked to supply two acts of demonstrative imagination: first, to get from Steps 11–12 to the unasserted (circumscribed:inscribed<sphere:cone); second, to unpack the "alternately" at Step 13 to mean (circumscribed:sphere<inscribed:cone).

Each act of imagination is in itself manageable, a trivial omission. This is like reading the chess column: the diagram of the position is given, and the first one or two moves are clearly "read" inside the diagram. But at some stage most readers begin to lose touch. It is one thing to imagine an operation on a *present* position, it is quite another to imagine an operation on an *imagined* position. Most chess readers would therefore at some point reach for their boards. Did Archimedes want the ancient readers to reach for their geometrical board – did he want his readers to *work* – so that, by working through the argument, they will all the more appreciate it?

This brings us back to one of our main themes: what does the reader take for granted? What does Archimedes take for granted? I offer footnotes with

"Archimedes' tool-box," that is, results *implied* by the arguments. Some of these are clearly assumed explicitly – for instance, *Elements* VI.16 – that if four magnitudes are in proportion, the rectangle contained by the extremes is equal to that contained by the means. But some other results may be just all too obvious for Archimedes: for instance, the extension of proportion-theory to inequalities of ratios (the assumption, for instance, that if a:b>c:d, then a:c>b:d).

The text signals varying degrees of obviousness. Some arguments are spelled out; some are skipped completely. For example, take the corollary, and the multiplication table. Step 1 implies 3*2=6; Step 3 implies (an equivalent of) 6/4=1.5, as does Step 12. So the multiplication table is indeed taken for granted. However, it is not completely implicit – the corollary works hard to state, explicitly, the numbers six and four. One is tempted to conclude, therefore, that the multiplication table was not as directly accessible to the Greek reader as it is to the modern reader.

/35/

The surface of the figure inscribed inside the segment of the sphere[330] is equal to a circle, whose radius is equal, in square, to the <rectangle> contained by: one side of the polygon inscribed in the segment of the great circle, and <by> the <line> equal to: all the <lines> parallel to the base of the segment, with the half of the segment's base.

Let there be a sphere and in it a segment whose base <is> the circle around the <diameter> AH [. Let a figure be inscribed inside it[331] – as has been said – contained by conical surfaces], and <let there be> a great circle AHΘ, and an even-sided polygon AΓEΘZΔH, (without the side AH),[332] and let a circle be taken – Λ – whose radius is equal, in square, to the <rectangle> contained by the side AΓ and by: all the <lines> EZ, ΓΔ, and also the half of the base, that is AK; it is to be proved that the circle is equal to the surface of the figure.

(a) For let the circle M be taken, whose radius is, in square, the <rectangle> contained by the side EΘ and <by> the half of EZ; (1) so the circle M will then be equal to the surface of the cone whose base <is> the circle around EZ, and <whose> vertex <is> the point

---

[330] This expression, "the figure inscribed inside the segment of the sphere" is an extension of "the figure inscribed inside the sphere," with the rotation now not of a polygon, but of a segment of a polygon.

[331] "Inside it"="inside the segment" (*not* inside the *sphere*; this is seen from Greek genders).

[332] "Without:" i.e. for the purposes of counting the number of sides and getting an even number AH is ignored.

Θ.[333] (b) And let also another <circle> be taken, N, whose radius is equal, in square, to the <rectangle> contained by EΓ and <by> the half of EZ, ΓΔ taken together; (2) this will be equal to the surface of the cone between the parallel planes at EZ, ΓΔ.[334] (c) And let another circle be taken similarly, Ξ, whose radius is, in square, the <rectangle> contained by: AΓ, and <by> the half of ΓΔ, AH taken together; (3) now, this, too, is equal to the conical surface between the parallel planes at AH, ΓΔ.[335] (4) Now, all the circles will be equal to the whole surface of the figure, and their radii will be equal, in square, to the <rectangle> contained by one side, AΓ, and <by> the <line> equal to: EZ, ΓΔ and the half of the base, AK. (5) But the radius of the circle Λ, too, was equal in square to the same area; (6) therefore the circle Λ will be equal to the circles M, N, Ξ; (7) so that <it will be equal> to the surface of the inscribed figure, too.

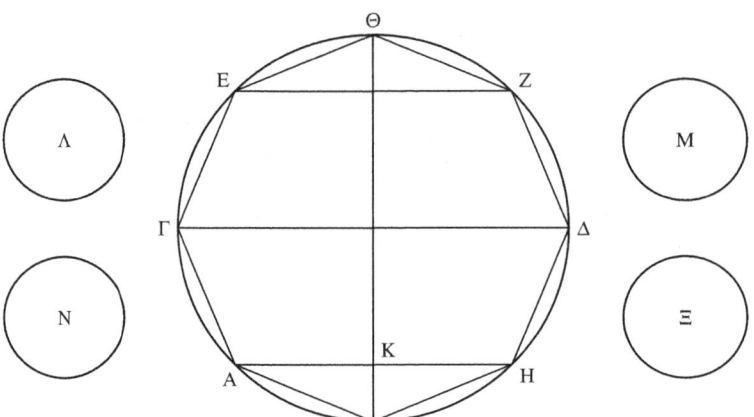

I.35
Codex C has the circles Λ, M (equal to each other) greater than the circles N, Ξ (equal to each other), which may be preferable. Codex B has the circle Λ greater than the remaining three small circles (equal to each other). Heiberg of course arranges all four circles by size.

TEXTUAL COMMENTS

The setting-out includes the passage: "[. Let a figure be inscribed inside it – as has been said – contained by conical surfaces]." The original makes for strange Greek syntax, because of the absence of any connector. (This is represented by my comma immediately after the square brackets.) So this could be a scholion incorporated lamely into the text. Heiberg further notes that this sentence would be more natural following the construction of the inscribed polygon, a sentence later: so, Heiberg suggests, the scholion was in fact originally meant to be based on the sentence following it.

However, Heiberg goes on to note, transplanting the clause alone will not help: what becomes of the reference "it" in this clause? And where would the

---

[333] *SC* I.14.   [334] *SC* I.16.   [335] *SC* I.16.

reference to "the figure" (later referred to in the proposition) come from, if this clause is indeed assumed to be a scholion? Finally, then, Heiberg offers the following hypothesis: that Proposition 35 as a whole ought to have followed Proposition 36 (which, as we shall see immediately, is another construction-proposition, similar to 23 and 28 above). Then the structure 36/35 becomes similar to the sequence 23/24 from earlier in the book, and the references of so-called Proposition 35 are settled through those of Proposition 36.

This is a coherent story: originally, the sequence was 36–35; then it was confused by some copier; 35 became uneasily elliptic as a consequence, and the scholion was added; later on it was incorporated into the text.

On the other hand, even so, Proposition 35 as it stands now in the text remains strange. Proposition 24 – although immediately following 23 (as well as will later be the case with Propositions 37–8) – does contain a construction of the figure, and does not just take it for granted. Thus one is tempted to offer a story alternative to Heiberg's: that the difference between 35 on the one hand, and 24, 37/38 on the other hand, is precisely that Proposition 35 comes *before* its construction-proposition. Because he has not yet constructed the figure explicitly, Archimedes refers to it in a half-hearted, vague way (notice, for instance, that the diameter around which the figure should be rotated is not constructed at all in 35! It does occur in the diagram, where the polygon is filled in as well: the diagram, and not the text, furnishes the continuity with earlier propositions). Whichever hypothesis we take, then, the relation between 35 and 36 is problematic, on both textual and substantial levels.

GENERAL COMMENTS

The nature of the final "chapter 6"

We have now reached the final "chapter" of the book, and it is something of an odd one out.

In architectonic terms, Archimedes made an important decision in delaying the preparatory material concerning segments of spheres until this stage. The alternative would have been to have first all the preparatory material (for spheres and for segments of spheres alike), and then proceed to apply the preparations in a brief sequence of concluding propositions. Instead, he chose to deal with the sphere first, and then to devote a substantial part of the work (about a quarter) to segments of spheres.

The truth is that everything would be an anti-climax following Proposition 34, and Archimedes faces a real difficulty. The propositions on segments of spheres logically depend upon the propositions on spheres, and therefore must follow them. On the other hand, segments are less interesting than spheres, and therefore must be an anti-climax. The solution adopted by Archimedes was to put a brave face on it and to behave as if segments were just as interesting. This proposition, however, is essentially identical to 24, and the sense of repetition now is inevitable.

Ambiguous language, its significance and its practical resolution

Take for instance the first sentence of the setting-out: "Let there be a sphere and in it a segment, whose base <is> the circle around the <diameter> AH." There is a consistent ambiguity in this work between "segment" as two dimensional or three dimensional, so that those two meanings can be told apart only by contextual clues. Here this leads to something of a clash: the expression "let there be a sphere and in it . . ." is well known with the completion "a great circle." Hence, the "segment" is first read as two dimensional, on analogy from the two-dimensional great circle. However, the text continues to give the base of the segment as a circle – hence the segment must be three dimensional. The interest of this is in how untroubled Archimedes seems to be with such potential ambiguities. The references of objects are not seen as problematic – and, typically, the identity of AH can be settled through the diagram – if by the diagram alone. It is through such ambiguities, taken with such lack of concern, that we draw a sense of Archimedes' mathematics as non-symbolic, as focused on the geometrical substance rather than the verbal form.

A central case of ambiguity in this proposition is the very expression "the figure inscribed inside the segment of the sphere," appearing in the enunciation. What is this? We have become used to "the figure inscribed inside the sphere." But – unless this was meant to follow Proposition 36 – the text gives no direct clue to what "inscribed inside the *segment*" means. We can only understand it through the diagram and through a mathematical intuition of the kind of object to make sense in this construction. Yet I do not think, finally, that this is an argument for putting 35 after 36. It is an argument for showing how much Archimedes relies upon analogy, rather than upon definition, in developing his conceptual scheme.

In general, the reader is expected to fill in details, based on his or her growing understanding of the mathematical situation. For instance the polygon is said to be "even-sided," no more; that it is equilateral, too, we assume on the basis of preceding propositions, formulaic language, and the diagram. I give two further elementary examples. First, for an example of the role of formulaic language in this process of disambiguation in practice, take the expression ". . . and an even-sided polygon AΓEΘZΔH, (without the side AH)." The position of the clause "(without the side AH)" is awkward. Something like "the polygon AΓEΘZΔH, having the number of its sides, excluding AH, even" would have made better Greek and mathematical sense. But "an even-sided polygon" is a formula, not to be broken: because it is read as a formula, then, we understand that normal syntax can be broken, so as not to break the formula. We thus reapply the clause to the relevant part, based on our understanding of the local mathematical language. Second, for an example of the role of the lettered diagram in the same process, take the lettering of the circle and of the polygon: AHΘ as against AΓEΘZΔH. The circle is counter-clockwise; the polygon is clockwise. Such counter directions may be used (here and elsewhere in Greek mathematics) to differentiate further between figures whose end-points are similar. "Direction" is a further meaning in the diagram, introduced to it by lettering, and which can be employed for clarifying distinctions between objects.

/36/

(a) Let a sphere be cut by a plane <not passing> through the center, (b) and <let there be> in it a great circle, AEZ, cutting, in right <angles>, the cutting plane, (c) and let a polygon, equilateral and even-angled (without the base AB), be inscribed inside the segment ABΓ. (d) So similarly to the earlier <propositions>, if the figure is carried in a circular motion (ΓZ remaining fixed), (1) the angles Δ, E, A, B will be carried along circles whose diameters <are> ΔE, AB, (2) while the sides of the segment[336] <will be carried> along a conical surface, (3) and there will be the resulting solid figure, contained by conical surfaces, having, <as> base, a circle whose diameter <is> AB, and Γ <as> vertex. (4) So similarly to the earlier <propositions>,[337] (5) it will have the surface smaller than the surface of the segment[338] (which contains it); (6) for their same limit – <namely, the limit> of the segment <=the spherical segment> and of the figure <= the figure contained by conical surfaces> is in a plane – <namely,> the circumference of the circle whose diameter <is> AB, (7) and the surfaces are both concave in the same direction, (8) and one is contained by the other.[339]

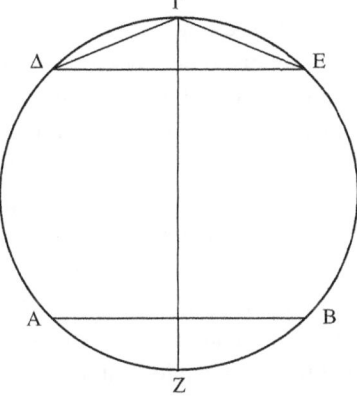

I.36
Heiberg has transposed AB (and the remainder following) to above the diameter, and has introduced straight lines AΔ, BE. See general comments.

TEXTUAL AND GENERAL COMMENTS

Heiberg intensely disapproved of this proposition. Added to his suggestion that the order of 35/36 should be reversed, he red-inks here the following:

---

[336] "Segment" means, in this context, "polygon."

[337] The reference, here as in Step d, is vague. The most obvious candidate in both cases for "earlier <propositions>" is the two construction-propositions 23, 28, but here the reference could equally be to all the earlier uses of Post. 4.

[338] Here, "segment" means "segment of sphere."   [339] Post. 4.

"even-angled" in Step c instead of the customary "even-sided;" "segment,"
"conical surface" in Step 2 instead of "polygon," "conical surfaces" respectively; and the awkward grammar of Step 6. Besides, he considers the whole
sequence of Steps 4–8 mathematically superfluous. Perhaps, this proposition
is especially mutilated; or alternatively, we see here Archimedes at his most
absent-minded, in a proposition of secondary importance.

The diagrams here contain a brilliant contrivance (indeed, one hopes it is
authorial): by not filling in the lines of the polygon between ΔE and AB, the
diagram does not affirm any specific number of sides. This is the diagrammatic
equivalence of a modern expression such as "and let its sides be B . . . EΓ,
ΓΔ, . . . A." The "blank" parts of the polygon in the diagram are equivalent to
the three dots in our written convention. (Heiberg's intended correction of the
diagram, introducing straight lines AΔ, BE, spoils the effect.)

/ 37 /

The surface of the figure inscribed in the segment of the sphere is
smaller than the circle whose radius is equal to the <line> drawn from
the vertex of the segment on the circumference of the circle which is
<the> base of the segment.

Let there be a sphere and in it a great circle ABEZ, and let there
be a segment in the sphere, whose base <is> the <circle> around the
diameter AB, [and let the said figure be inscribed inside it, and <let
a> polygon <be inscribed> inside the segment of the circle], and the
rest the same, with the diameter of the sphere being ΘΛ, and with ΛE,
ΘA being joined; and let there be a circle M (let its radius be equal to
AΘ); it is to be proved that the circle M is greater than the surface of
the figure.

(1) For the surface of the figure has been proved equal to a circle,
whose radius is equal, in square, to the <rectangle> contained by: EΘ,
Eut. 266   and <by> EZ, ΓΔ, KA;[340] (2) but the <rectangle contained> by EΘ
and <by> EZ, ΓΔ, KA has been proved equal to the <rectangle> con-
Eut. 266   tained by EΛ, KΘ;[341] (3) and the <rectangle contained> by EΛ, KΘ
is smaller than the <square> on AΘ [(4) for <it is> also <smaller>
than the <rectangle contained by> ΛΘ, KΘ];[342] (5) so it is obvious
that the radius of the circle which is equal to the surface of the figure,

---

[340] *SC* I.35.   [341] *SC* I.22, further interpreted by *Elements* VI.16.

[342] See Eutocius for the argument, which works (in this order) through *Elements*
III.31, VI.4, VI.17, and III.15.

is smaller than the radius of the <circle> M;[343] (6) therefore it is clear that the circle M is greater than the surface of the figure.[344]

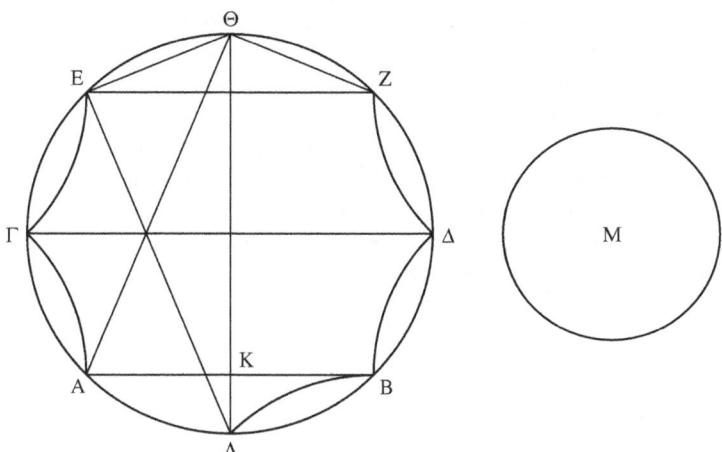

I.37
Codices BDG, followed by Heiberg, have straight lines instead of arcs in the polygon. Codices BD add a straight line AΛ; Heiberg removed the line ΛB, further transposing the line AB (and the remainder following) to above the diameter. Codex C is fragmentary here and the bottom of the main circle is lost.

TEXTUAL COMMENTS

In the setting-out, the bracketed passage offended Heiberg by being too similar to the passage, also bracketed by him, in the setting-out of Proposition 35. He also notes that the words "and the rest the same," following the bracketed passage, could follow nicely the imperative "let there be," from before the bracketed passage, adding some further plausibility to the brackets. Of course, none of this is conclusive.

Step 4 seems like a very brief summary of Eutocius' commentary (in which case, it would have to be very late). Alternatively, Eutocius' commentary could have been a spelling-out of this brief step. I discuss this in a footnote to Eutocius' commentary.

GENERAL COMMENTS

Another look at the history of "the obvious"

SC I.22 proved that, given a certain construction, a proportion holds: "The drawn <lines> and half the base have to the height of the segment the same ratio, which the <line> joined from the diameter of the circle to the side of the polygon <has> to the side of the polygon." This is used in Step 2 of this proposition: "The <rectangle contained> by EΘ and by EZ, ΓΔ, KA has been proved equal to the <rectangle> contained by EΛ, KΘ."

Using a modern notation, we may say that while Proposition 22 proved that a:b::c:d, Step 2 recalls this result in the form a*d=b*c.

---

[343] Setting-out: the radius of M is equal to AΘ.   [344] *Elements* XII.2.

The equivalence between two forms of this type is proved in *Elements* VI.16, so Archimedes is walking on safe ground. But it should be noticed that, at first glance, it seems as if, for Archimedes, Proposition 22 proved the claim of Step 2 *directly* – as if, for him, VI.16 were so central, that a:b::c:d and a*d=b*c seem not as two separate (but mutually derivable) claims, but as two notational variations of the very same claim. This was not the case for Eutocius, who comments explicitly, now, on the equivalence between the two expressions. What counts as a derivation, then, and what as a notational variation? This is a theme in the history of mathematics, and we see here a case where such questions are differently answered, by different authors. It is fair to suggest, perhaps, that commentators take more time with their arguments, and therefore would see more closely the fine-grained structure of derivations: less would be "obvious" to them.

/38/

The figure inscribed in the segment, contained by conical surfaces, with the cone having a base the same as the figure, and, <as> vertex, the center of the sphere, is equal to the cone having a base equal to the surface of the figure, and a height <equal> to the perpendicular drawn from the center of the sphere on one side of the polygon.

For let there be a sphere and in it a great circle and a segment smaller than a semicircle, ABΓ, and a center E, and let a polygon[345] (even-sided, without AΓ) be inscribed inside the segment ABΓ, similarly to the earlier <propositions>, and let the sphere, carried in a circular motion (BA remaining fixed)[346] make some figure contained by conical surfaces, and, upon the circle around the diameter AΓ, let a cone be set up having, <as> vertex, the center, and let a cone be taken, K, having a base equal to the surface of the figure, and a height <equal> to the perpendicular drawn from the center E on one side of the polygon; it is to be proved that the cone K is equal to: the contained figure, with the cone AEΓ.

(a) So also, let cones be set up on the circles around the diameters ΘH, ΔZ, having, <as> vertex, the point E; (1) now indeed, the solid rhombus HBΘE is equal to a cone, whose base is equal to the surface of the cone HBΘ, while <its> height <is equal> to the perpendicular drawn from E on HB,[347] (2) and the remainder contained by the surface between the parallel planes at HΘ, ZΔ and <by> the conical

---

[345] "Equilateral" is omitted.

[346] The letter A, confusingly, is used twice in the circle. See textual comments.

[347] *SC* I.18.

<surfaces> ZEΔ, HEΘ is equal to a cone, whose base is equal to the surface between the parallel planes at HΘ, ZΔ, while <its> height <is equal> to the perpendicular drawn from E on ZH.[348] (3) Again, the remainder contained by: the surface between the parallel planes at ZΔ, AΓ, and <by> the conical <surfaces> AEΓ, ZEΔ is equal to a cone, whose base is equal to the surface between the parallel planes at ZΔ, AΓ, while <its> height <is equal> to the perpendicular drawn from E on ZA;[349] (4) now, the said cones[350] will be equal to: the figure, together with the cone AEΓ. (5) And they have a height equal to the perpendicular drawn from E on one side of the polygon,[351] (6) and bases equal to the surface of the figure AZHBΘΔΓ; (7) but the cone K, too, has the same height and a base equal to the surface of the figure; (8) therefore the cone is equal to the said cones. (9) But the said cones were proved equal to the figure and the cone AEΓ. (10) Therefore the cone K, too, is equal to the figure and the cone AEΓ.

/Corollary/

From this it is obvious that the cone having: <as> base, the circle whose radius is equal to the <line> drawn from the vertex of the segment to the circumference of the circle which is <the> base of the segment;[352] and a height equal to the radius of the sphere – is greater than the inscribed figure with the cone; (1) for the cone mentioned above[353] is greater than the cone equal to: the figure, together with the cone having, <as> base, the base of the segment, and the vertex at the center, (2) that is, <it is greater> than the <cone> having the base equal to the surface of the figure, and a height <equal> to the perpendicular drawn from the center on one side of the polygon;[354] (3) for both the base is greater than the base [(4) for this has been proved][355] (5) and the height <is greater> than the height.[356]

---

[348] *SC* I.20.   [349] *SC* I.20.   [350] I.e. the three cones introduced in Steps 1–3.

[351] All these perpendiculars are equal through the equilaterality of the polygon, as well as *Elements* III.14.

[352] E.g. with the diagram of this proposition, the line AB.

[353] A rather confusing expression (there are so many cones mentioned above!). The intended reference is to the complicated cone introduced in the first sentence of the corollary.

[354] This step spells out Step 1 in terms of the equality proved at *SC* I.38.

[355] *SC* I.37.

[356] This is because the radius is greater than the perpendicular drawn from the center on an internal line, a result that is systematically taken for granted in this book, e.g. in Proposition 13, Step 44 (not directly a result proved in the *Elements*, but related to results such as *Elements* III.15).

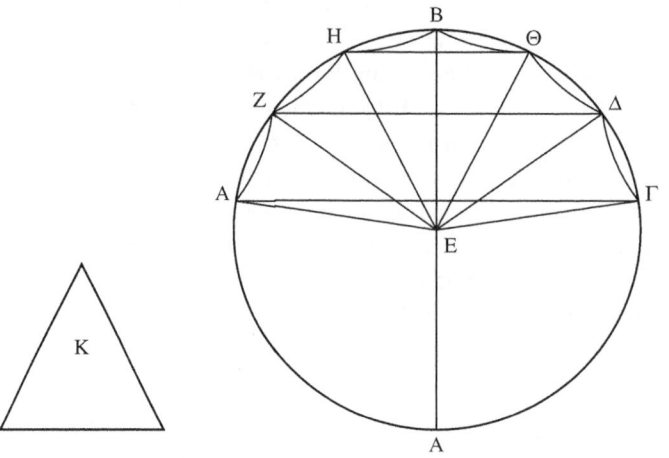

I.38
Codex B, followed by Heiberg, has straight lines instead of arcs in the polygon.
Codices GH introduce a straight line AB.
Codex A does not have the bottom A, and the letter is supplied on the basis of a faint trace in codex C as well as the text (see textual comments). Codex B, followed by Heiberg, has introduced the letter Λ instead.

TEXTUAL COMMENTS

The text refers in the setting out to the line BA, clearly intending the diameter drawn down from the vertex B. There is no letter at the other end of this diameter: it was almost certainly omitted by scribal error. Heiberg thought the omitted letter was Λ, not A, and that an obvious scribal error is responsible for the A in the manuscripts. In this he may be right, particularly since the letter A is already taken up by another point on the circle – while the letter Λ itself is free and immediately follows alphabetically upon the last otherwise used letter, K. For reasons similar to those expressed in the textual comments to I.3, I keep, without any certainty, the manuscripts' reading.

Heiberg had a difficulty with Step 4 of the corollary, whose words drive a wedge, as it were, between 3 and 5 (so that Heiberg wished to remove it); but Greek can be very tolerant in such matters, and no decision can be made.

GENERAL COMMENTS

Archimedean carelessness as regards detail

First, notice how the enunciation and definition of goal have "the figure *with* the cone" (*sun*), while Step 4 of the proof has "the figure *together with* the cone" (*meta*), Step 9 has "the figure *and* the cone" (*kai*) and Step 10 has "*both* the figure *and* the cone" (*te . . . kai*). The language of addition is anarchic, here as elsewhere in Greek mathematics.

The following is mathematically more significant. The proposition begins with the words: "The figure inscribed in the segment . . ." Now, that the segment is that of a sphere (and not of a circle) is by now a known aspect of the practice. More interesting, however, is the fact that the proposition is true, in fact, only for a *particular* type of segment – namely only those smaller than a hemisphere. So why is that not said in the enunciation? Is this textual corruption, which made us lose a qualification Archimedes himself had made? (Not, I would think, the most plausible explanation.) Does this represent a reliance upon the diagram?

(But, generally speaking, the diagram is not supposed to give information as regards *sizes*.) One is tempted to see here, once again, a certain willingness on Archimedes' part to let the readers sort out any residual ambiguities for themselves. The basic thrust of the argument, at this stage of the book, is sufficiently clear for such details not to make a difference.

/39/

Let there be a sphere and in it a great circle, ABΓ, and let <a segment> (which the <line> AB cuts), smaller than a semicircle, be cut, and <let there be> a center, Δ, and let the <lines> AΔ, AB be joined from the center Δ to the <points> A, B, and, around the resulting sector,[357] let a polygon[358] be circumscribed, and around it a circle. (1) So it <=the circle around the circumscribed polygon> will have the same center as the circle ABΓ. (a) So if the polygon, carried in a circular motion (EK remaining fixed), returns to the same <position>, (2) the circumscribed circle will be carried along a surface of a sphere, (3) and the angles of the polygon will draw circles, whose diameters (being parallel to AB) join the angles of the polygon, (4) and the points at which the sides of the polygon touch the smaller circle draw circles (in the smaller sphere),[359] whose diameters will be the <lines> joining the touching points (being parallel to AB), (5) and the sides will be carried along conical surfaces, (6) and there will be the circumscribed figure, contained by conical surfaces, whose base <is> the circle around the <diameter> ZH; (7) so the surface of the said figure is greater than the surface of the smaller segment, whose base is the circle around the <diameter> AB.

(b) For let tangents be drawn: AM, BN; (8) therefore they will be carried along a conical surface, (9) and the figure resulting from the polygon AMΘE<Λ>NB[360] will have the surface greater than the segment of the sphere whose base <is> the circle around the diameter AB [(10) for they have the same limit in one plane, <namely> the circle around the diameter AB, (11) and the segment is contained by the figure].[361] (12) But the surface of a cone, that has resulted from

Eut. 266

---

[357] N.B.: "sector," rather than the "segment" we were used to. (The difference is that a sector is contained by a circumference and two radii, while a segment is contained by a circumference and a chord.)

[358] Not just any polygon: the reference, by now formulaic, is to the equilateral, even-sided polygon.

[359] We learn from this step of the existence of another, *smaller* sphere.

[360] I.e.: the solid figure resulting from the circular motion of this polygon.

[361] Post. 4.

ZM, HN, is greater than the <surface> that has resulted from MA, NB; (13) for ZM is greater than MA ((14) for it is subtended by a right <angle>),[362] (15) while NH <is greater> than NB, (16) and when this is <the case>, the surface is then greater than the surface [(17) for these have been proved in the lemmata].[363] (18) So it is clear that the surface of the circumscribed figure, too, is greater than the surface of the segment (of the smaller sphere).

Eut. 266

/Corollary/

And it is obvious that the surface of the figure circumscribed around the sector is equal to a circle, whose radius is, in square, the <rectangle> contained by: one side of the polygon, and <by> all the <lines> joining the angles of the polygon and yet again the half of the base of the said polygon [(1) for the figure drawn by the polygon is inscribed inside the segment of the greater sphere][364] [(2) and this is clear through what has been written above].

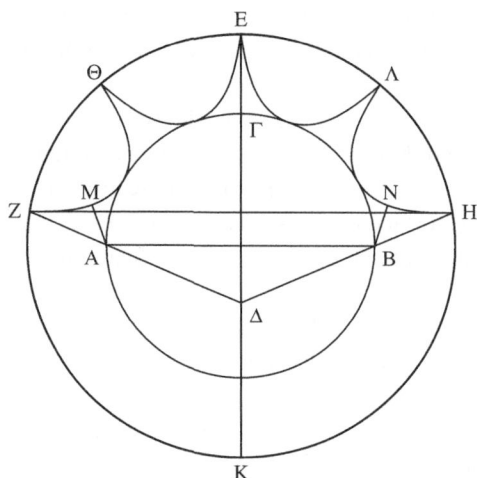

TEXTUAL COMMENTS

The main proposition is a construction-proposition, similar to 23, 28, 36 above. The construction, however, is so well understood by now, that it is presented not

I.39
Codex B, followed by Heiberg, has straight lines instead of arcs in the polygon.
Codex B, followed by Heiberg, has the center of the circles, in a sense correctly, at Δ. It appears that Codex C had the center passing through the line AB, while the copies of A were divided: codices DH, too, have the center passing through the line AB, while codices EG have the center passing through the line ZH. (The two lines, indeed, are hard to distinguish.) Codex A has omitted Γ (reinstated by Codex D). It is impossible to say if codex C had it or not.

---

[362] *Elements* III.18 for the right angle; the move from Step 14 to 13 is based on *Elements* I.19, I.32.

[363] Referring to Eutocius.

[364] The claim of the corollary holds through *SC* I.35.

as a series of imperatives ("let this be done . . ."), but as a series of assertions (Steps 2–5: "so this will be the case . . ."). As a result, the argument following the construction (Step 8 onwards) "feels" like an extra argument, something added on to the main piece of argument. The structure therefore is very backwards-looking, and the main connector is "for." Heiberg disliked backwards-looking justifications, and therefore bracketed Steps 10–11 and 14, I believe for no real reason. Step 17, on the other hand, must refer to Eutocius.

In the corollary, Heiberg saw it as obvious that the "this" in Step 2 refers to the main claim of the corollary, not to Step 1, hence he bracketed Step 1 (which seemed to him to stand in the way). Then he went on to say (this begins to be Stalinist, really) that Step 2 was suspect, as well – presumably because it is a backwards-looking argument. Current scholarship tends to be more lenient.

GENERAL COMMENTS

Backwards-looking structure and its significance for how the proposition is to be read

As mentioned in the textual comments, the proposition is remarkable for its backwards-looking structure. It is instructive to see that the structure is much more difficult to "read" than standard, forwards-looking structures. In such standard structures, two assumptions are broadly true:

1. The assertions required for any argument were made prior to the assertion that provides the conclusion for the argument. That is, there is no need to look beyond a step, when checking its legitimacy.
2. If an argument is made that does not use a certain assertion made earlier in the proof, this implies that the earlier assertion is likely not to be used any further. That is, there is no need to go much earlier when checking the legitimacy of a given step.

The combination of these two rules yields the result that one knows exactly which assertions one might use while checking any given step of the argument – generally speaking, one might use the small set of assertions made in the immediately previous steps.

A backwards-looking structure violates both rules: (1) the argument for Step 12, for instance, consists of Steps 13, 15, 16; (2) the argument for Step 18 uses both Steps 9 and 12, although, after they were made, the argument moved on to the long and independent sequence 13–17. The result is that, when checking the legitimacy of steps in this proposition, one needs to have not a local acquaintance with the steps made in the immediate vicinity of each individual step, but a global acquaintance with the overall thrust of the argument. This is natural, given that, in fact, we are supposed to be persuaded now not so much locally, one step leading to another, but globally, through our wider familiarity, at this stage of the book, with the procedures of such construction-propositions. We have reached the point where "Sphere and Cylinder" procedures, themselves, have become elementary for us.

/ 40 /

The surface of the figure circumscribed about a sector is greater than a circle, whose radius is equal to the <line> drawn from the vertex of the segment[365] to the circumference of the circle which is <the> base of the segment.

For let there be a sphere and on it[366] a great circle, ABΓΔ, and a center E, and let the polygon ΛKZ be circumscribed around the sector, and let a circle be circumscribed around it <=around the polygon>, and let a figure come to be, as before, and let there be a circle N, whose radius is equal, in square, to the <rectangle> contained by: one side of the polygon, and <by>: all the <lines> joining <the angles>, with the half of KΛ.[367] (1) But the said area <=the rectangle> is equal to the <rectangle contained> by MΘ and ZH[368] [((2) which <=ZH> is, in fact, <the> height of the segment of the greater sphere)[369] (3) for this has been proved].[370] (4) Therefore the radius of the circle N is equal, in square, to the <rectangle> contained by MΘ, HZ. (5) But HZ is greater than ΔΞ [((6) which is <the> height of the smaller segment)[371] ((a) for if we join KZ, (7) it will be parallel to ΔA.)[372] (8) But AB is parallel to KΛ, too,[373] (9) and ZE is common; (10) therefore the triangle ZKH is similar to the triangle ΔAΞ.[374] (11) And ZK is greater than AΔ; (12) therefore ZH is greater than ΔΞ, too],[375] (13) while MΘ is equal to the diameter, ΓΔ[376] [(b) for – if EO is joined[377] – (14) since MO is

---

[365] "The segment:" here meaning a surface, namely the reference is to the segment of the *surface of the sphere*, which is the external boundary of the *solid sector of the sphere*.

[366] See textual comments.

[367] I.e. one side of the rectangle is equal to the side of the polygon; the other side, more complicatedly, is equal to the sum of: all the lines joining the angles of the polygon, plus the half of KΛ.

[368] *SC* I.22 together with *Elements* V.16.

[369] This is essentially based on unpacking the diagram, besides being an application of *SC* I.22.

[370] Step 3 is probably in support of Step 1, not of Step 2. See textual comments.

[371] Again, this is essentially based on the diagram.

[372] Can be obtained in several ways, e.g. *Elements* VI.2: the line ΔA cuts the radii ZE, KA proportionally (for it leaves as remainders of both – the two being equal radii of the greater circle – an equal radius of the smaller circle); hence it is parallel to the base ZK.

[373] The same as the preceding step.   [374] *Elements* I.29.

[375] *Elements* VI.4, V.14 (extended to proportion-inequality).

[376] The same as Step 6 of Proposition 30.

[377] Note that the identity of O as the touching point of the tangent MZ will be taken for granted in the following.

equal to OZ,[378] (15) while ΘE <is equal> to EZ,[379] (16) therefore EO is parallel to MΘ;[380] (17) therefore MΘ is twice EO.[381] (18) But ΓΔ, too, is twice EO;[382] (19) therefore MΘ is equal to ΓΔ], (20) and the <rectangle contained> by ΓΔ, ΔΞ is equal to the <square> on AΔ;[383] (21) therefore the surface of the figure KZΛ is greater than the circle, whose radius is equal to the <line> drawn from the vertex of the segment to the circumference of the circle which is <the> base of the segment, (<namely, the circumference of> the <circle> around the diameter AB); (22) for the circle N is equal to the surface of the figure circumscribed around the sector.[384]

Eut. 267 *(margin)*

/Corollary 1/

So also: the figure circumscribed around the sector, with the cone whose base <is> the circle around the diameter KΛ and whose vertex <is> the center, is then equal to a cone, whose base is equal to the surface of the figure, and whose height <is equal> to the perpendicular drawn from the center on the side [(1) which <=the perpendicular> is equal to the radius of the sphere;[385] (2) for the figure circumscribed about the sector is inscribed inside the segment of the greater sphere, whose center is the same; (3) so the claim is clear from what has been written above].[386]

/Corollary 2/

And from this it is obvious, that the circumscribed figure, with the cone,[387] is greater than a cone: having, <as> base: the circle, whose radius is equal to the <line> drawn from the vertex of the segment of the smaller sphere to the circumference of: the circle, which is <the> base of the segment; and <whose> height <is equal> to the radius;[388] (1) for the cone equal to the figure with the cone, will have the base

---

[378] *Elements* III.18, 3.   [379] Radii of the same circle.

[380] *Elements* VI.2.   [381] *Elements* VI.4.

[382] Diameter and radius in the same circle.

[383] *Elements* III.31 (angle at A is right), VI.8 Cor. (so ΓΔ:AΔ::AΔ:ΔΞ), VI. 17 (so ΓΔ*ΔΞ=ΔA²).

[384] *SC* I.39 Cor. I.

[385] "The radius of the sphere:" referring to the smaller sphere.   [386] *SC* I.38.

[387] "The cone" here is the cone having the same base as the figure, with its vertex at the center.

[388] "Radius" here is "the radius *of the smaller sphere*."

greater than the said circle,[389] (2) and the height equal to the radius of the smaller sphere.[390]

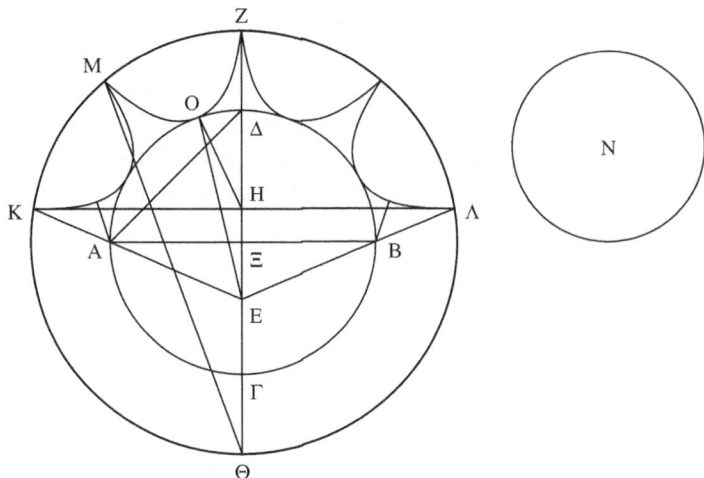

I.40
Codex B, followed by Heiberg, has straight lines instead of arcs in the polygon.
Codex B, followed by Heiberg, has introduced a line KZ. See general comments. Heiberg further removed the line OH, as well as the small lines extending from the points A, B.
The position of the center is confused and perhaps irrecoverable: codices BG have it, in a sense correctly, at E, codex D has it below E, codex E has it at H, codex H has it at Ξ (so, it seems, codex C). There is an error on codex E with the line "AΔ," failing to make it reach the terminus A; instead, it settles on the intersection KΛ/inner circle.   Codex D (and B) has the circle N to the left of the main circle.   Codices DEH have the circle N aligned more with the center of the main circle, rather than higher. I follow codices G4 in a sort of *lectio difficilior*.   Codex A introduced a curious Σ at the intersection KΛ/MΘ (not copied by Codex B).   Codices EH4 omit Γ (perhaps so Codex A). Codex 4 has N instead of H.   Codex H omits the letters Θ, Γ, N as well as the line MΘ.

TEXTUAL COMMENTS

Two minor points to begin with. First, the setting-out starts with ". . . let there be a sphere and on it a great circle ABΓΔ and a center E." Exciting: "on" (*epi*) instead of the customary "in" (*en*). Is this textual corruption or real lexical variation? But, if corruption, why should this particular one take place? (Note that, as mere textual corruption, this is not negligible, since the case of the noun needs to change with the different preposition.) Possibly, the "on" refers to both circle and *center*, so the different preposition actually could mean something new and different from what we are used to.

Second, note that in the first corollary, Heiberg's doubts are relatively speaking well founded: the argument of Steps 1–3 adds little, and Step 3, in particular, is a typical scholion. It is very instructive therefore to see that whoever inserted Step 3 into the corollary did not yet have a numbered text (otherwise he would not say "from what is written above," referring as far back as Proposition 38!).

Now to the main difficulty, namely the three series of backwards-looking justifications in the main proposition: Steps 2–3, 6–12, 14–19.

First, consider the relations between the three: Steps 2 and 6 are parallel assertions, relating a similar content; so are the sequences 7–12, 14–19. Second, note that if these are interpolations, at least 7–12, 14–19 must have antedated Eutocius (whose comment to Step 21 assumes a text similar to ours, i.e. one in which the difficulty is to piece together 4, 5, 13, 20 and 22, clearly because of the absence of present-day Steps 7–12, 14–19).

---

[389] *SC* I.40 (plus Cor. 1 above).

[390] The same as Step 1 above. Also necessary for the argument is Interlude, recalling *Elements* XII.14.

Now, Steps 2–3, in themselves, are to the point. Indeed, Step 3 does indeed refer to 1, not to 2 (so this might be a reason to doubt the authenticity of Step 2), but Step 2 is necessary for the applicability of Step 3 as an argument supporting Step 1. The only question is whether Archimedes ever gave any such cross-references (Heiberg seems to have doubted this).

Step 6 can perhaps be seen as a lame attempt to copy Step 2. In fact it is difficult to see the role of Step 6 in its actual place. Its only possible function is as an argument for 5[391] (but note then the absence of a connective signaling such a function). But whoever inserted it for such a purpose could not have had Steps 7–12 in front of him. So if Steps 6–12 are the combined result of two separate interpolations, 6 must have come before 7–12. Both could not be genuine; therefore the second *must* be interpolated (and 6 remains problematic). This is helpful from another point of view: Steps 5 and 13 are related by the connectors *men . . . de*, expressed in my translation by the "while" at the start of 13. Greek can suffer such breaks between connected clauses, but it is better to do without them. Finally, note the absence of the line ZK from the diagrams in the manuscripts. In short, the bracketing of Steps 7–12, at least, seems very well founded.

The line EO is indeed present in the diagrams (as well as the redundant HO!), but then it could not just be "imagined:" the letter O would not come into existence without the passage 14–19. This passage, indeed, is problematic in that it argues, in detail, for what is the equivalent of Step 6 of Proposition 30. There, that step was probably left without any argument (and at most was given the briefest of arguments, Step 7). So it would seem that some (the same?) hand inserted both Steps 7–12 and Steps 14–19 (and so we can do without the peculiar conditionals, Steps a and b). Assuming we still have Step 6, the resulting structure is brief, but not impossibly opaque: 5 is argued for by 6, 13 is already known, and 20 is based on the most basic of tools. And we should expect brevity, for the end is now in sight.

GENERAL COMMENTS

Limits on visualization

The line KZ is interesting in its absence from the diagram. This is seen not only in the manuscripts' tradition of the diagram, but also in the text itself, where the expression of Step a, "for if we join KZ . . ." suggests that the line is a virtual object, whose possibility is entertained but, in actuality, is not drawn. Why is that? The answer seems to be quite simple. Since the diagram follows the convention of representing the polygon with circular arcs, it is impossible to draw the line KZ without creating phantom, misleading intersections with the lines KM, MZ. (Note, indeed, that in all diagrams with circular-arcs polygon, the space between the circle and the circular-arcs polygon is always

---

[391] Meaning perhaps: "the height of the smaller segment is also in itself smaller:" imprecise, but within the acceptable limits of imprecision in this treatise.

empty, preventing such meaningless intersections.) Indeed, the very convention of drawing a polygon with circular arcs is apparently driven by considerations of visualization: with such pseudo-polygons, visual resolution of circle and polygon is made easier. In other words, we see a certain trade-off in visualization: by making one feature of the configuration easier to visualize, another is made more difficult to visualize, with the result that visualization is dropped altogether in this particular case and an unvisualized, virtual object is used instead. The configuration is simply too complicated. Visualization – even in highly visual Greek geometry – has its physical limits.

So-called "corollaries," and the structure of the proposition

Heiberg's titles, "Corollary 1," "Corollary 2," are in this case particularly misleading. "Corollary 1" is not a corollary at all, since it is not a result made obvious by this proposition: it becomes evident from the construction, which is common to Propositions 39 and 40, and if anything it is based on Proposition 38, not on Proposition 40 (indeed, it does not even begin with the words "so it is obvious that . . ."). Rather than a corollary, this is an argument *within* the "main proposition." The proposition supplies two main proofs. First, that of what Heiberg calls "Proposition 40," then that of what Heiberg calls "Corollary 2" – and which is based on "Proposition 40," together with what Heiberg calls "Corollary 1." Notice also that in Step 21, both references to lettered objects are, as it were, afterthoughts: ". . . therefore the surface of the figure KZΛ is greater than the circle, whose radius is equal to the <line> drawn from the vertex of the segment to the circumference of the circle which is <the> base of the segment (<namely, the circumference of> the <circle> around the diameter AB)." In other words, Step 21 is already on the general, letter-less level of the two "corollaries." There is no clear break between the main proposition and the so-called "corollaries," but a smooth sequence. Essentially, brevity has reached the point where standard proofs are no longer distinguishable from so-called "corollaries." (This brevity, indeed, becomes much more marked if we excise the passages 7–12, 14–19.)

Related to this complicated structure – of main proof followed by a secondary proof – is the absence of definition of goal in this proposition (no "it is to be proved that . . ."): the general statement of the goal is left till the end of the proof, and this is because the double goal of this proof cannot be revealed before the first proof is completed.

/ 41 /

Again, let there be a sphere and in it a great circle, and a segment smaller than a semicircle, <namely> ABΓ, and a center Δ, and let an even-angled polygon be inscribed inside the sector ABΓ, and let a <polygon> similar to it be circumscribed, and let the sides be parallel to the sides, and let a circle be circumscribed around the circumscribed polygon, and similarly to the earlier <propositions>, let the circles make, as they are carried in a circular motion (HB remaining fixed), figures contained by conical surfaces;[392] it is to be proved that the surface of the circumscribed figure has to the surface of the inscribed figure a ratio duplicate of the side of the circumscribed polygon to the side of the inscribed polygon; and the figure with the cone has a triplicate ratio of the same.[393]

(a) For let there be the circle M, whose radius is equal, in square, to the <rectangle contained> by: one side of the circumscribed polygon, and <by> all the <lines> joining the angles and yet again the half of EZ; (1) so the circle M will be equal to the surface of the circumscribed figure.[394] (b) So let there be taken also the circle N, whose radius is equal, in square, to the <rectangle> contained by: one side of the inscribed polygon, and <by> all the <lines> joining the angles, with the half of AΓ; (2) so this, as well, will be equal to the surface of the inscribed figure.[395] (3) But the said areas <=rectangles> are to each other, as the <square> on the side EK to the <square> on the side AΛ[396] [(4) and therefore as the polygon to the polygon, the circle M to the circle N];[397] (5) now, it is obvious that the surface of the circumscribed figure, too, has to the surface of the inscribed figure a duplicate ratio of EK to AΛ [(6) and the same <ratio> which the polygon <has>, too <to the polygon>].[398]

Eut. 267

(c) Again, let there be a cone, Ξ, having a base equal to the <circle> M, and, <as> height, the radius of the smaller sphere; (7) so this cone is equal to the circumscribed figure, with the cone whose base <is> the circle around EZ, and <whose> vertex <is> Δ.[399] (d) And let there be another cone, O, having a base equal to the <circle> N, and <as> height, the perpendicular drawn from Δ on AΛ; (8) so this, as well, will be equal to the inscribed figure, with the cone, whose base <is> the circle around the diameter AΓ, and <whose> vertex <is>

---

[392] "Figures" in the plural: one inscribed, the other circumscribed.

[393] That is, taking the two "figures with the cone," one based on the circumscribed figure, the other on the inscribed figure, the circumscribed has to the inscribed the triplicate ratio of the side to the side.

[394] *SC* I.39 Cor.   [395] *SC* I.35.

[396] Essentially the same as Step 17 of Proposition 32 (though there the comparison is between lines, not between areas). Here, as there, see Eutocius.

[397] The polygons are to each other as the squares on the sides (*Elements* VI.20). The areas mentioned in Step 3 – which are as the squares on the sides – are also equal to the circles M, N (Steps 1, 2).

[398] See textual comments on the structure of the Steps 3–6.

[399] *SC* I.40 Cor. 1.

Eut. 268 the center Δ;[400] (9) For these have been all demonstrated above. (10)
And [since] it is: as EK to the radius of the smaller sphere, so AΛ to
Eut. 268 the perpendicular drawn from the center, [Δ], to AΛ,[401] (11) and it was
proved that as EK to AΛ, so the radius of the circle M to the radius of
the circle N[402] [(12) and the diameter to the diameter]; (13) therefore
it will be: as the diameter of the circle which is <the> base of Ξ, to
the diameter of the circle which is <the> base of O, so the height of
the cone Ξ to the height of the cone O[403] [(14) therefore the cones are
similar].[404] (15) Therefore the cone Ξ has to the cone O a triplicate
ratio of the diameter to the diameter;[405] (16) now, it is obvious that
the circumscribed figure with the cone, too, has to the inscribed figure,
with the cone, a triplicate ratio of EK to AΛ.

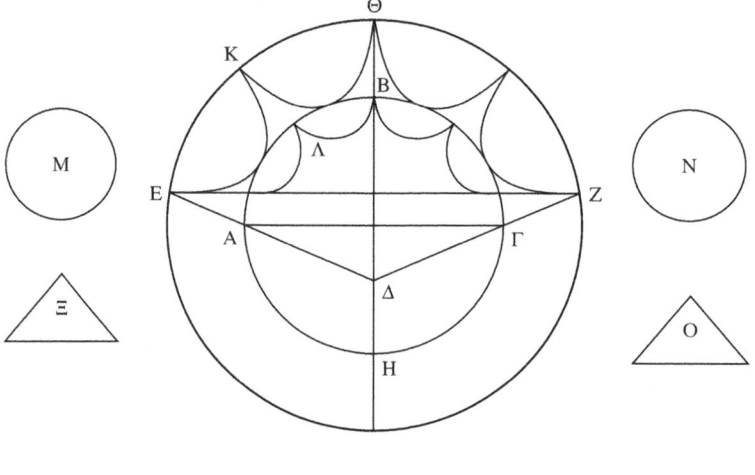

I.41
Codex B, followed by
Heiberg, has straight
lines instead of arcs in
the polygon.
Codices BG, followed
by Heiberg, put the
center of the circles, in
a sense correctly, at Δ;
Codex E puts it on the
line EZ; codex D puts it
below Δ. Codex C,
however, like Codex H,
puts it on the line AΓ,
which I follow here.
Codex G has the
triangles a little tilted to
the right.   Codices
EG omit B.

---

[400] *SC* I.38.

[401] See Eutocius, who uses *Elements* III.18, VI.4. The argument is in principle similar to that of Step 3 above (and so to Step 17 in Proposition 32), which all seem to
rely on the implicit rule "homologous lines in similar polygons are in the ratio of the
sides of the polygons" (here, the polygons are triangles, and the homologous lines are
perpendiculars).

[402] Eutocius derives this from Step 4 above (together with *Elements* XII.2).

[403] Structure of the argument: (10) (EK:radius (smaller sphere)::AΛ:perpendicular
from Δ on AΛ). With implicit alternation: (10*) (EK:AΛ::radius (smaller sphere): perpendicular from Δ on AΛ)). Then (11) (EK:AΛ::radius of M:radius of N). So, the implicit result of (10*) and (11) is (11*) (radius (smaller sphere):perpendicular from Δ on
AΛ::radius of M:radius of N). Then, (12) (diameter:diameter::radius:radius). The implicit result now: (12*) (radius (smaller sphere):perpendicular from Δ on AΛ::diameter
of M:diameter of N). Now, with Step c: (radius (small sphere)=height of Ξ), (diameter
of M=diameter of base of Ξ), and Step d: (perpendicular from Δ on AΛ=height of O),
(diameter of N=diameter of base of O), and all four terms of the implicit result (12*)
are now substituted. Step 13 is the result of this substitution.

[404] *Elements* XI. Def. 24.   [405] Interlude, recalling *Elements* XII.12.

TEXTUAL COMMENTS

Again, let us start from the less complicated issues. First, in Step 10, Heiberg brackets two words. This is because these words do not occur inside Eutocius' quotation in his commentary. This is an argument about which I have been skeptical throughout. Further, Steps 12, 14 may both be (mathematically sound) interpolations made to clarify the argument. Step 12 also "feels" strange, as Step 11 does not lead to it easily. But such clumsiness can be authorial, too. Hence no decision is possible concerning those two steps.

So far, "normal" cases where there is very little to say either way. Steps 4, 6 are much more bizarre.

This is not so much because those steps do not follow directly from the preceding assertions. As can be seen, e.g. in the note to Step 13, Archimedes often leaves much of the argument implicit (and, following Heiberg himself and omitting Step 12, the argument leading to 13 would become even more elliptic). But in fact the steps *are* worrying, and one can see why Heiberg chose them as an example of pieces of text that ought to be bracketed. The problem is that they do not lead anywhere. Steps can be either goals in themselves (i.e. specified in the definition of goal), or steps on the route to other goals. The two goals of this proposition are attained at Steps 5 and 16, respectively. Steps 4 and 6 are relevant for neither conclusion. True, Eutocius derived Step 11 from Step 4, but Step 4 does not lead *directly* to Step 11, and the combination of Steps 1, 2 and 3 offers other, better ways to get to Step 11. As for Step 6, it is completely useless for the development of the proposition.

But why should a scholiast, then, insert these steps? Or is this a case of textual corruption – did we lose something more meaningful? But it is difficult to see what this could have been.

Another hypothesis: Archimedes here prepares material which will be used for the following proposition? But then – can the relation between different propositions be so intimate? But of course we must remember that the division into "propositions" was not so clear cut in the original (which does not number them). Perhaps we should think of the entire sequence of this last, "chapter 6," as coming close to being a single proposition?

And another – very difficult, but tempting, hypothesis: that Archimedes, in a hurry, piles up relations in the first part of the proof, *since he does not know in advance what exactly he will require in the second part of the proof*. This proposition, then, would become a first draft. Let us bear this hypothesis in mind as we go on reading the remainder of the treatise.

GENERAL COMMENTS

A sketchy proposition, with objects sketchily described

How heavy, that "again" sighed at the very start of this proposition!

For indeed, whether a draft or not, the proposition is a hurried passage through well-known territory. The thought of Proposition 32 is repeated but, so that he can abbreviate, Archimedes uses his special tool, that of construction-propositions (Proposition 32, unlike 41, was a "standard" proposition): this allows him to squeeze together the construction and the arguments. The result is an interesting hybrid, a construction-proposition – with a definition of goal. As a consequence, we have here a new twist on an important theme of relation between the general and the particular. When we reach the definition of goal, following a very sketchy construction, there are very few letters to go around, referring to the diagram. Instead, the definition of goal uses no letters, and is phrased in completely general language: to prove the particular result, for the diagram constructed here, is therefore seen as tantamount to proving the general conclusion. Once again, then, innocent variations of the discourse – here motivated by sheer laziness, it seems – have important logical consequences.

The same sketchiness of reference to objects is seen elsewhere. For example, almost at the very start, we find a surprising expression: "... let an even-angled polygon be inscribed inside the sector ΑΒΓ:" notice the deviant word "sector" (instead of "segment"); and the qualification "without the base," dropped. Then Step a: "For let there be the circle M ..." instead of the more standard "for let there be a circle, M." This is comparable to the use of "again" at the start of this proposition. In short, the etiquette of mathematical practice, requiring us to announce formally each new *débutant* at the mathematical *monde* – that is, the *ad hoc* universe created for the individual proposition – is eroded by the familiarity of the domain of discourse. We know them all.

/ 42 /

The surface of every segment of a sphere smaller than a hemisphere is equal to a circle, whose radius is equal to the <line> drawn from the vertex of the segment to the circumference of the circle which is <the> base of the segment of the sphere.

Let there be a sphere and a great circle in it, ΑΒΓ, and a segment in it, smaller than a hemisphere, whose base <is> the circle around the <diameter> ΑΓ (being <=the circle around ΑΓ> in right <angles> to the circle ΑΒΓ), and let a circle be taken, Z, whose radius is equal to AB. So it is required to prove that the surface of the segment ΑΒΓ is equal to the circle Z.

(a) For if not, let the surface be greater than the circle Z, (b) and let the center Δ be taken, (c) and let <lines>, joined from Δ to Α, Γ, be produced; (1) and, there being two unequal magnitudes – the surface of the segment and the circle Z – (d) let an equilateral and even-angled polygon be inscribed inside the sector ΑΒΓ, (e) and let another similar to it be circumscribed, (f) so that the circumscribed has to the inscribed a smaller ratio than the surface of the segment of the sphere to the circle Z,[406] (g) and the circle being carried in a circular motion, just as before, (2) there will be two figures contained by conical surfaces, of which

---

[406] *SC* I.6.

one <is> circumscribed, and the other inscribed (3) and the surface of the circumscribed figure will be to the <surface> of the inscribed, as the circumscribed polygon to the inscribed; (4) for each of the ratios is duplicate the <ratio>, which the side of the circumscribed polygon has to the side of the <inscribed> polygon.[407] (5) But the circumscribed polygon has to the inscribed a smaller ratio than the surface of the said segment to the circle Z, (6) and the surface of the circumscribed figure is greater than the surface of the segment;[408] (7) therefore, also, the surface of the inscribed figure is greater than the circle Z; (8) which is impossible; (9) for the surface of the said figure[409] has been proved to be smaller than the circle of this kind.[410]

Eut. 268

(h) Again, let the circle be greater than the surface, (i) and let similar polygons be circumscribed and inscribed, (j) and let the circumscribed have to the inscribed a smaller ratio than <that> which the circle has to the surface of the segment.[411] (10) Therefore the surface is not greater than the circle Z.[412] (11) And it was proved, that neither <is it> smaller;[413] (12) therefore equal.

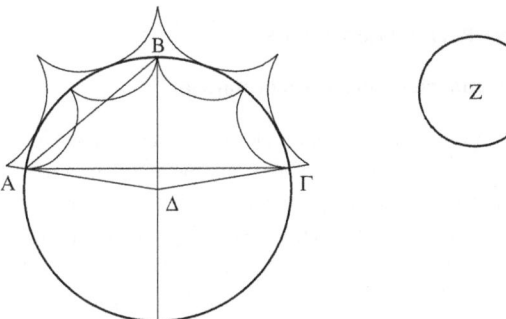

I.42
Codex B, followed by Heiberg, has straight lines instead of arcs in the polygon.
Codices DG, followed by Heiberg, omit the straight line AB; codex E (like Heiberg) has a straight line BΓ.
Codex D has Ξ instead of Z.

---

[407] See Eutocius for a derivation of this claim from the *enunciation* of the preceding proposition. (But is this not identical with the "superfluous" result of the preceding proposition, Step 6 – which would then provide that step with a meaning?)

[408] *SC* I.39.

[409] Literally, "the said surface of the figure." The reference is to the surface of the inscribed figure.

[410] *SC* I.37. "A circle *of this kind*:" a rare expression, referring to the circle Z through a general (but periphrastic) description.

[411] *SC* I.6. The argument may go on like this (following Heiberg): given Step 3 we have (circumscribed figure:inscrib figure::circumscribed polygon:inscribed polygon), while the construction at Step j specifies that (circumscribed polygon:inscribed polygon<circle Z:surface of the segment). So we immediately get (circumscribed figure:inscribed figure<circle Z:surface of the segment) – from *Elements* V.11, extended to inequality. However, "surface of the segment" and "inscribed figure" are one and the same thing, so (circumscribed figure:inscribed figure<circle Z:inscribed figure), hence, from *Elements* V.8 (circumscribed figure <circle Z), in direct contradiction to *SC* I.40.

[412] The opposite of what needs affirming. See textual comments.

[413] Wrong: it was proved that it is not *greater*.

TEXTUAL COMMENTS

The second part of the proof is extremely interesting. It has two main features: first, all the argument is missing (see note to Step j). Second, the conclusion is false (Step 10 should have read "Therefore circle Z is not greater than the surface," and then Step 11 falls into place, as well).

First the missing argument. Either it represents a lacuna – a gap created by a failure in textual transmission – or, alternatively, it was intentionally left out by Archimedes: as it were, an exercise for the reader. But then, probably, Eutocius would have solved such an exercise: the lack of commentary on this part of the proof seems to imply that Eutocius' text was fuller. It is probably a lacuna, then. If so, then the false conclusion might be interpreted as a feeble attempt to fill in such a lacuna. Once again, the false conclusion could well be an authorial slip of the pen. Archimedes does not generally commit *such* errors, but the end of the book is drawing near. Finally: if, in spite of Eutocius' lack of commentary, this part of the proof would be interpreted as an exercise left for the reader – would it not be especially fitting if Archimedes intentionally left here a *false solution*?

GENERAL COMMENTS

Meager diagrams, rich verbal formulations

The concluding set of propositions, 42–4 are similar to the previous set of conclusions (Propositions 33–4) in that they have some very meager diagrams. It is typical that even the auxiliary circle – usually from high up in the alphabet – is here called Z. Even so, there is an unused letter, E. (Probably the gap in the alphabetical sequence is meant to stress the independence of Z.) There is hardly any diagrammatic activity inside the sphere itself. It seems that Archimedes wishes to make his conclusions as diagram-independent as possible and, in this way, as general-seeming and simple as possible. The conclusions rely, much more, upon a rich verbal formulation.

So I believe it is no accident that the line BΓ does not appear in the diagrams in the manuscripts. The proof is not so much about BΓ, as about the baroque object – "the <line> drawn from the vertex of the segment to the circumference of the circle which is <the> base of the segment of the sphere." Correlated with the meager diagram, then, is a certain attempt at richness of formulation. That the segment is smaller than a hemisphere is stated now for the first time (partly to force the distinction between this proposition and the next one). Also, consider the setting-out: "(being <=the circle around AΓ> in right <angles> to the circle ABΓ)" – a precise qualification, remarkable in its absence so far.

A more complicated case is Step c: "... and let <lines>, joined from Δ to A, Γ, be produced ..." Now, "joining and producing" is a recognized operation in Greek mathematics. A line is from point X to point Y ("joined") then it is extended to beyond point Y ("produced"). We have not yet had this basic operation in this treatise: lines were joined to the limit of the sphere, which often functions as the limit of the mathematical universe. Here, something takes place beyond this limit, on an external polygon. The twist is the following: the external polygon is not yet constructed at Step c, mainly since Archimedes avoids a detailed diagrammatic, lettered construction (from Step c onwards, the proposition is almost letter-less: notice the description of the circle Z at Step 9, "of this kind" – a very untypical circumlocution, meant to avoid designation by letters). The end result is that Step c is very ambiguous in context, since there is no indication as to why the lines are "produced," and to where.

Little overarching structure of the proposition

While some of the local formulations in this proposition are fuller than usual, the overall structure of the proof is freer than in earlier comparable propositions. The proposition builds upon an acquaintance with both subject matter and procedure. Most importantly, there is no "for if it is unequal, it is either greater or smaller. First, let it be greater . . ." Instead, Step a goes directly to assume one inequality, while Step h assumes the other. There is no architectonic, governing the two: the reader is by now expected to supply this architectonic. Compare also the construction at Step i: "circumscribed and inscribed" instead of the standard "inscribed and circumscribed." The reversal of order points to the unreality of the action. No one is meant actually to circumscribe and inscribe – we just imagine the actions (and therefore we may imagine them in any order we please, disregarding the natural order of actual constructions).

Perhaps, then, even if Archimedes did give a proof for the second part, we may see him as somewhat impatient at this point of the argument? Possibly, then, he gave no more than a very truncated proof of the second part: which would explain its total disappearance from the manuscript tradition? But this is a speculation.

/ 43 /

And even if a segment is greater than a hemisphere, similarly, its surface is equal to a circle whose radius will be equal to the <line> drawn from the vertex to the circumference of the circle which is <the> base of the segment.

(a) For let there be a sphere (b) and in it a great circle, (c) and let it be imagined cut by a plane <which is> right to the <plane> at AΔ, (d) and let the ABΔ be smaller than a hemisphere, (e) and let <there be> a diameter BΓ, at right <angles> to AΔ, (f) and let BA, AΓ be joined from B, Γ to A, (g) and let there be the circle E, whose radius is equal to AB, (h) and the circle Z, whose radius is equal to AΓ, (i) and the circle H, whose radius is equal to BΓ; (1) therefore, also, the circle H is equal to the two circles E, Z.[414] (2) But the circle H

---

[414] *Elements* III.31: angle BAΓ is right. *Elements* I.47: Pythagoras' theorem. *Elements* XII.2: circles proportional to squares on diameters.

is equal to the whole surface of the sphere [(3) since each is four times the circle around the diameter ΒΓ],[415] (4) and the circle Ε is equal to the surface of the segment ΑΒΔ [(5) for this has been proved in the case of the smaller than a hemisphere];[416] (6) therefore the remaining circle Ζ is equal to the surface of the segment ΑΓΔ, (7) which, in fact, is greater than a hemisphere.

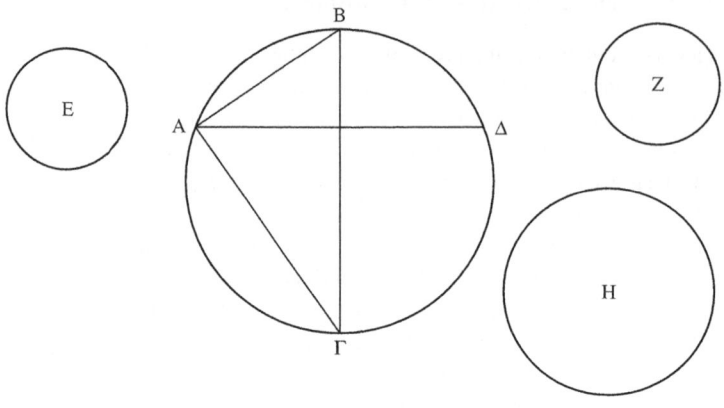

I.43
Codex C is not preserved for this diagram. Codices GH4 (as well as B) have Z aligned exactly above H. I follow codices DE in the position of Z, in a sort of *lectio difficilior*. Codices GH have E a little greater than Z, while Codex B has Z a little greater than E. Codex D has Ξ instead of Z.

TEXTUAL COMMENTS

There are two brief backwards-looking justifications, each worrying in a different way. Step 3 is difficult to comprehend, since the reference of "each" is unclear (the word "each" appears in the feminine gender, referring presumably to "surfaces," but the "surface of the circle H" was never mentioned as such: what was mentioned was "circle," in the masculine). As for Step 5, it simply seems redundant: why refer so explicitly to what has been proved so recently? Both difficulties are real – but of course these statements could still be made by the author. In particular, if this proposition is understood as a direct continuation of the preceding one, then the reference at Step 5 becomes less surprising (such "it has been proved" statements do occur occasionally *inside* propositions, referring to results that were proved considerably earlier in the same argument). I follow this up in the general comment that follows.

GENERAL COMMENTS

Relations between propositions become even more direct

Here is a new departure in terms of inter-propositional dependence. Proposition 43 is another *case*, complementing Proposition 42. The two are related in the

---

[415] *SC* I.33 (the surface of the sphere is four times the great circle), Step i (the radius of the circle H is twice the radius of the great circle), *Elements* XII.2 (circles are to each other as the squares on their diameter, i.e. the circle H is four times the great circle, too). The "each" refers to the two "surfaces:" that of the sphere, and that of the circle H.

[416] Step g, and then *SC* I.42.

same way in which the two parts of Proposition 41, for instance, are related. In Proposition 41, the first part proved the ratio between two-dimensional figures, and the second, *based on the first*, proved the ratio between three-dimensional figures. Here, Proposition 42 proves the equality for segments smaller than hemispheres, while Proposition 43, *based on Proposition 42*, proves the equality for segments greater than hemispheres.

This dependence is stressed at the start of the enunciation by the words "and even if . . .", which make the very enunciation depend on the preceding proposition. Further, notice the use of the future tense: ". . . its surface is equal to a circle, whose radius *will be* equal . . ." I.e., the circle constructed in Proposition 43 is not a *real* circle, actually constructed: it is a virtual circle, a spin-off of the real circle of the original Proposition 42. (See also the verb "imagine" in Step c.) Similarly, there is no definition of goal, the proposition understood to be covered still by the same goal as the preceding one.

This said, the proposition does have its own identity. For one thing, it does have a separate diagram – the essential mark of a proposition. The reasoning employed here is not a mere adaptation of an earlier piece of reasoning, and the overall result is very elegant. Archimedes did well, one may say, to keep this proposition at least partly independent. Having it as an after-thought would have added heaviness to Proposition 42, while depriving Proposition 43 of its inherent elegance.

/44/

Every sector of a sphere is equal to a cone having a base equal to the surface of the segment of the sphere at the sector, and a height equal to the radius of the sphere.

Let there be a sphere and in it a great circle, ABΔ, and a center, Γ, and a cone having, <as> base, the circle equal to the surface at the circumference ABΔ, and a height equal to BΓ; it is to be proved that the sector ABΓΔ is equal to the said cone.

(a) For if not, let the sector be greater than the cone, (b) and let the cone Θ be set out, as has been said; (1) so there being two unequal magnitudes, the sector and the cone Θ, (c) let two lines be found, Δ, E ((d) – and Δ greater than E) (e) and let Δ have to E a smaller ratio than the sector to the cone,[417] (f) and let two lines be taken, Z, H, (g) so that Δ exceeds Z and Z <exceeds> H and H <exceeds> E, <all> by an equal <difference>,[418] (h) and let an equilateral

---

[417] *SC* I.2.

[418] Δ is used twice in this diagram, once standing on a point on the sector, once again standing on a line defined by the main proportion-inequality of the proposition. See general comments.

and even-angled polygon be circumscribed around the plane sector of the circle, (i) and let <a polygon> similar to it be inscribed, (j) so that the side of the circumscribed has to the <side> of the inscribed a smaller ratio than that which Δ has to Z,[419] (k) and similarly to the earlier <propositions>, with the circle being carried in a circular motion, (1) let there come to be two figures contained by conical surfaces; (2) therefore the circumscribed <figure>, with the cone having, <as> vertex, the point Γ, has to the inscribed, with the cone, a ratio triplicate of that which the side of the circumscribed polygon has to the side of the inscribed.[420] (3) But the side of the circumscribed <polygon> has <to the side of the inscribed> a smaller ratio than Δ to Z. (4) Therefore the said solid figure <=the circumscribed, with the cone> will have <to the inscribed, with the cone> a smaller ratio than triplicate the <ratio> of Δ to Z. (5) But Δ has to E a greater ratio than triplicate the <ratio> of Δ to Z;[421] (6) therefore the solid figure circumscribed about the sector has to the inscribed figure a smaller ratio than that which Δ has to E. (7)

Eut. 269

But Δ has to E a smaller ratio than the solid sector to the cone Θ; (8) <therefore> the figure circumscribed about the sector, too, <has> to the inscribed, <a smaller ratio than the solid sector to the cone Θ>.[422] (9) And alternately;[423] (10) but the circumscribed solid figure is greater than the segment;[424] (11) therefore the figure inscribed in the sector, too, is greater than the cone Θ; (12) which is impossible;[425] (13) for, in the <propositions> above it has been proved smaller than a cone of this kind[426] [(14) that is, than a <cone> having, <as> base, a circle whose radius is equal to the line joined from the vertex of the segment to the circumference of the circle which is <the> base of the segment; and, <as> height, the radius of the sphere; (15) and this is the said cone Θ; (16) for it has: <as> base, a circle equal to the surface of the segment, (17) that is <equal> to the said circle,[427] (18) and a height equal to the radius of the sphere]; (19) therefore the solid sector is not greater than the cone Θ.

---

[419] *SC* I.4.   [420] *SC* I.41.

[421] See Eutocius' commentary on Step 9 of Proposition 34.

[422] See textual comments.

[423] *Elements* V.16, with the result (circumscribed figure:solid sector<inscribed figure:cone Θ).

[424] "Segment:" should have been "sector."

[425] The structure of the argument is new: instead of deriving a proportion inequality, and asserting that the *proportion* is impossible (ultimately because it will yield an impossible inequality), Archimedes here takes the proportion as if it is, in itself, acceptable, and then derives from it the impossible equality – only then asserting the impossibility.

[426] *SC* I.38 Cor.

[427] That the two descriptions of the circle coincide (the circle equal to the surface of the sector, and the circle whose radius is the line drawn from the vertex of the segment to the base), is asserted in *SC* I.42.

(l) So, again, let the cone Θ be greater than the solid sector. (m) So, again, similarly, let Δ have to E (being greater than it) (n) a smaller ratio than that which the cone has to the sector,[428] (o) and similarly, let Z, H be taken, so that the differences are the same, (p) and let the side of the even-angled polygon circumscribed around the plane sector have to the <side> of the inscribed a smaller ratio than that which Δ has to Z[429] [(q) and let the solid figures around the sector come to be]; (20) now, we shall similarly prove that the solid figure circumscribed around the sector has to the inscribed a smaller ratio than that which Δ has to E,[430] (21) and than that which the cone Θ has to the sector[431] [(22) so that the sector, too, has to the cone a smaller ratio than the solid inscribed in the segment[432] to the circumscribed].[433] (23) But the sector is greater than the figure inscribed inside it;[434] (24) therefore the cone Θ is greater than the circumscribed figure; (25) which is impossible [(26) for it has been proved that the cone of this kind is smaller than the figure circumscribed around the sector];[435] (27) therefore the sector is equal to the cone Θ.

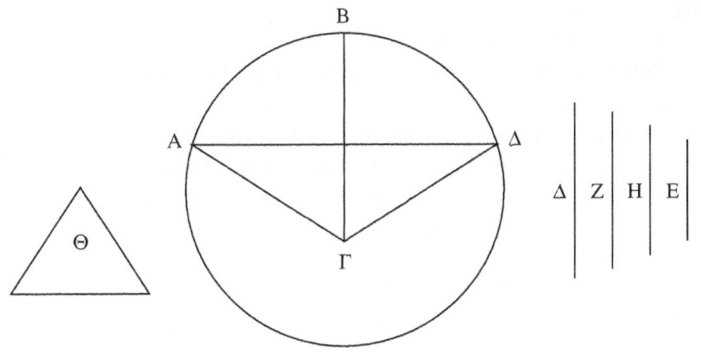

Archimedes' On Sphere and Cylinder 1.

I.44
Codex C is not preserved for this diagram.   Codices BG, followed by Heiberg, put the center of the circle, in a sense correctly, at Γ.
Codex E has the bases of the four lines at the same level, arranged by their tops, as in the thumbnail.   Codex D has the triangle between the circle and the four lines.
Codex E has Θ instead of E, K instead of Z.

---

   [428] *SC* I.2.   [429] *SC* I.4.

   [430] To complete this piece of argument, Steps 1–6 can be repeated here without any modification (so that the "similarly" claim is clearly justified).

   [431] Notice that the subject of the sentence is carried over from Step 20 to Step 21, so that Step 21 asserts, effectively: "[T]he solid figure circumscribed around the sector has to the inscribed a smaller ratio than that which the cone Θ has to the sector."

   [432] Again, "segment" should be "sector."

   [433] Bearing in mind that Step 21 asserts that (figure circumscribed:figure inscribed<cone Θ:sector), extending *Elements* V.7 to inequalities, we derive (sector:cone Θ<figure inscribed:figure circumscribed), which is what Step 22 asserts. An implicit result is that, with an extension of *Elements* V.16, we may derive (sector:figure inscribed<cone Θ:figure circumscribed). This is what the following steps refer to.

   [434] *SC* I.38 Cor., I.42.   [435] *SC* I.40 Cor.2.

TEXTUAL COMMENTS

The text of Step 8 stands in the manuscripts as

ἤ	τὸ περιγεγραμμένον τῷ	τομεῖ σχῆμα	πρὸς	τὸ	ἐγγεγραμμένον
e	*to perigegrammenon toi*	*tomei schema*	*pros*	*to*	*engegrammenon*
"or" / "than"	"the circumscribed about the	sector figure	to		the inscribed"

Obviously there is something wrong with the start of this step. Heiberg suggests a lacuna, and completes the step to read (additions underlined): "Therefore the solid sector has to the cone Θ a greater ratio than the figure circumscribed about the sector to the inscribed."

This is elegant and in itself plausible, since the ending of Step 7, "the solid sector to the cone Θ," in the Greek, becomes, with Heiberg's interpolation, an homoeoteleuton, neatly accounting for the lacuna. However, Eutocius has a completely different lemma: "Therefore the circumscribed solid has to the inscribed a smaller ratio than the solid sector to the cone Θ."

Eutocius often deviates from Archimedes' words, but it is much more difficult to have Eutocius speak of a "smaller ratio" where Archimedes' text had a "greater ratio." Indeed, there is no reason for Archimedes to switch now into "greater ratio:" Steps 6, 7 have "smaller ratio."

On the other hand, if Archimedes becomes very condensed, he can easily write, as I suggest: (7) But Δ has to E a smaller ratio than the solid sector to the cone Θ; (8) <therefore> the figure circumscribed about the sector to the inscribed, too <*scilicet* has a smaller ratio than the solid sector to the cone Θ>.

I still need the "too," and therefore I suggest that the η at the start of the sentence as it stands is a simple scribal error for a twirled κ, shorthand for καί, "and, too" (very common shorthand: see Heiberg's introduction to Vol. III., p. XI). The picture emerging is that of a hasty, abbreviating Archimedes; more of this below.

Now to the usual problems. Steps 14–18 are very suspicious indeed. Archimedes concludes Step 13 with "a cone of this kind," and Step 14 starts with "namely . . ." Why would Archimedes be as elliptic as he is at the end of Step 13, to restart Step 14 immediately with an expansion? On the other hand, Step 14 is very natural for a scholion. The following Steps 15–18 depend on Step 14, and so the entire passage goes together. It is a valid scholion, which most readers will require, given the extremely cryptic construction of the cone at Step b.

Next, Heiberg brackets Step q because it speaks of figures "around" the sector, whereas in fact one of the figures is not *around*, but *inside* the sector. Once again, however, this may be Archimedes being brief.

Heiberg's remaining bracketings – Steps 22, 26 – are the regular type of brackets based on an indefinite sense of "mathematical propriety," and as usual this is problematic. But before taking leave of Heiberg in this book, let us remember him as the great man he was, and enjoy an example of his many brilliant emendations, at Step o. The manuscripts have: "*ωστε ειναι τας δυο πλευρας τας αυτας" ("*so that the two sides are the same"), and Heiberg corrects into: "ωστε ειναι τας διαφορας τας αυτας" ("so that the differences are the same"). One needs to have an instinctive grasp of the Greek language and of the scribal practice, to make such emendations – which cannot possibly be wrong. Heiberg had an instinctive ear for Archimedes. It is just because he had this immediate, oracular grasp of the original, that he dared to interfere with it so much; now, of course, we prefer to keep the original readings whenever possible. But that we have a text of Archimedes at all is something we owe to Heiberg's powers as a medium.

And now let us take our leave of Archimedes, as well. The medieval (Byzantine? Ancient?) scribe took leave like this:

Archimedes' On Sphere and Cylinder 1
Archimedes himself did not have this in his own text (surely he did not call this Book 1!). The letter to Dositheus ended with the last step of the last proposition. An anti-climax, typical of Greek mathematics. The hidden rhetoric is that nothing can be more beautiful than this, unadorned mathematical text.

GENERAL COMMENTS

Haste

Haste is the most noticeable feature of this proposition. Start at Step b. Strictly speaking, this cone should have been constructed already in the setting-out (not for nothing is it required to *set out* the cone!). Archimedes forgot, simply, to letter the cone in the setting-out proper: an instance of haste. Further, in Steps 10, 22, "segment" is used where the right word is "sector." And now to the end of the proposition, which is even more abrupt than the start: there is no worked through conclusion of the double *reductio* argument (what we miss, between Steps 26/27, are *26: "so the cone is not greater than the sector," and **26: "but it was proved, that neither is it smaller").

Now review Steps c–g in the construction. First, four lines are set out. Second, they appear inside the text in their alphabetical order (*not* their order of size: note that the function of Z, H is to be arithmetical means between Δ, E). This is even though, in the diagrams, the four lines do appear in the order of size.

Conspicuously, the four lines occupy a mid-position, alphabetically, between the letters of the sphere and those of the cone. This, however, is badly handled: Θ, the letter of the cone, was chosen without careful preparation, so not enough "alphabetic room" has been left for the four letters of the lines. Some squeezing-in was made necessary, so that Δ was doubled up functioning as both a point on the sphere and a line. We have seen such duplications already in Propositions 3 and 38 above. There, however, the evidence for the reduplication was relatively weak, and Heiberg simply avoided it. Here the textual evidence is compelling and Heiberg had to keep the manuscripts' reading. In fact, it may well be that Heiberg was right and that this is the only proposition in the treatise where such duplication takes place – it is, indeed, something Greek mathematicians do usually avoid.

The sequence of events could have been roughly like this. First, the unlettered diagram had been drawn (note the size sequence of lines). Second, the diagram was lettered, roughly speaking as the text was written (note the alphabetical order of the lines in their appearance in the text). There was, however, no planning of the distribution of letters – and no later proofreading. Another case of haste, then.

Finally, note that, in the same rush, Archimedes did not care to give a separate proof for the case of a segment greater than a hemisphere. This was left, as it were, as an exercise for Dositheus; and Eutocius would be rightly perplexed in his commentary to *SC* II.2, Steps 33–4, which assume such a proof.

Why this haste? We can imagine anything, and I like to imagine Archimedes in the Syracusean spring – the season of spheres and cylinders. The weather improves daily, and the ship to Alexandria, where Dositheus awaits Archimedes' letters, is about to set sail.

ON THE SPHERE AND THE CYLINDER, BOOK II

/Introduction/

Archimedes to Dositheus: greetings

Earlier you sent me a request to write the proofs of the problems, whose proposals[1] I had myself sent to Conon; and for the most part they happen to be proved[2] through the theorems whose proofs I had sent you earlier: <namely, through the theorem> that the surface of every sphere is four times the greatest circle of the <circles> in it,[3] and through <the theorem> that the surface of every segment of a sphere is equal to a circle, whose radius is equal to the line drawn from the vertex of the segment to the circumference of the base,[4] and through <the theorem> that, in every sphere, the cylinder having, <as> base, the greatest circle of the <circles> in the sphere, and a height equal to the diameter of the sphere, is both: itself, in magnitude,[5] half as large again as the sphere; and, its surface, half as large again as the surface of the sphere,[6] and through <the theorem> that every solid sector is equal to the cone having, <as> base, the circle equal to the surface of the segment of the sphere <contained> in the sector, and a height equal to the radius of the sphere. Now, I have sent you those theorems and problems that are proved through these theorems <above>, having proved them in this book. And as for those that are found through some other theory,

---

[1] *Protasis*: see general comments.

[2] "Prove" and "write" use the same Greek root.

[3] *SC* I.33.   [4] *SC* I.42–3.

[5] The words "in magnitude" refer to what we would call "volume" (to distinguish from the following assertion concerning "surface").

[6] *SC* I.34 Cor.

<namely:> those concerning spirals, and those concerning conoids, I shall try to send quickly.[7]

Of the problems, the first was this: Given a sphere, to find a plane area equal to the surface of the sphere. And this is obviously proved from the theorems mentioned already; for the quadruple of the greatest circle of the <circles> in the sphere is both: a plane area, and equal to the surface of the sphere.

TEXTUAL COMMENTS

Analogously to the brief sequel to the postulates in the first book, so here, again, the introductory material ends with a brief unpacking of obvious consequences. Assuming that Archimedes' original text did not contain numbered propositions, there is a sense in which this brief unpacking can count as "the first proposition:" it is the first argument. It is also less than a proposition, in the crucial sense that it does not have a diagram. This liminal creature, then, helps mediate the transition between the two radically distinct portions of text – introduction and sequence of propositions.

The propositions probably did not possess numberings; the books certainly did not. It is perfectly clear that the titles of treatises, let alone their arrangement as a consecutive pair, are both later than Archimedes. (It is interesting to note that the same arrangement is present in both the family of the lost codex A, and the Palimpsest, even though the two codices differ considerably otherwise in their internal arrangement.) As for Archimedes, he simply produced two unnamed treatises, with obvious continuities in their subject matter, as well as differences in their focus, that he himself spells out in this introduction. There is no harm in referring to them – as the ancients already did – as "First Book on Sphere and Cylinder" or "Second Book on Sphere and Cylinder." We should take this, perhaps, as our own informal title, akin to the manner in which philosophers sometimes refer to "Kant's First Critique," etc.

GENERAL COMMENTS

Practices of mathematical communication

In this introduction, rich in references to mathematical communication, we learn of several stages in the production of a treatise by Archimedes.

First comes the "proposal" – my translation of the Greek word *protasis*. Now, this word came to have a technical sense, first attested from Proclus' *Commentary to Euclid's Elements* I: that part of the proposition in which the general enunciation is made. It is not very likely, however, that this technical sense is what Archimedes himself already had in mind here: in the later, Proclean sense, a *protasis* has meaning only when accompanied by other, non-*protasis* parts

---

[7] A reference to *SL, CS*. (To appear in Volume II of this translation.)

of the proposition, and clearly Archimedes had sent only the *protasis*. What could that be, then? Literally, *protasis* is "that which is put forward," and one sense of the word is "question proposed, problem" – in other words, a puzzle. It was such puzzles, then, that Archimedes sent Conon. ("All right, I give up," came back Dositheus' reply.)

Next comes the proof. As noted in n. 2, "prove" uses the same Greek root as "write," *graph*. This is also closely related to terms referring to the figure (which is a kata*graph*e, or a dia*gram*ma), so we see a nexus of ideas: writing down, drawing figures, proving; all having to do with translating an idea in the mind of a mathematician to a product that is part of actual mathematical communication – answering the three *sine qua non* conditions of Greek mathematical communication – written, proved, drawn.

What is the relation between the idea in the mind of the mathematician and the idea in actual mathematical communication? Archimedes' references to results he already seems to have in some sense – from *SL* and *CS* – are especially tantalizing. Why does he promise to send them "quickly?" He probably knows how all those theorems and problems are proved – for otherwise he would not send out the puzzles concerning them. So why not send them straight away? Perhaps he was still busy proofreading them. (If so, the morass of inconsistent style and abbreviated exposition we know so well by now from Book I, is what Archimedes can show *after* the proofreading stage!) Or perhaps, all Archimedes had, prior to "sending" to Dositheus, were notes – stray wax tablets with diagrams that he alone could interpret as solutions for intricate problems.

Or perhaps, he does not have a perfect grasp on the proofs, yet? "I have sent you those theorems and problems that are proved through these theorems <above>, having proved them in this book. And as for those that are found through some other theory..." Things, then, are either *proved through theorems* or *found through theory*. Perhaps, "theory" (a cognate of "theorem," roughly referring, in this context, to the activity of which "theorems" are the product) is a more fuzzy entity, comprising a bundle of unarticulated bits of mathematical knowledge present to the mathematician's mind. Perhaps, it is such knowledge – and not explicitly written down proofs – which is active in the mathematical *discovery*?

Leaving such speculations aside, we ought to focus not on the stage of mathematical *discovery*, but on the stage of mathematical *communication*. The decisive verb in this introduction is not "discovery," not even "prove," but, much more simply, "send." It is the act of sending which gives rise to a mathematical treatise. In this real sense, then, it was the ancient mathematical community – and not the ancient mathematicians working alone – who were responsible for the creation of Greek mathematical writing.

/ 1 /

The second was: given a cone or a cylinder, to find a sphere equal to the cone or to the cylinder.

Eut. 270   Let a cone or a cylinder be given, A, (a) and let the sphere B be
equal to A,[8] (b) and let a cylinder be set out, ΓΖΔ, half as large again
as the cone or cylinder A, (c) and <let> a cylinder <be set out>,
half as large again as the sphere B, whose base is the circle around
the diameter ΗΘ, while its axis is: ΚΛ, equal to the diameter of the
sphere B;[9] (1) therefore the cylinder E is equal to the cylinder K. [(2) But
the bases of equal cylinders are reciprocal to the heights];[10] (3) therefore
as the circle E to the circle K, that is as the <square> on ΓΔ to the
<square> on ΗΘ[11] (4) so ΚΛ to ΕΖ. (5) But ΚΛ is equal to ΗΘ [(6)
for the cylinder which is half as large again as the sphere has the axis
equal to the diameter of the sphere, (7) and the circle K is greatest of
the <circles> in the sphere];[12] (8) therefore as the <square> on ΓΔ to
the <square> on ΗΘ, so ΗΘ to ΕΖ. (d) Let the <rectangle contained>
by ΓΔ, ΜΝ[13] be equal to the <square> on ΗΘ; (9) therefore as ΓΔ
to ΜΝ, so the <square> on ΓΔ to the <square> on ΗΘ,[14] (10) that
is ΗΘ to ΕΖ, (11) and alternately, as ΓΔ to ΗΘ, so (ΗΘ to ΜΝ) (12)
and ΜΝ to ΕΖ.[15] (13) And each of <the lines> ΓΔ, ΕΖ is given;[16]

---

[8] We are not explicitly told so, but we are to proceed now through the method of
analysis and synthesis, in which we assume, at the outset, that the problem is solved – in
this case, that we have found a sphere equal to the given cone or cylinder. We then use
this assumption to derive the way by which a solution may be found.

[9] This construction is a straightforward application of *SC* I.34 Cor., as explained in
Steps 6–7.

[10] *Elements* XII.15. This is recalled in the interlude of the first book, but no such
reference needs to be assumed in this, second book, and in general I shall not refer in
this book to the interlude of the first book.

[11] *Elements* XII.2.   [12] *SC* I.34 Cor.

[13] ΓΔ is given, and it is therefore possible (through *Elements* I.45) to construct a
parallelogram on it – therefore also a rectangle – equal to a given area, in this case
equal to the square on ΗΘ. It is then implicit that ΜΝ is *defined* as the second line in a
rectangle, contained by ΓΔ, ΜΝ, which is equal to the square on ΗΘ.

[14] Compare VI.1, ". . . parallelograms which are under the same height are to one
another as the bases," and then the square on ΓΔ and the rectangle contained by ΓΔ,
ΜΝ can be conceptualized as lying both under the height ΓΔ, with the bases ΓΔ, ΜΝ respectively (so ΓΔ:ΜΝ::the square on ΓΔ:the rectangle contained by ΓΔ, ΜΝ); and then
the rectangle contained by ΓΔ, ΜΝ has been constructed equal to the square on ΗΘ.

[15] A complex situation. We have just seen (Steps 9–10) that A. ΓΔ:ΜΝ::ΗΘ:ΕΖ,
which, "alternately" (*Elements* V.16), yields B. ΓΔ:ΗΘ::ΜΝ:ΕΖ. On the other hand,
the construction at Step d, together with *Elements* VI.17, yields C. ΓΔ:ΗΘ::ΗΘ:ΜΝ.
Archimedes starts from A, and then says, effectively, "(Step 11:) alternately C (Step 12:)
and B." This is very strange: the "alternately" should govern B, not C. Probably Step 11
should be conceived as if inside parenthesis – which I supply, as an editorial intervention
in the text, in Step 11.

[16] I.e., they are determined by the "given" of the problem, namely the cone or cylinder
A (see Step b in the construction). Note, however, that they are given only as a couple.
Both together determine a unique volume, but they may vary simultaneously (the one

(14) therefore HΘ, MN are two mean proportionals between two given lines, ΓΔ, EZ; (15) therefore each of <the lines> HΘ, MN are given.[17]

So the problem will be constructed[18] like this:

So let there be the given cone or cylinder, A; so it is required to find a sphere equal to the cone or cylinder A.

Eut. 272

(a) Let there be a cylinder half as large again as the cone or cylinder A,[19] whose base is the circle around the diameter ΓΔ, and its axis is the <axis> EZ, (b) and let two mean proportionals be taken between ΓΔ, EZ, <namely> HΘ, MN, so that as ΓΔ is to HΘ, HΘ to MN and MN to EZ, (c) and let a cylinder be imagined, whose base is the circle around the diameter HΘ, and its axis, KΛ, is equal to the diameter HΘ. So I say that the cylinder E is equal to the cylinder K.

(1) And since it is: as ΓΔ to HΘ, MN to EZ, (2) and alternately,[20] (3) and HΘ is equal to KΛ [(4) therefore as ΓΔ to MN, that is as the <square> on ΓΔ to the <square> on HΘ,[21] (5) so the circle E to the circle K],[22] (6) therefore as the circle E to the circle K, so KΛ to EZ. [(7) Therefore the bases of the cylinders E, K are reciprocal to the heights]; (8) therefore the cylinder E is equal to the cylinder K.[23] (9) But the cylinder K is half as large again as the sphere whose diameter is HΘ; (10) therefore also the sphere whose diameter is equal to HΘ, that is B,[24] (11) is equal to the cone or cylinder A.

---

growing, the other diminishing in reciprocal proportion) without changing that volume. To say that "each of them is given" is, then, misleading. We may in fact derive a solution for the problem, regardless of how we choose to set the cone ΓZΔ, since what we are seeking is for *some* cylinder or cone satisfying the equality: one among the infinite family of such cylinders and cones, their bases and heights reciprocally proportional.

[17] A single mean proportional of A, C is a B satisfying A:B:B:C. Two mean proportionals satisfy A:B::B:C::C:D, where B and C are the two mean proportionals "between" A and D.

[18] Greek "*sunthesetai*," "will be synthesized." The word belongs to the pair *analysis/ synthesis*, perhaps translatable as "deconstruction/construction," literally something like "breaking into pieces," "putting the pieces together." As we saw above, Archimedes (as is common in Greek mathematics) did not introduce in any explicit way his *analysis*; but the *synthesis* is introduced by an appropriate formula.

[19] See Eutocius for this problem, which is essentially relatively simple (it requires one of several propositions from *Elements* XII, e.g. XII.11 or 14).

[20] *Elements* V.16, yielding the unstated conclusion: ΓΔ:MN::HΘ:EZ.

[21] From *Elements* V. Deff. 9–10, and the stipulation that the lines ΓΔ, HΘ, MN, EZ are in continuous proportion (which is an equivalent way of saying that HΘ, MN are two mean proportionals between ΓΔ, EZ).

[22] *Elements* XII.2.   [23] *Elements* XII.15.

[24] This sphere B – the real requirement of the problem – has not been constructed at all at the synthesis stage. Archimedes offers two incomplete arguments that only taken together provide a solution to the problem. See general comments to this and following problems, for the general question of relation between analysis and synthesis.

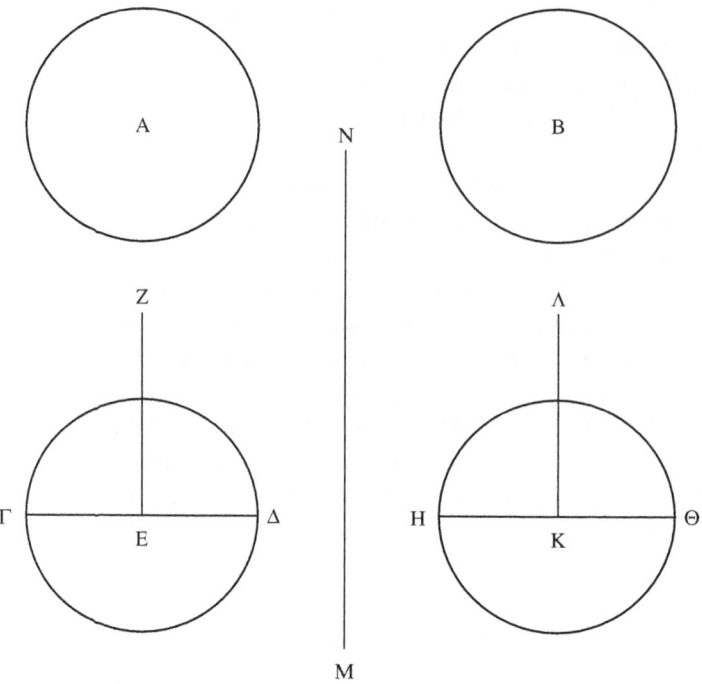

TEXTUAL COMMENTS

Heiberg brackets Step 2 in the analysis, as well as the related Step 7 in the synthesis, presumably for stating what are relatively obvious claims: but this being the very beginning of the treatise, we may perhaps imagine Archimedes being more explicit than usual. Steps 6–7 in the analysis, on the other hand, are very jarring, in repeating, in such close proximity, the claim of Step c: they seem most likely to be a scholion to Step 5, interpolated into the text.

Steps 4–5 in the synthesis are more difficult to explain. They make relevant and non-obvious claims. They are problematic only in that their connector is wrong: the "therefore" at the start of Step 4 yields the false expectation, that the claim of Steps 4–5 taken together is somehow to be derived from the preceding steps. I can not see why this mistaken connector should not be attributed to Archimedes, as a slip of the pen.

GENERAL COMMENTS

Does "analysis" *find* solutions?

The pair of analysis and synthesis is a form of presenting problems, whose intended function has been discussed and debated ever since antiquity. In the comments to this book, I shall make a few observations on the details of some arguments offered in this form.

A basic question is whether the analysis in some sense "finds" the solution to the problem. In this problem, the solution can be seen quite simply (arguably,

II.1
Codex A had the slightly different lay-out of the thumbnail (clearly the difference is that codex C has two columns of writing in the page, while codex A probably had only one: with wider space available, A adopted a shorter arrangement. Late ancient writing would tend to have two columns, answering to the narrow column of the papyrus roll, hence I prefer the layout of C).   Codices DH4 do not have the point M extending to below the lower circles, perhaps representing codex A. Once again, I follow codex C.   Codices DG had KΛ greater than EZ.   Codex G had the two circles A, B (equal to each other) greater than the circles ΓEΔ, HKΘ (also roughly equal to each other); circle HKΘ somewhat lower than circle ΓEΔ.   Codex 4 permutes M/N.

the problem is simpler than the synthesis/analysis approach makes it appear), and it is therefore a useful case for answering this question.

We may conceive of the problem of finding a sphere equal to a given cone or cylinder, as that of transformation: we wish to transform the cone or cylinder into a sphere. Consider a cylinder. Given any cylinder, we may transform it into a "cubic" cylinder (where the diameter of the base equals the height), by conserving (new circle):(old circle)::(old height):(new height). (This is not a trivial operation, and it already calls for two mean proportionals, involving as it does a proportion with both lines and areas.) The sphere obtained inside this "cubic" cylinder would be, following *SC* I.34 Cor., $\frac{2}{3}$ the cylinder itself. We may therefore enlarge this new sphere by a factor of $\frac{3}{2}$, by enlarging its diameter by a factor of $\sqrt[3]{\frac{3}{2}}$. This new sphere, with its new diameter, would now be the desired sphere; but it is obviously simpler to enlarge the original cylinder by a factor of $\frac{3}{2}$ (no need to specify how, but the simplest way is by enlarging its height by the same factor, following *Elements* XII.14). Then all we require to do is to transform this new, enlarged cylinder into a "cubic" cylinder, which is done through two mean proportionals.

Thus the solution to the problem has two main ideas. One is to use *SC* I.34 Cor. to correlate a sphere and a "cubic" cylinder; the second is to make this correlation into an equality, by enlarging the given cylinder in the factor $\frac{3}{2}$. The second idea is an *ad-hoc* construction, which does not emerge in any obvious way out of the conditions stated by the problem. And indeed, it is not anything we derive in the course of the analysis: to the contrary, this is a stipulated construction, occurring as Step b of the analysis. Thus this second aspect of the solution clearly is not "found" by the analysis.

But neither is the first one. To begin with, the main idea is derived not from the analysis process, but from *SC* I.34 Cor. itself. But this obvious observation aside, it should be noticed that the idea of using two mean proportionals – arguably, the most important point of the analysis – is, once again, not a direct result of the analysis as such. Once again, it has to be stipulated into the analysis by an *ad-hoc* move – that of Step d, where the line MN is stipulated into existence (with several further manipulations, this line yields the two mean proportionals). Nothing in the analysis necessitates the introduction of this line, which was inserted into the proposition, just like the auxiliary half-as-large cylinder, because Archimedes already knew what form the solution would make.

In other words: in this case, there is nothing "heuristic" about analysis. Here we see analysis not so much a format for finding solutions, but a format for presenting them.

/ 2 /

Every segment of the sphere is equal to a cone having a base the same as the segment, and, <as> height, a line which has to the height of the segment the same ratio which: both the radius of the sphere and the height of the remaining segment, taken together, have to the height of the remaining segment.[25]

Let there be a sphere, in it a great circle whose diameter is AΓ, and let the sphere be cut by a plane, <passing> through the <points> BZ, at right <angles> to AΓ, and let Θ be center, and let it be made: as ΘA, AE taken together to AE, so ΔE to ΓE, and again let it be made: as ΘΓ, ΓE taken together to ΓE, so KE to EA, and let cones be set up on the circle around the diameter BZ, having <as> vertices the points K, Δ; I say that the cone BΔZ is equal to the segment of the sphere at Γ, while the <cone> BKZ <is equal> to the at the point A.

(a) For let BΘ, ΘZ be joined, (b) and let a cone be imagined, having, <as> base, the circle around the diameter BZ, and, <as> height, the point Θ, (c) and let there be a cone, M, having, <as> base, a circle equal to the surface of the segment of the sphere, BΓZ ((1) that is, <a circle> whose radius is equal to BΓ),[26] and a height equal to the radius of the sphere; (2) so the cone M will be equal to the solid sector BΓΘZ; (3) for this has been proved in the first book.[27] (4) And since it is: as ΔE to EΓ, so ΘA, AE taken together to AE, (5) it will be dividedly: as ΓΔ to ΓE, so ΘA to AE,[28] (6) that is ΓΘ to AE,[29] (7) and alternately, as ΔΓ is to ΓΘ, so ΓE to EA,[30] (8) and compoundly, as ΘΔ to ΘΓ, ΓΔ to AE,[31] (9) that is, the <square> on ΓB to the <square> on BE;[32] (10) therefore as ΔΘ to ΓΘ, the <square> on ΓB to the <square> on BE. (11) But ΓB is equal to the radius of the circle M, (12) and BE is radius of the circle around the diameter BZ; (13) therefore as ΔΘ to ΘΓ, the circle M to the circle around the diameter BZ.[33] (14) And ΘΓ is equal to the axis of the cone M; (15) therefore as ΔΘ to the axis of the cone M, so the circle M to the circle around the diameter BZ; (16) therefore the cone having, <as> base, the circle M, and, <as> height, the radius of the sphere, is equal to the solid rhombus BΔZΘ [(17) for

Eut. 306

---

[25] Every plane cutting through a sphere divides it into two segments. One is taken as *the segment*; the other, then, is taken as *the remaining segment*. There are thus four leading lines in this proposition. Three of them are: height of the segment (S); radius of the sphere (R); height of the remaining segment (S′). (Note that one of S/S′ is greater than R, and the other is smaller, e.g. S′>R>S, except the limiting case, where the two segments are each a hemisphere and S′=R=S.) The fourth line is the height of the constructed cone (C), which is here defined as C:S::(R+S′):S′.

[26] *SC* I.42.   [27] *SC* I.44.   [28] *Elements* V.17.

[29] Both ΘA and ΓΘ are radii in the sphere. The implicit result of Steps 5–6 is: ΓΔ:ΓE::ΓΘ:AE. Step 7 refers to this implicit result.

[30] *Elements* V.16.   [31] *Elements* V.18.

[32] *Elements* VI.8 Cor., VI.20 Cor.2; for details, see Eutocius.

[33] *Elements* XII.2.

this has been proved in the lemmas of the first book.[34] Or like this: (18) since it is: as $\Delta\Theta$ to the height of the cone M, so the circle M to the circle around the diameter BZ, (19) therefore the cone M is equal to the cone, whose base is the circle around the diameter BZ, while <its> height is $\Delta\Theta$; (20) for their bases are reciprocal to the heights.[35] (21) But the cone having, <as> base, the circle around the diameter BZ, and, as height, $\Delta\Theta$, is equal to the solid rhombus BΔZΘ].[36] (22) But the cone M is equal to the solid sector BΓZΘ; (23) therefore the solid sector BΓZΘ, too, is equal to the solid rhombus BΔZΘ. (24) Taking away as common the cone, whose base is the circle around the diameter BZ, while <its> height is EΘ; (25) therefore the remaining cone BΔZ is equal to the segment of the sphere BZΓ.

And similarly, the cone BKZ, too, will be proved to be equal to the segment of the sphere BAZ.

(26) For since it is: as $\Theta\Gamma$E taken together to ΓE, so KE to EA, (27) therefore dividedly, as KA to AE, so $\Theta\Gamma$ to ΓE;[37] (28) but $\Theta\Gamma$ is equal to ΘA;[38] (29) and therefore, alternately, it is: as KA to AΘ, so AE to EΓ;[39] (30) so that also compoundly: as KΘ to ΘA, AΓ to ΓE,[40]

Eut. 306

(31) that is the <square> on BA to the <square> on BE.[41] (d) So again, let a circle be set out, N, having the radius equal to AB; (32) therefore it is equal to the surface of the segment BAZ.[42] (e) And let [the] cone N be imagined, having the height equal to the radius of the

Eut. 307

sphere; (33) therefore it is equal to the solid sector BΘZA; (34) for this is proved in the first <book>.[43] (35) And since it was proved: as KΘ to ΘA, so the <square> on AB to the <square> on BE, (36) that is the <square> on the radius of the circle N to the <square> on the radius of the circle around the diameter BZ, (37) that is the circle N to the circle around the diameter BZ,[44] (38) and AΘ is equal to the height of the cone N, (39) therefore as KΘ to the height of the cone N, so the circle

Eut. 308

N to the circle around the diameter BZ; (40) therefore the cone N, that is the <solid> sector BΘZA (41) is equal to the figure BΘZK.[45] (42)

---

[34] The reference could be to *Elements* XII.14, 15.   [35] *Elements* XII.15.

[36] Can be derived from *Elements* XII.14.   [37] *Elements* V.17.

[38] Both are radii. The implicit result of Steps 27–8, taken up by Step 29, is KA:AE::ΘA:EΓ.

[39] *Elements* V.16.   [40] *Elements* V.18.

[41] Steps 26–31 follow precisely Steps 4–9, and therefore see note to Step 9 (the required Euclidean material: *Elements* VI.8 Cor., VI.20 Cor.2).

[42] *SC* I.43.   [43] *SC* I.44. But see Eutocius' comments.   [44] *Elements* XII.2.

[45] The figure intended is a cone out of which another smaller cone has been carved out. See Eutocius for the argument. It is essentially identical to that of Step 16 above, applying *Elements* XII.14, 15 with the difference that here we subtract, rather than add, cones.

Let the cone, whose base is the circle around BZ, while <its> height is EΘ, be added <as> common; (43) therefore the whole segment of the sphere ABZ is equal to the cone BZK; which it was required to prove.

/Corollary/

And it is obvious that a segment of a sphere is then, generally, to a cone having the base the same as the segment, and an equal height, as: both the radius of the sphere and the perpendicular of the remaining segment, taken together, to the perpendicular of the remaining segment; (44) for as ΔE to EΓ, so the cone ΔZB, (that is the segment BΓZ),[46] (45) to the cone BΓZ.[47]

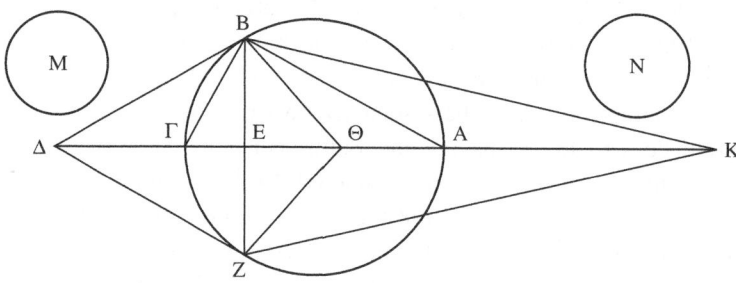

With the same laid down: <to prove> that the cone KBZ, too, is equal to the segment of the sphere BAZ.

(f) For let there be a cone, N, having, <as> base, [the] <surface> equal to the surface of the sphere, and, <as> height, the radius of the sphere; (46) therefore the cone is equal to the sphere [(47) for the sphere has been proved to be four times the cone having, <as> base, the great circle, and, <as> height, the radius.[48] (48) But then, the cone N, too, is four times the same, (49) since the base is also <four times> the base,[49] ((50) and the surface of the sphere is <four times> the greatest of the <circles> in it)].[50] (51) And since it is: as ΘA, AE taken together to AE, ΔE to EΓ, (52) dividedly and alternately: as ΘΓ to ΓΔ, AE to EΓ.[51] (53) Again, since it is: as KE to EA, ΘΓE taken together to ΓE, (54) dividedly and alternately: as KA to ΓΘ, that is to ΘA,[52] (55) so

II.2
Codex A had the two smaller circles projecting more to the left and the right of the main figure – see comments to previous diagram.   Codex D, followed by Heiberg, has moved Θ to coincide with the center of the circle. Codex E omits line ΓB.

---

[46] Proved in the preceding proposition.   [47] *Elements* XII.14.
[48] *SC* I.34.   [49] And then apply *Elements* XII.11.
[50] *SC* I.33.   [51] *Elements* V.17, 16.   [52] Both radii.

AE to EΓ,[53] (56) that is ΘΓ to ΓΔ.[54] (57) And compoundly;[55] (58)

Eut. 308

and AΘ is equal to ΘΓ; (59) therefore as KΘ to ΘΓ, ΘΔ to ΔΓ, (60) and the whole KΔ is to ΔΘ, as ΔΘ to ΔΓ,[56] (61) that is as KΘ to ΘA; (62) therefore the <rectangle contained> by ΔK, ΘA is equal to the <rectangle contained> by ΔΘK.[57] (63) Again, since it is: as KΘ to ΘΓ, ΘΔ to ΓΔ, (64) alternately;[58] (65) and as ΘΓ to ΓΔ, AE

Eut. 309

was proved to be to EΓ; (66) therefore as KΘ to ΘΔ, AE to EΓ; (67) therefore also: as the <square> on KΔ to the <rectangle contained> by KΘΔ, the <square> on AΓ to the <rectangle contained> by AEΓ.[59] (68) And the <rectangle contained> by KΘΔ was proved equal to the <rectangle> contained by KΔ, AΘ; (69) therefore as the <square> on KΔ to the <rectangle contained> by KΔ, AΘ, that is KΔ to AΘ,[60] (70) the <square> on AΓ to the <rectangle contained> by AEΓ, (71) that is to the <square> on EB.[61] (72) and AΓ is equal to the radius of the circle N; (73) therefore as the <square> on the radius of the circle N to the <square> on BE, that is the circle N to the circle around the diameter BZ,[62] (74) so KΔ to AΘ, (75) that is KΔ to the height of the cone N; (76) therefore the cone N, that is the sphere, (77) is equal to the solid rhombus BΔZK.[63] [(78) Or like this; therefore[64] it is: as the circle N to the circle around the diameter BZ, so ΔK to the height of the cone N; (79) therefore the cone N is equal to the cone, whose base is the circle around the diameter BZ, while <its> height is ΔK; (80) for their bases are reciprocal to the heights.[65] (81) But this cone[66] is equal to the solid rhombus BKZΔ;[67] (82) therefore the cone N, too,

---

[53] *Elements* V.17, 16.

[54] The implicit result of Steps 54–6 is KA:ΘA::ΘΓ:ΓΔ. It is from this that Step 57 starts.

[55] *Elements* V.18. The result of this operation is not spelled out. It would be (KA+ΘA:ΘA::ΘΓ+ΓΔ:ΓΔ), or (KΘ:ΘA::ΘΔ:ΓΔ). Step 58 refers to this implicit result.

[56] *Elements* V.16, 18; see Eutocius.   [57] *Elements* VI.16.

[58] *Elements* V.16. I.e. KΘ:ΘΔ::ΘΓ:ΓΔ.

[59] The derivation from Step 66 to Step 67 implies a general result in geometrical proportion-theory that is not provided in the *Elements* (Archimedes either refers to a lost result, or takes it here for granted). See Eutocius' proof, which uses *Elements* V.7, 18, 21, VI.1.

[60] *Elements* VI.1.

[61] *Elements* VI.8 Cor. The implicit result of Steps 69–71 is (KΔ:AΘ::(sq. AΓ):(sq. EB)). Steps 72–4 further manipulate this implicit proportion.

[62] *Elements* XII.2.

[63] *Elements* XII.14–15. the following passage explicates this.

[64] The "therefore" means that we are taking our cue from Steps 73–5, so as to reach Steps 76–7 by another route.

[65] *Elements* XII.15.   [66] The last cone mentioned in Step 79.

[67] Can be derived from *Elements* XII.14.

that is the sphere, (83) is equal to the solid rhombus BZKΔ].[68] (84) Of which,[69] the cone BΔZ was proved equal to the segment of the sphere BΓZ; (85) therefore the remaining cone BKZ is equal to the segment of the sphere BAZ.

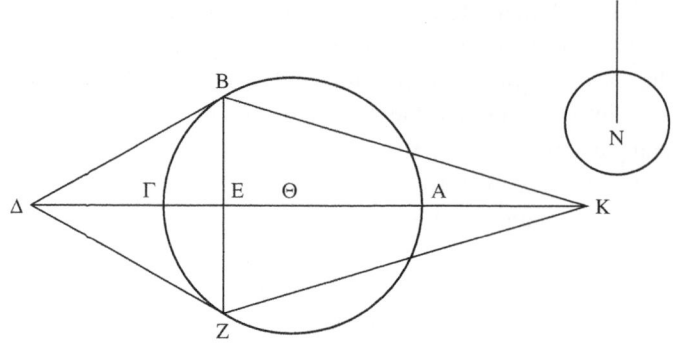

II.2
In codex A, the relation KA>ΔΓ is much more pronounced (see previous comments). Codex D has the line extend not upwards, but downwards from the circle N.   Codex A and all its copies had M instead of N. It is possible (no more) that codex C had the same mistake.

TEXTUAL COMMENTS

In the setting-out, "a plane, <passing> through the <points> BZ, at right <angles> to AΓ," I keep the manuscripts' reading against Heiberg (who follows Nix), with the geometrically curious "points" (instead of the expected "line." In Greek, this is the difference between plural and singular, τῶν and τῆς).

Steps 47–50 are silly and, what clinches the matter, the particle in 48, ἀλλὰ μήν (which I translate, rather lamely, "but then") is never used elsewhere by Archimedes. Steps 78–83 seem to come from a similar source, perhaps the same interpolator (though this cannot be proved).

Now to the glaring textual difficulty of this proof. There are two separate arguments for the equality of the greater segment to the cone BKZ: Steps 26–43, and Steps 46–85. Since the first, but not the second, is very closely modeled on the proof for the smaller segment, it is possible to imagine that the first proof was added by a less competent mathematician, who simply extended the proof for the smaller segment to the case of the greater segment. The introduction of the second proof, "With the same laid down: <to prove> that the cone KBZ too, is equal to the segment of the sphere BAZ," is bizarre as it stands in the sequence of text as we read it right now, but if we remove the first proof then this becomes a natural way for Archimedes to introduce this extended proof. Having given a proof for the smaller segment, he now goes on to give a proof for the greater segment. So the whole of Steps 26–43 is perhaps to be bracketed (this, incidentally, will help explain why there are no minor interpolations in the sequence 26–43). Needless to say, had Heiberg

---

[68] Rounding back to Steps 76–7.

[69] Namely, the sphere and the solid rhombus.

bracketed Steps 26–43 I would probably have found something nice to say about them.

Heiberg was clearly at his most clement here. I am amazed that he did not bracket Step 3, "for this has been proved in the first book," as well as the similar Step 34. At least the reference to the "first book" cannot be authentic (is this how one refers to previous letter?). True, Greek βιβλίον may mean as little as "roll," but the word "first" instead of, say, "previous," is damning. For similar reasons, Heiberg is certainly right in bracketing Steps 17–21.

The title "corollary" has of course no original manuscript authority. It was probably the mistake of inserting this title that caused Heiberg to fail to understand the wider structure of the text, as if the main text and the so-called "corollary" were totally independent; hence Heiberg's failure to bracket Steps 26–43.

GENERAL COMMENTS

The two cones and the generality of the argument

There is a special complication regarding generality here. Why does one need *two* cones, proving for the two segments? Clearly the expectation is that the two cases (smaller or greater than a hemisphere) will be qualitatively different, calling for a different argument. The generality of each of the arguments stops short of being applicable to the other case. The line BZ acts as a barrier, as it were, blocking the transmission of results (which are, however, directly transmittable to any other sphere with a similar configuration).

But how do we tell which of the two segments is which, *by the construction itself*? How do we know – without referring to the diagram – which case we are dealing with at each stage? If we cannot, in what sense can the two cases be said to be *qualitatively* distinct?

Now, there is a qualitative difference between the proof for the smaller segment and the *first proof* offered for the greater segment: Step 24 (smaller segment) takes a cone away; Step 42 (greater segment) adds a cone. However, although the *second* proof for the greater segment is so fantastically complex and, at its surface structure, quite distinct from the proof for the smaller segment, it is in fact not a proof for a greater segment at all, but completely general. Steps 46–85 nowhere use the specific character of the segment, as greater than a hemisphere. Of course, Step 24 still implies that the original segment is a smaller segment, so, to the extent that the definition of goal governing Steps 46–85 sets them in opposition to Steps 1–25, those Steps 46–85 have to apply to the greater segment. Yet Steps 46–85 would apply, as a matter of logic, regardless of what kind of segment was taken at Steps 1–25.

I suggest that the second proof is Archimedes' own, perhaps (as mentioned in the textual comments) the only proof "for the greater segment" offered by Archimedes himself. So in fact Archimedes does not give a proof for the greater segment at all. He gives a proof for the smaller segment (that with a very minor modification can cover the greater segment, too), and then goes on to give a completely general proof, that *if* the assertion is true for one segment, it will also be true for the remaining one – no matter which segment we start with! Apparently Archimedes valued this generality enough to go through the length of Steps 46–85; some later editor preferred the more direct case-by-case approach of Steps 26–43.

It may be of course that Archimedes gave the more difficult proof of Steps 46–85, because he realized that *SC* I.44, as it stands, does not support the claim of Step 33, necessary for the argument in Steps 26–43 (since *SC* I.44, as it stands, deals only with segments smaller than a hemisphere). But I doubt this. The enunciation of *SC* I.44 is completely general, for any sector; and Archimedes knew that the claim of *SC* I.44 holds completely generally: the fact that the generalized proof for *SC* I.44 was left implicit should have made no difference. But this returns us to the basic philosophical question: why were we allowed to leave the second case implicit in *SC* I.44, whereas here, in *SC* II.2, the second case is proved separately? What are the criteria for a genuine *case*? Perhaps the criteria for what counts as a *case* are to be externally motivated: in *SC* I.44, Archimedes is in a hurry, towards the end of the book; here the argument develops more leisurely, the book having just begun, and cases are taken with greater care.

The operation of "imagination:" the border between the conceptual and spatial

The construction furnishes us with a new handle on the operation of imagination: "(b) and let a cone be imagined, having, <as> base, the circle around the diameter BZ . . . (c) and let there be a cone, M, having, <as> base, a circle equal to the surface of the segment of the sphere . . ." Why is the cone on BZ imagined, while M is taken to *be*? If anything, M requires a bolder act of imagination (given that it is represented solely by a circle)! It seems that imagination is required only when it is necessary to furnish a full spatial object, participating in the geometrical configuration. Imagination is a spatial, not a conceptual act. The purely conceptual cone M need not be imagined – it is beyond the pale of imagination, it exists not in geometrical space but in the verbal universe of proportions and propositions. The actual cone on BZ is manipulated in the spatial world, and therefore it needs to be imagined there.

But of course the point is precisely that this border – between the visual and the conceptual – can be so easily crossed. This trespassing is one of the keys to Archimedes' magic. Consider the following pair of tricks:

We start from Step 4, $\Delta E : E\Gamma :: \Theta A + AE : AE$. Now a rapid series of acts:

First trick: Step 5. The ratio $\Delta E : \Gamma E$ is implicitly reinterpreted as $\Gamma\Delta + \Gamma E : \Gamma E$ (and so the "dividedly" operation bites). That is, a spatial decomposition enters inside a proportion. With this implicit reinterpretation and the verbal manipulation of the "dividedly," we get $\Gamma\Delta : \Gamma E :: \Theta A : AE$.

Second trick: Step 6. The ratio $\Theta A : AE$ is converted to the ratio $\Gamma\Theta : AE$, based on the fact that both ΘA, $\Gamma\Theta$ are radii. That is, a spatial reidentification enters inside a proportion. So, implicitly, $\Gamma\Delta : E\Gamma :: \Gamma\Theta : AE$.

Now, with the purely verbal manipulation of "alternately," we get Step 7, ΔΓ:ΓΘ:: ΓΕ:ΕΑ.

Compare now the starting point, Step 4, and – so rapidly evolving from it! – Step 7:

(4) ΔΕ:ΕΓ:: ΘΑ+ΑΕ:ΑΕ, (7) ΔΓ:ΓΘ:: ΓΕ:ΕΑ.

The terms of the proportion have mutated beyond recognition, in a sequence of surprising combinations of the conceptual and the spatial. It is from such rapid successions of tricks that Archimedes' proofs take off.

/ 3 /

The third problem was this: to cut the given sphere by a plane, in such a way that the surfaces of the segments will have to each other a ratio the same as the given <ratio>.[70]

(a) Let it come to be, (b) and let there be a great circle of the sphere, AΔBE, (c) and its diameter AB, (d) and let a right plane be produced, <in right angles> to AB,[71] (e) and let the plane make a section in the circle AΔBE, <namely> ΔE, (f) and let AΔ, BΔ be joined.

(1) Now since there is a ratio of the surface of the segment ΔAE to the surface of the segment ΔBE, (2) but the surface of the segment ΔAE is equal to a circle, whose radius is equal to AΔ,[72] (3) and the surface of the segment ΔBE is equal to a circle, whose radius is equal to ΔB,[73] (4) and as the said circles to each other, so the <square> on AΔ to the <square> on ΔB,[74] (5) that is AΓ to ΓB,[75] (6) therefore a ratio, of AΓ to ΓB, is given;[76] (7) so that the point Γ is given.[77] (8) And ΔE is at right <angles> to AB; (9) therefore the plane <passing> through ΔE, too, is <given> in position.

Eut. 309
Eut. 310

So it will be constructed like this: (a) Let there be a sphere, whose great circle is ABΔE and <whose> diameter is AB, (b) and <let> the given ratio <be> the <ratio> of Z to H, (c) and let AB be cut at Γ, so that it is: as AΓ to BΓ, so Z to H, (d) and let the sphere be cut by a plane <passing> through Γ at right <angles> to the line AB, (e) and

---

[70] I.e., we are given a sphere and a ratio, and we are required to cut the sphere so that the surfaces will have the given ratio. Archimedes' own formulation slightly obscures this, since the given ratio is mentioned as an afterthought. See general comments.

[71] In itself this does not say much. The idea is for the plane to be right to the great circle that passes through AB.

[72] *SC* I.43.   [73] *SC* I.42.   [74] *Elements* XII.2.

[75] See Eutocius. This is essentially from *Elements* VI.8 Cor.

[76] It is the *same* as a given ratio.   [77] See Eutocius, who uses *Data* 7, 25, 27.

let ΔE be a common section,[78] (f) and let AΔ, ΔB be joined, (g) and let two circles be set out, Θ, K – Θ having the radius equal to AΔ, K having the radius equal to ΔB; (1) therefore the circle Θ is equal to the surface of the segment ΔAE,[79] (2) while K <is equal to the surface> of the segment ΔBE; (3) for this has been proved in the first book.[80] (4) And since the <angle contained> by AΔB is right,[81] (5) and ΓΔ is a perpendicular, (6) it is: as AΓ to ΓB, that is Z to H (7) the <square> on AΔ to the <square> on ΔB,[82] (8) that is the <square> on the radius of the circle Θ to the <square> on the radius of the circle K, (9) that is the circle Θ to the circle K,[83] (10) that is the surface of the segment ΔAE to the surface of the segment of the sphere ΔBE.

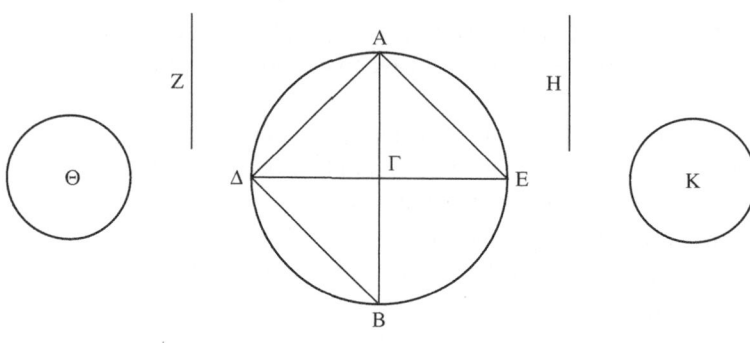

TEXTUAL COMMENTS

It is remarkable that Heiberg brackets nothing here. Of course he ought to have bracketed Step 3 in the synthesis, for reasons explained in regard to Steps 3, 34 in the previous proposition.

Step 4 in the synthesis, "and since the <angle contained> by AΔB is right," appears in the manuscripts as "and since the <angle contained> by AΔB is *given.*" That the manuscripts cannot be right is clear, but the mistake is interesting, because this is possibly authorial. It is not a natural scribal mistake, since the actual word "given" does not appear here very often. The concept, however, is *mathematically* important to the proposition, and therefore the mistake is more likely to issue from a mathematician: an Archimedean slip of the pen?

II.3
Codex C is not preserved for the diagram. By analogy with II.1, it might be suggested that it could have the two smaller circles nearer the main circle, more underneath the two lines.
Codex A has omitted line AΔ, perhaps drawing line AE by mistake instead.
Codex E aligns the two circles even higher up and away from the main circle, while codex D adopts the rather different arrangement of the thumbnail.
Codices BD have Z greater than H.
Codex E has N instead of H.   Codex 4 has Δ instead of A.

---

[78] "Common section:" of the plane mentioned at Step d, and of the great circle mentioned at Step a.
[79] *SC* I.43.   [80] *SC* I.42.   [81] *Elements* III.31.
[82] See Eutocius (the same as Step 5 in the analysis above).
[83] *Elements* XII.2.

GENERAL COMMENTS

Enumerating problems and the structure of the book

The first few words, "the third problem was this:," are a second-order intervention, going back to the introduction. Similar comments are made at the
enunciation of the first proposition, and further back, at the end of the introduction itself. This is the last such second-order intervention: from now on, the
style reverts to pure mathematical presentation. To a certain extent, Archimedes
uses this brief title to create a continuity between the introduction and the main
text: starting from pure introduction, we move to a series of propositions, the
first explicitly connected with the introduction, the latter becoming pure propositions. Thus Archimedes somehow manages to bridge this, the main stylistic
divide of Greek mathematical writing. Another effect of those brief titles is to
stress the nature of the treatise: it is very much an *ad-hoc* compilation, a set of
independent solutions. It is arranged not according to an internal deductive or
narrative order, but simply according to a list of problems that it tackles one
by one. It is thus very different from the first book, with its clear goal and its
playful indirect route of obtaining that goal. Instead of a large-scale narrative
structure, this treatise is a sequence of independent tours-de-force, each having
its own separate character.

The strange nature of "being given"

The logic of "being given" combines here with the logic of analysis and synthesis, with an interesting result.

In a problem, the parameters for the problem itself – the objects defining the
problem – must of course be given, simply so that the problem may be stated.
A problem is always about doing something, *something else being given*. Thus
one is given, in the statement of this problem, both a sphere and a ratio. In
the analysis, however, one starts from the assumption of the problem being
solved. What do we have then? A sphere, cut in such a way that its surfaces
satisfy a given ratio. But what does this tell us? The given ratio, in a sense,
is not geometrically significant. We do not do anything with the fact that the
ratio is given, since there is nothing we *can* do with this: a ratio which is given
is no different from any other ratio, its givenness endows it with no specific
geometrical properties. All the given ratio does, is to supply us with a suitable
ending point for the analysis process.

Thus, when Archimedes starts the analysis with the words "let it come to be,"
we are left asking – "let *what* come to be?" That the surfaces are to each other
as . . . as *what*? This – and here is the beauty of the situation – is immaterial. All
we need to know is that the ratio of the surfaces has this effectively meaningless
property, of being *given*. Hence also the interesting Step 1 of the analysis: "Now
since there is a ratio of the surface of the segment $\triangle AE$ to the surface of the
segment $\triangle BE$." This step, at face value, asserts nothing for, in the context, it
can be directly assumed that all pairs of objects of the same kind have some ratio
between them. Still, givenness being empty of special geometrical meaning, it

is very natural for Archimedes to state not that the ratio is *given*, but that it *is* – as it were, an allowed member of the universe of discourse. The ratio is "on the table."

/4/

To cut the given sphere so that the segments of the sphere have to each other the same ratio as the given.

Let there be the given sphere, $AB\Gamma\Delta$; so it is required to cut it by a plane so that the segments of the sphere have to each other the given ratio.

(a) Let it be cut by the plane $A\Gamma$. (1) Therefore the ratio of the segment of the sphere $A\Delta\Gamma$ to the segment of the sphere $AB\Gamma$ is given. (b) And let the sphere be cut through the center, and let the section be a great circle, $AB\Gamma\Delta$,[84] (c) and <let its> center be K, (d) and <its> diameter ΔB, (e) and let it be made: as $K\Delta X$ taken together to ΔX, so PX to XB,[85] (f) and as KBX taken together to BX, so ΛX to $X\Delta$,[86] (g) and let $A\Lambda$, $\Lambda\Gamma$, AP, $P\Gamma$ be joined; (2) therefore the cone $A\Lambda\Gamma$ is equal to the segment of the sphere $A\Delta\Gamma$,[87] (3) while the <cone> $AP\Gamma$ <is equal> to the $AB\Gamma$;[88] (4) therefore the ratio of the cone $A\Lambda\Gamma$ to the cone $AP\Gamma$ is given, too. (5) And as the cone to the cone, so ΛX to XP [(6) since, indeed, they have the same base, the circle around the diameter $A\Gamma$];[89] (7) therefore the ratio of ΛX to XP is given, too. (8) And through the same <arguments> as before, through the construction, as $\Lambda\Delta$ to $K\Delta$, KB to BP (9) and ΔX to XB.[90] (10) And since it is: as PB to BK, $K\Delta$ to $\Lambda\Delta$,[91] (11) compoundly, as PK to KB, that is to $K\Delta$,[92] (12) so $K\Lambda$ to $\Lambda\Delta$;[93] (13) and therefore

Eut. 310

Eut. 310

---

[84] Any plane cutting through the center will produce a great circle; the force of the clause is to provide this great circle with its letters. (Note further that it is by now taken for granted that this cutting plane, producing the great circle, is at right angles to the plane $A\Gamma$.)

[85] Defining the point P. (K, Δ, B are defined by the structure of the sphere, X is taken to be defined through the make-believe of the analysis.)

[86] Analogously defining the point Λ.

[87] *SC* II.2.   [88] *SC* II.2.   [89] *Elements* XI.14.

[90] Translating the letters appropriately between the diagrams, the claims made here can be seen to be equivalent to *SC* II.2, Step 29 (=Step 8 here), Steps 7–8, 29 (=Step 9 here). There is the standard problem that interim conclusions are not asserted in general terms, and are therefore more difficult to carry over from one proposition to another, hence Archimedes' *explicit* reference in Step 8. Also, see Eutocius.

[91] *Elements* V.7 Cor.   [92] Both KB and KΔ are radii.

[93] *Elements* V.18.

the whole PΛ is to the whole KΛ as KΛ to ΛΔ;[94] (14) therefore the
<rectangle contained> by PΛΔ is equal to the <square> on ΛK.[95]

Eut. 311 (15) Therefore as PΛ to ΛΔ, the <square> on KΛ to the <square>
on ΛΔ.[96] (16) And since it is: as ΛΔ to ΔK, so ΔX to XB, (17) it will
be, inversely and compoundly: as KΛ to ΛΔ, so BΔ to ΔX[97] [(18)
and therefore as the <square> on KΛ to the <square> on ΛΔ, so
the <square> on BΔ to the <square> on ΔX. (19) Again, since it is:
as ΛX to ΔX, KB, BX taken together to BX, (20) dividedly, as ΛΔ to

Eut. 311 ΔX, so KB to BX].[98] (h) And let BZ be set equal to KB; ((21) for it is
clear that it will fall beyond P)[99] [(22) and it will be: as ΛΔ to ΔX,

Eut. 311 so ZB to BX; (23) so that also: as ΛΔ to ΛX, BZ to ZX].[100] (24) And
since <the> ratio of ΛΔ to ΛX is given, (25) therefore <the> ratio

Eut. 312 of PΛ to ΛX is given as well.[101] (26) Now, since the ratio of PΛ to
ΛX is combined of both: the <ratio> which PΛ has to ΛΔ, and <that

Eut. 316 which> ΛΔ <has> to ΛX,[102] (27) but as PΛ to ΛΔ, the <square>
on ΔB to the <square> on ΔX,[103] (28) while as ΛΔ to ΛX, so BZ
to ZX. (29) Therefore the ratio of PΛ to ΛX is combined of both: the
<ratio> which the <square> on BΔ has to the <square> on ΔX, and

Eut. 317 <the ratio which> BZ <has> to ZX. (i) And let it be made: as PΛ to
ΛX, BZ to ZΘ.[104] (30) And <the> ratio of PΛ to ΛX is given; (31)
therefore <the> ratio of ZB to ZΘ is given as well. (32) And BZ <is>
given; (33) for it is equal to the radius; (34) therefore ZΘ is given as
well.[105] (35) Also, therefore, the ratio of BZ to ZΘ is combined of
both: the <ratio> which the <square> on BΔ has to the <square>
on ΔX, and <that which> BZ <has> to ZX. (36) But the ratio BZ to

---

[94] As Eutocius explains very briefly, we have, as an implicit result of Steps
11–12, (PK:KΛ::KΛ:ΛΔ), from which can be derived, through *Elements* V.12,
(PK+KΛ:KΔ+ΛΔ::KΛ:ΛΔ) – if we have a:b::c:d, we can derive (a+c):(b+d)::c:d.

[95] *Elements* VI.17.

[96] This could be derived directly from Step 13, through *Elements* VI.20 Cor.

[97] *Elements* V.7 Cor., 18.   [98] *Elements* V.17.

[99] See Eutocius. The result derives from the assumption that ABΓ is the smaller
segment.

[100] *Elements* V.12.

[101] A complex claim in the theory of proportions. See Eutocius, who uses *Elements*
V. 7 Cor., 19 Cor., and *Data* 1, 8, 22, 25, 26.

[102] The operation of "composition of ratios" was never fully clarified by the Greeks:
see Eutocius for an honest attempt. It can be connected with what we would understand
as "multiplication of fractions." (The ratio a:f is composed, as it were, from two ratios b:c,
d:e that satisfy (b:c)*(d:e)=a:f – whatever this multiplication and this equality actually
mean. The simplest case is the one here, a:c composed of a:b and b:c.)

[103] As Eutocius shows, this can be derived from Steps 15 and 17. See textual comments.

[104] Defining the point Θ.   [105] *Data* 2.

ZΘ is combined of both: the <ratio> of BZ to ZX, and of the <ratio> of ZX to ZΘ. [(37) Let the <ratio> of BZ to ZX be taken away <as> common];[106] (38) remaining, therefore, it is: as the <square> on BΔ, that is a given[107] (39) to the <square> on ΔX, so XZ to ZΘ, (40) that is to a given. (41) And the line ZΔ is given.

Eut. 317

(42) Therefore it is required to cut a given line, ΔZ, at the <point> X and to produce: as XZ to a given <line> [<namely> ZΘ], so the given <square> [<namely> the <square> on BΔ] to the <square> on ΔX.

This, said in this way – without qualification – is soluble only given certain conditions,[108] but with the added qualification of the specific characteristics of the problem at hand[109] [(that is, both that ΔB is twice BZ and that ZΘ is greater than ZB – as is seen in the analysis)], it is always soluble;[110] and the problem will be as follows:

Given two lines BΔ, BZ (and BΔ being twice BZ), and <given> a point on BZ, <namely> Θ; to cut ΔB at X, and to produce: as the <square> on BΔ to the <square> on ΔX, XZ to ZΘ.

And these <problems>[111] will be, each, both analyzed and constructed at the end.[112]

---

[106] We have (translating the composition of ratios into anachronistic notation): (35) BZ:ZΘ=((sq. BΔ):(sq. ΔX))*(BZ:ZX), (36) BZ:ZΘ=(BZ:ZX)*(ZX:ZΘ). From which of course we can derive, ((sq. BΔ):(sq. ΔX))*(BZ:ZX)=(BZ:ZX)*(ZX:ZΘ). Archimedes now (37) takes away the common term (BZ:ZX) and derives (38–9) the proportion (sq. BΔ):(sq. ΔX)::(ZX:ZΘ).

[107] It is a square on the given diameter of the sphere.

[108] "Soluble only given certain conditions:" is literally, in the Greek, "has a *diorismos*." *Diorismos* is a technical term, meaning (in this context), limits under which a problem is soluble. What Archimedes says is that, when the last statement following the analysis is stated as a general problem, where the given lines and square may vary freely – so that they may be any given lines and area whatsoever – some combinations will prove to be insoluble.

[109] Literally, "with the addition of the problems at hand." The Greek for "problem" (*problema*) is wider in meaning than our modern mathematical sense, and can mean, as it does here, "*specific characteristics* of a problem."

What Archimedes means is that the specific given square and line of the problem of *SC* II.4 make the problem possible. They are not just any odd square and line. The given square is uniquely determined by one of the given lines, namely by ΔZ. It is the square on two thirds the line ΔZ. The remaining given line, ZΘ, is not uniquely determined by the given line ΔZ, but it has a boundary: it is less than a third of ΔZ. So with these specific determinations and limits, the problem can always be solved ("always" – i.e. no matter where Θ falls on the line BZ). For all of this, see Eutocius.

[110] Literally, "it does not have a *diorismos*."

[111] I.e. both the unqualified and the qualified problem.

[112] Do not reach for the end of the treatise: this promised appendix vanished from the tradition of the *SC*. See, however, the extremely interesting note by Eutocius.

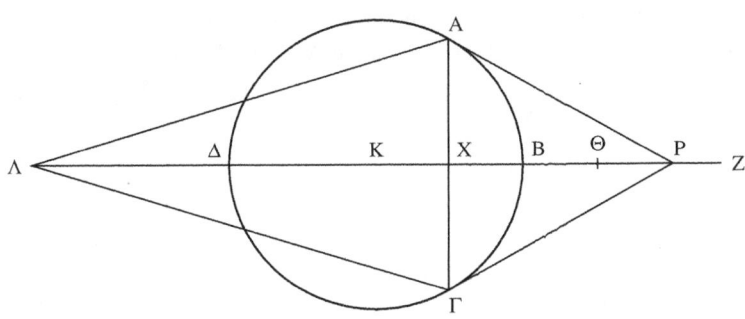

II.4
Codex C is not
preserved for this
diagram.

So the problem will be constructed like this:

Let there be the given ratio, the <ratio> of Π to Σ (greater to smaller), and let some sphere be given and let it be cut by a plane <passing> through the center, and let there be a section <of the sphere and the plane, namely> the circle ABΓΔ, and let BΔ be diameter, and K center, and let BZ be set equal to KB, and let BZ be cut at Θ, so that it is: as ΘZ to ΘB, Π to Σ, and yet again let BΔ be cut at X, so that it is: as XZ to ΘZ, the <square> on BΔ to the <square> on ΔX, and, through X, let a plane be produced, right to the <line> BΔ; I say that this plane cuts the sphere so that it is: as the greater segment to the smaller, Π to Σ.

Eut. 344

(a) For let it be made, first as KBX taken together to BX, so ΛX to ΔX, (b) second as KΔX taken together to XΔ, PX to XB, (c) and let ΑΛ, ΛΓ, ΑΡ, ΡΓ be joined; (1) so through the construction (as we proved in the analysis), the <rectangle contained> by PΛΔ will be equal to the <square> on ΛK,[113] (2) and as KΛ to ΛΔ, BΔ to ΔX;[114] (3) so that, also: as the <square> on KΛ to the <square> on ΛΔ, the <square> on BΔ to the <square> on ΔX. (4) And since the <rectangle contained> by PΛΔ is equal to the <square> on ΛK [(5) it is: as PΛ to ΛΔ, the <square> on ΛK to the <square> on ΛΔ],[115] (6) therefore it will also be: as PΛ to ΛΔ, the <square> on BΔ to the <square> on ΔX, (7) that is XZ to ZΘ. (8) And since it is: as KBX taken together to BX, so ΛX to XΔ, (9) and KB is equal to BZ, (10) therefore it will also be: as ZX to XB, so ΛX to XΔ; (11) convertedly, as XZ to ZB, so XΛ to ΛΔ;[116] (12) so that also, as ΛΔ to ΛX, so BZ to ZX.[117] (13) And since it is: as PΛ to ΛΔ, so XZ to ZΘ, (14)

Eut. 344

and as ΔΛ to ΛX, so BZ to ZX, (15) and through the equality in the perturbed proportion, as PΛ to ΛX, so BZ to ZΘ;[118] (16) therefore

---

[113] Step 14 in the analysis. [114] Step 17 in the analysis.

[115] Step 15 in the analysis. [116] *Elements* V. 19 Cor.

[117] *Elements* V.7 Cor.

[118] *Elements* V.23. To explain the expression: to move from A:B::C:D and B:E::D:F to conclude that A:E::C:F is to have an argument *from the equality*. Here the premises

also: as ΛX to XP, so ZΘ to ΘB.[119] (17) And as ZΘ to ΘB, so Π to Σ; (18) therefore also: as ΛX to XP, that is the cone ΑΓΛ to the cone ΑΡΓ,[120] (19) that is the segment of the sphere ΑΔΓ to the segment of the sphere ΑΒΓ,[121] (20) so Π to Σ.

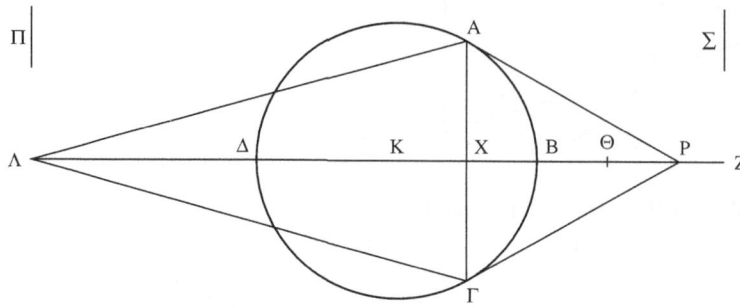

II.4
Codex C is not preserved for this diagram.   Codex D has positioned the two lines Π, Σ to the two sides of the main figure, and has made Σ greater than Π.

TEXTUAL COMMENTS

Heiberg's bracketed passages (Steps 6, 18–20, 22–3, 37 and bits of 42 in the analysis, a few bits of the interlude between analysis and synthesis, and Step 5 in the synthesis) are not trivial, but are still relatively moderate given the size of the proposition. All of them, with the exception of Step 6 in the analysis (a fairly obvious, so also suspect, backward-looking justification), are bracketed because of some tensions they create when read together with Eutocius' commentary. They either state what Eutocius seems to prove separately from Archimedes, or their text disagrees with Eutocius' quotations. As usual, I doubt if such tensions are at all meaningful. Thus this fiendishly complicated proposition seems to be in relatively good textual order, which is not at all a paradox: its intricacies are such to deter the scholiast.

The text refers, in the interlude between analysis and synthesis, to an appendix to the work. This appendix was lost to the main lines of transmission, it is absent from all the extant manuscripts, and was initially unknown to Eutocius. After what he implies was a long search, Eutocius was capable of finding some vestiges of this appendix, apparently in some text totally independent of the *On the Sphere and the Cylinder*. For all of this, see Eutocius.

---

are not the direct sequence A–B–E and C–D–F, but rather A:B::C:D and B:E::F:C. The second sequence is not C–D–F, but F–C–D, and the conclusion is accordingly A:E::F:D. This then is a *perturbed proportion*. (None of those labels is very instructive, but they are established by tradition, and are enshrined in our text of Euclid.) Also, see Eutocius.

[119] For instance: From (15) PΛ:ΛX::BZ:ZΘ, get PΛ:XP::BZ:ΘB (*Elements* V.19 Cor.), hence XP:PΛ::ΘB:BZ (*Elements* V.7 Cor.) which, with (14) again, yields the conclusion XP:ΛX::ΘB:ZΘ (*Elements* V.22). Applying *Elements* V.7 Cor. again, we get the desired conclusion: (15) ΛX:XP::ZΘ:ΘB.

[120] *Elements* XI.14.   [121] *SC* II.2.

GENERAL COMMENTS

A suggestion on the function of analysis in a complex solution

As noted in the textual comments, this proposition is very complicated. The complexity, however, does not stem from any deep insight gained by the proposition. The complex construction required to solve the problem is the result of a direct manipulation, through proportion theory, of the reduction of sphere segments to cones, provided in *SC* II.2. Thus the solution is in a way less than completely satisfactory: the baroque construction has no deep motivation, and stands in contrast to the extremely simple statement of the problem. Essentially, this is because Archimedes' tools here, geometrical proportions, were designed to state in clear, elegant form relations in plane geometry. Archimedes cleverly reduces the three-dimensional curvilinearity of spheres into the line segments along ΛZ, but the solid nature of the problem remains irreducible, in the form of cumbersome, non-obvious proportions. (It might perhaps be suggested that the search for ways of dealing with non-planar geometric relations, in the same elegance available for plane geometry, ultimately led to the emergence of modern mathematics.)

One way in which the solution may appear more satisfactory is, quite simply, by prefacing the synthetic solution by an analysis. The purpose of the analysis, I suggest, may be in this case a sort of *apology* for the synthesis. The analysis shows how the parameters of the problem force the author to solve the problem in this particular way and no other, and in this way make this complicated solution appear a bit more "natural." It is almost as if, to make the synthesis appear less cumbersome than it is, Archimedes prefaces it by an even more cumbersome analysis, so that, by comparison, the synthesis appears to be straightforward.

At any rate, once again: there is no reason to believe that the synthesis was *discovered* by following the analysis. It is instructive to note that the points Z, Θ appear in their natural alphabetic order in the synthesis and not the analysis, suggesting that the analysis might have been written by Archimedes only after the synthesis was already written. At any rate, the main ideas behind this solution are very clear – and have nothing to do with the method of analysis and synthesis. The solution is motivated by the desire to transform solid relations into linear relations. To do this, the relation between the segments of spheres is transformed into a relation between cones (which are then easy to translate into lines, with the tools provided in the *Elements*). Thus the main idea of the proof is simply *SC* II.2 which – crucially – was *not* offered in synthesis and analysis form. Why? Because, as a theorem, it called for no apology. Put simply: when you state the truth, its ugliness is no shame. Ugliness is a shame only (as in a problem) when you choose it among infinitely many other options.

The use of interim results

As mentioned already in n. 90 above, Steps 8–9 in the analysis show us the difficulty which arises with interim results. *SC* II.2 had reached a number of interim proportions, which were stepping-stones for further argumentation. Here the same stepping-stones are required. However, Archimedes' way of referring to them is extremely mystifying: "And through the same <arguments> as before, through the construction, as ΛΔ to ΚΔ, ΚΒ to ΒΡ, and ΔΧ to ΧΒ." (Note that the word "construction" refers not to the drawing of the diagram, but to the verbal stipulation made concerning the ratios obtaining in this proposition.) This opaque form of reference is due to a combination of two reasons. First, the stepping-stones were not enshrined at any enunciation. They were not goals in themselves, to be proved in the most general way, and hence they were never stated in general form and apart from a reference to diagrammatic letters. Second, the lettering of the two propositions, *SC* II.2 and 4, differs (although they both deal with exactly the same position). This is typical of the practice of Greek mathematics, where, at the end of each proposition, the "deck of cards is reshuffled," letters being re-assigned to the diagram according to many local factors (especially the order in which those letters are introduced into the texts of the different propositions). As a consequence, there is no specific statement Archimedes can refer to: the general statement of the interim results was never enunciated, while the particular statement was not given in a form usable in this context. All Archimedes has to refer to is the assertion: "and therefore, alternately, it is: as ΚΑ to ΑΘ, so ΑΕ to ΕΓ" – Step 29 in *SC* II.2 – which has no bearing at all on *SC* II.4 (where, for instance, there is not even an Ε!).

It is interesting that Archimedes did not solve this difficulty by allowing a further, interim lemma, expressed as a general enunciation. It is typical of this treatise, that the focus is throughout on the problems themselves. Once again: this is not a gradually evolving, self-sufficient treatise, like the previous book, but a series of solutions to certain striking problems, with only a very few theorems mentioned only where absolutely necessary. This is most obvious with the lemma mentioned here in the interlude between the analysis and the synthesis: perhaps the most striking result in this book, it was delegated to an appendix, set apart from the main work, and perhaps consequently lost from the main manuscript tradition.

Finally, note that, once again, we see that Archimedes does not have the tools required for making explicit references of any kind: quite simply, the propositions are not numbered, so that all he can refer to is the vague "same as before" – which could be anywhere in the treatise. Indeed, the vestigial system of numbering used in this treatise refers to problems alone: *SC* II.2, a theorem, escapes, as it were, Archimedes' coarse net.

/ 5 /

To construct <a segment of a sphere> similar to a given segment of a sphere and, the same , equal to another given .

Let the two given segments of a sphere be ΑΒΓ, ΕΖΗ, and let the circle around the diameter ΑΒ be base of the segment ΑΒΓ, and <its> vertex the point Γ, and <let> the <circle> around the diameter ΕΖ be base of the ΕΖΗ, and <its> vertex the point Η; so it is required to find a segment of a sphere, which will be equal to the segment ΑΒΓ, and similar to the <segement> ΕΖΗ.

(a) Let it be found and let it be the ΘΚΛ, and let its base be the circle around the diameter ΘΚ, and <its> vertex the point Λ. (b) So let there also be circles[122] in the spheres: ΑΝΒΓ, ΘΞΚΛ, ΕΟΖΗ, (c) and their diameters, at right <angles> to the bases of the segments: ΓΝ, ΛΞ, ΗΟ, (d) and let Π, Ρ, Σ be centers (e) and let it be made: as ΠΝ, ΝΤ taken together to ΝΤ, so ΧΤ to ΤΓ, (f) and as ΡΞ, ΞΥ taken together to ΞΥ, so ΨΥ to ΥΛ, (g) and as ΣΟ, ΟΦ taken together to ΟΦ, so ΩΦ to ΦΗ, (h) and let cones be imagined, whose bases are the circles around the diameters ΑΒ, ΘΚ, ΕΖ, their vertices the points Χ, Ψ, Ω.

(1) So the cone ΑΒΧ will be equal to the segment of the sphere ΑΒΓ, (2) and the <cone> ΨΘΚ <will be equal> to the ΘΚΛ, (3) and the <cone> ΕΩΖ to the ΕΗΖ; (4) for this has been proved.[123] (5) And since the segment of the sphere ΑΒΓ is equal to the segment ΘΚΛ, (6) therefore the cone ΑΧΒ is equal to the cone ΨΘΚ, as well [(7) and the bases of equal cones are reciprocal to the heights];[124] (8) therefore it is: as the circle around the diameter ΑΒ to the circle around the diameter ΘΚ, so ΨΥ to ΧΤ. (9) And as the circle to the circle, the <square> on ΑΒ to the <square> on ΘΚ;[125] (10) therefore as the <square> on ΑΒ to the <square> on

Eut. 345 ΘΚ, so ΨΥ to ΧΤ. (11) And since the segment ΕΖΗ is similar to the segment ΘΚΛ,[126] (12) therefore the cone ΕΖΩ, as well, is similar to the cone ΨΘΚ [(13) for this shall be proved];[127] (14) therefore it is:

Eut. 346 as ΩΦ to ΕΖ, so ΨΥ to ΘΚ. (15) And <the> ratio of ΩΦ to ΕΖ is given;[128] (16) therefore <the> ratio of ΨΥ to ΘΚ is given as well. (h) Let the <ratio> of ΧΤ to Δ be the same; (17) and ΧΤ is given; (18)

Eut. 347 therefore Δ is given as well. (19) And since it is: as ΨΥ to ΧΤ, that is the <square> on ΑΒ to the <square> on ΘΚ, (20) so ΘΚ to Δ,[129] (i) let the <rectangle contained> by ΑΒ, ς be set equal to the <square> on ΘΚ;[130] (21) therefore it will also be: as the <square> on ΑΒ to the <square> on ΘΚ, so ΑΒ to ς.[131] (22) But it was also proved: as the

[122] By "circles," Archimedes refers here to *great* circles.

[123] *SC* II.2.   [124] *Elements* XII.15.   [125] *Elements* XII.2.

[126] The assumption of the analysis (Step a).

[127] See Eutocius (to whom the reference probably points).

[128] This is obvious since the segment itself is given. See Eutocius for a detailed exposition.

[129] Step h, and then *Elements* V.16 (see Eutocius, who comments on this, probably not because there is any need to remind the readers of the existence of *Elements* V.16 but because of the difficult structure of Steps 19–20).

[130] I.e. the line ς is determined by this Step i to satisfy the equality rect. ΑΒ,ς =sq.ΘΚ.

[131] *Elements* VI.1.

Eut. 347   <square> on AB to the <square> on ΘK, so ΘK to Δ,[132] (23) and alternately: as AB to ΘK, so ς to Δ.[133] (24) And as AB to ΘK, so ΘK to ς [(25) through the <fact> that the <square> on ΘK is equal to the <rectangle contained> by AB, ς];[134] (26) therefore as AB to ΘK, so ΘK to ς and ς to Δ. (27) Therefore ΘK, ς are two means in a continuous proportion between two given <lines, namely> AB, Δ.

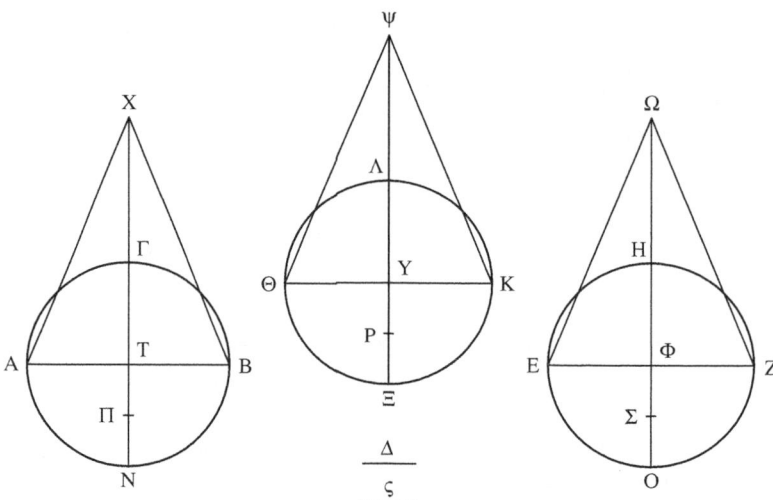

II.5
Codex C is not preserved for this diagram. However, the next diagram is preserved in C, and since the two are geometrically identical in A, I shall assume they are identical in C as well, and proceed to treat the next diagram of C as a source for this diagram, as well. Codex A has the somewhat different arrangement of the thumbnail (Δ to the left of ς), for which see previous comments. DE4 have ΣT instead of T (!), while D4 have MΦ instead of Φ (!). Clearly something went wrong in codex A, in those areas.   Codex E may have Δ instead of ς.

So the problem will be constructed like this: (a) Let the ABΓ be that to which it is required to construct an equal segment, (b) while EZH <is> that to which <it is required to construct> a similar , (c) and let ABΓN, EHZO be great circles of the spheres, and their diameters ΓN, HO, and centers Π, Σ, (d) and let it be made: as ΠN, NT taken together to NT, so XT to TΓ, (e) while as ΣOΦ taken together to OΦ, ΩΦ to ΦH; (1) therefore the cone XAB is equal to the segment of the sphere AΓB,[135] (2) while the <cone> ZΩE <is equal> to the EHZ.[136] (f) Let it be made: as ΩΦ to EZ, so XT to Δ, (g) and, between two given lines, <namely> AB, Δ, let two mean proportionals be taken, <namely> ΘK, ς, so that it is: as AB to ΘK, so KΘ to ς and ς to Δ, (h) and, on the <line> ΘK, let a segment of a circle be erected,[137] <namely> ΘKΛ, similar to the segment of circle EZH,

---

[132] An unstated result of Steps 19–20 above. The combination of 21–2 now yields the unstated result AB:ς::ΘK:Δ, which is the basis of the next step.

[133] See note to preceding step (as explained also by Eutocius). See general comments on the structure of the argument here.

[134] Step 24 derives from Step 25 through *Elements* VI.17.

[135] *SC* II.2.   [136] *SC* II.2.

[137] This "let . . . be erected" in my translation stands for the Greek word ἐφεστάσθω, a *hapax legomenon* in the Archimedean corpus (misspelled, too, in the manuscripts. A better translation might have been "let a segment of a circle be erekted").

(i) and let the circle be completed,[138] and let its diameter be ΛΞ, (j) and let a sphere be imagined, whose great circle is ΛΘΞK, and whose center is P, (k) and let a plane be produced through ΘK, right to ΛX; (3) so the segment of the sphere to the same side as Λ will be similar to the segment of the sphere EHZ, (4) since the segments of the circles were similar, too. (5) And I say that it is also equal to the segment of the sphere ABΓ. (l) Let it be made: as PΞ, ΞY taken together to ΞY, so ΨY to YΛ; (5) therefore the cone ΨΘK is equal to the segment of the sphere ΘKΛ.[139] (6) And since the cone ΨΘK is similar to the cone ZΩE,[140] (7) it is therefore: as ΩΦ to EZ, that is XT to Δ (8) so ΨY to ΘK; (9) and alternately and inversely;[141] (10) therefore as ΨY to XT, ΘK to Δ. (11) And since AB, KΘ, ς, Δ are proportional (12) it is: as the <square> on AB to the <square> on ΘK, ΘK to Δ.[142] (13) But as ΘK to Δ, ΨY to XT; (14) therefore also: as the <square> on AB to the <square> on KΘ, that is the circle around the diameter AB to the circle around the diameter ΘK,[143] (15) so ΨY to XT; (16) therefore the cone XAB is equal to the cone ΨΘK;[144] (17) so that the segment of the sphere ABΓ is equal to the segment of the sphere ΘKΛ, as well. (18) Therefore the ΘKΛ has been constructed, the same equal to the given segment ΑΓB and similar to another given, EZH.

Eut. 347

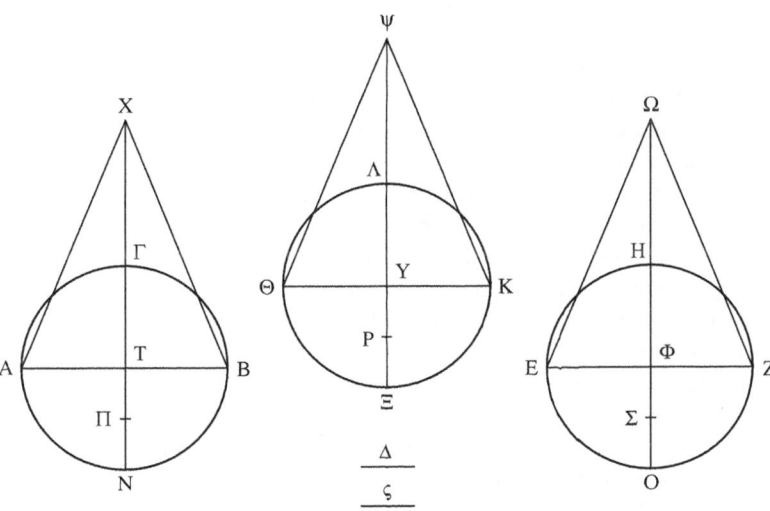

II.5 Second diagram See previous diagram for notes on the layout. Codex G changes somewhat the figure by raising the horizontal lines ATB, ΘYK, EΦZ to above the centers. Codex 4 has Π instead of Γ, codices E4 omit O.   Codex H has omitted K.

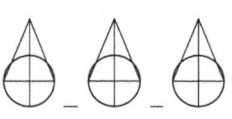

---

[138] I.e. the circle ΞKΛΘ.   [139] *SC* II.2.

[140] The same as Step 12 of the analysis; see discussion there.

[141] *Elements* V.16, 7 Cor.

[142] See Eutocius. Essentially, this is nothing more than *Elements* VI. 20 Cor. 2.

[143] *Elements* XII.2.   [144] *Elements* XII.15.

TEXTUAL COMMENTS

Once again, surprisingly little is problematic in this relatively complicated proposition. As I shall go on to note in the general comments, there is some mathematical reason to suspect Step 13 of the analysis. Heiberg further doubts Steps 7 and 25 of the analysis, on the usual inconclusive grounds of their being backwards-looking justifications, offering little mathematical insight. Other than this, the text seems clear and consistent.

GENERAL COMMENTS

Relying on informal intuitions

In Steps 11–12 of the analysis, Archimedes makes an interesting claim: "(11) And since the segment EZH is similar to the segment $\Theta K \Lambda$, (12) therefore the cone $EZ\Omega$, as well, is similar to the cone $\Psi\Theta K$." In other words, the segments determine their respective cones, up to similarity. This is nowhere proved in the text, though Step 13 goes on to add: "(13) for this shall be proved."

This may be a reference to another lost appendix (similar to the lost appendix, mentioned in the interlude between analysis and synthesis in Proposition 4), originally written by Archimedes himself. However, it is less likely that Archimedes would have gone to the trouble of furnishing such a separate appendix. This claim is interesting but, ultimately, it is relatively straightforward, and is certainly very far from the order of difficulty and originality of the lost appendix mentioned in Proposition 4. On the other hand, Step 13 is a natural way to refer to Eutocius' comment. Thus we may suggest that the claim of Steps 11–12 was directly intuited by Archimedes. This makes sense: he, after all, had invented the construction of the cones discussed here, in *SC* II.2, and so would have been aware of their features. Similar cones are such that the diameters of their bases, and their heights, are in proportion; similar segments of the sphere are such that the diameters of their bases, and their heights, are in proportion. The bases are shared between segments and cones; and so the claim of Steps 11–12 is that the heights of the cones discussed here are proportional to the heights of the segments of the sphere. This is true, because the height of the cone is the fourth term in a proportion where all other terms are sections of the diameter – all similarly affected, therefore, by enlarging or reducing this diameter (these three terms are, taking the present case of $\Omega\Phi$ as the height of the cone: $\Sigma O+O\Phi$, $O\Phi$, ΦH. By enlarging or reducing the sphere, the relative position of Φ is unaffected, i.e. everything dependent upon it is always the same fraction of the diameter). Now, Step 14 goes on as follows: "(14) therefore it is: as $\Omega\Phi$ to EZ, so ΨY to ΘK." Any rigorous, Eutocius-like derivation of Step 12 from Step 11 would necessarily have involved something equivalent to Step 14 and it therefore seems even more probable that Archimedes envisaged no such derivation. Furthermore, in Step 6 of the synthesis, the claim of Step 12 is already an obvious feature, requiring no argument. I therefore suggest that Archimedes understood implicitly the geometrical fact of the invariance of ratios between segments of the diameter, under enlargement and reduction of the sphere. And he may have felt that such intuitions, in this advanced context, called for no explicit proof.

Implicit, non-linear structures of argument

In the example of Steps 11–12, we have seen one sign of the relatively relaxed standards of explicitness in derivation, perhaps representing the advanced nature of the treatise. The structure of the argument in Steps 19–23 of the analysis is a further deviation from more standard, explicit practice; it is also a sign of things to come in this treatise.

Steps 19 and 20, taken together, yield a certain result. Archimedes does not state it explicitly (in itself, not an unprecedented practice). He then moves on to the independently argued Step 21, and only then invokes again the result of Steps 19–20, in Step 22. Then the combination of Steps 21 and 22 entails an implicit result (call it "22a"). Archimedes does not state 22a, but rather transforms it and states the explicit result that derives from this transformation, as Step 23 (see notes to the steps above).

Archimedes could easily have stated Step 22 immediately following Steps 19–20, then could bring in Step 21, derive 22a explicitly from Steps 21–2, and only then derive Step 23. This would have resulted in a linear structure, where the assumptions required are always those that were recently stated explicitly.

So there are two difficulties involved in Archimedes' actual structure. One difficulty is postponement of the use of a step: why not state Step 22 immediately? So, this postponement leads to a non-linear argument.[145] The other difficulty is implicitness of steps: why not state 22a explicitly? True, in this case both non-linearity and implicitness are very mild, but in their combination they already create a difficult structure – and we shall see much more of this in the following. Archimedes begins to demand that we have constantly at hand all the results stated so far (non-linearity) – more than this, he will demand that we have in mind results which were not stated explicitly, but were merely implied (implicitness). A new kind of audience is envisaged – if, indeed, a truly active audience is still being envisaged at all. More and more, it is as if we have been summoned to witness, passively, Archimedes' train of thought – that we are not truly expected to follow ourselves.

/ 6 /

Given two segments of a sphere (either of the same <sphere>, or not), to find a segment of a sphere that shall be similar to one of the given

---

[145] Another, much less disturbing though still a somewhat strange postponement, is the postponement of the use of Step 10 of the synthesis, till Step 13.

<segments>, while it shall have the surface equal to the surface of the other segment.

Let the given spherical segments be at the circumferences ABΓ, ΔEZ; and let that, to which it is required to find a similar , be the at the circumference ABΓ; and <let> that, to whose surface <it is required> to have an equal surface, <be> the at the <circumference> ΔEZ.

(a) And let it have come to be, and let the segment of the sphere KΛM be similar to the segment ABΓ – and let it have the surface equal to the surface of the segment ΔEZ, (b) and let the centers of the spheres be imagined, (c) and, through them, let planes be produced right to the bases of the segments, (d) and let sections <=of the planes> be, in the spheres: the great circles KΛMN, BAΓΘ, EZHΔ, (e) and in the bases of the segments: the lines KM, AΓ, ΔZ, (f) and let ΛN, BΘ, EH be diameters of the spheres, being at right <angles> to KM, AΓ, ΔZ, (g) and let ΛM, BΓ, EZ be joined.

(1) And since the surface of the segment of the sphere KΛM is equal to the surface of the segment ΔEZ, (2) therefore the circle, whose radius is equal to ΛM, is equal to the circle, whose radius is equal to EZ [(3) for the surfaces of the said segments were proved to be equal to circles, whose radii are equal to the <lines> joined from the vertices of the segments to the bases];[146] (4) so that MΛ, too, is equal to EZ.[147] (5)

Eut. 348   And since the KΛM is similar to the segment ABΓ (6) it is: as ΛP to PN, BΠ to ΠΘ;[148] (7) and inversely and compoundly: as NΛ to ΛP, so ΘB to BΠ.[149] (8) But also: as PΛ to ΛM, so BΠ to ΓB [(9) for the triangles are similar];[150] (10) therefore as NΛ to ΛM, that

Eut. 348   is to EZ, (11) so ΘB to BΓ. (12) And alternately;[151] (13) and <the> ratio of EZ to BΓ <is given>; (14) for each of the two is given[152] (15) therefore the ratio of ΛN to BΘ is also given. (16) And BΘ is given; (17) therefore ΛN is given as well;[153] (18) so that the sphere, too, is given.

---

[146] *SC* I.42–3.   [147] *Elements* XII.2.

[148] See Eutocius. The argument is similar in nature to that required in the preceding proposition for showing that the characteristic cones are proportional to the segments. (Steps 11–12 of the analysis.)

[149] *Elements* V.7 Cor., V.18.

[150] The same argument as that required for Step 6 above.

[151] *Elements* V.16. The unstated result of Steps 10–11 is NΛ:EZ::ΘB:BΓ, which, alternately, yields NΛ:ΘB::EZ:BΓ. This is the assertion made by Step 12.

[152] Step 13 derives from Step 14 by *Data* 13. As for Step 14, this is obvious in the same way as Steps 6 and 8 above. See also Eutocius, who uses *Elements* I.47 (besides the already well-known argument for the givenness of the parts of the diameter and of the base of the segment).

[153] *Data* 2.

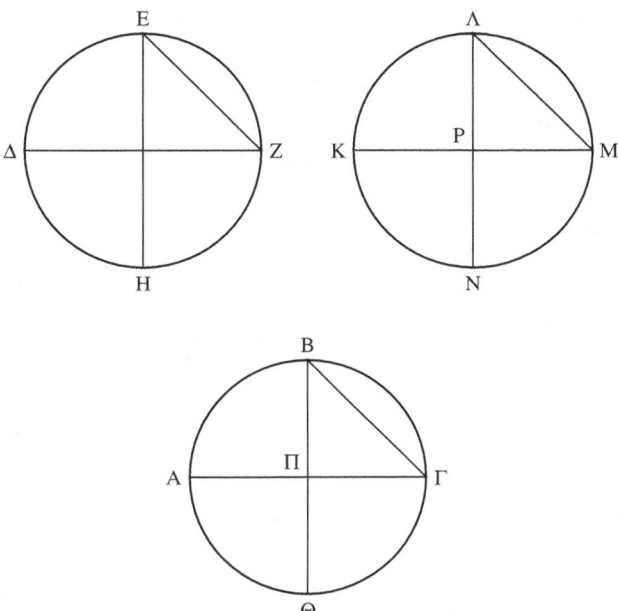

II.6
Codex A has the
somewhat different
arrangement of the
thumbnail, for which
see previous comments.
It also changes
somewhat the figure by
raising the horizontal
lines ΔZ, AΠΓ, KPM
to above the centers.
Perhaps, through a
lectio difficilior, this is
to be preferred to codex
C's use of diameters,
but since the use of
diameters is otherwise
so prevalent I prefer to
suspect that a slight
mistake in A's original
was exaggerated.

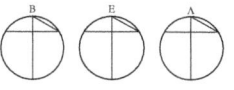

So it shall be constructed like this: Let there be the two given segments of the sphere, ABΓ, ΔEZ: ABΓ, to which <it> is required
<to find> a similar , and ΔEZ, to whose surface <it is
required> to have an equal surface, (a) and let the same be constructed
as in the analysis,[154] (b) and let it be made: as BΓ to EZ, so BΘ to
ΛN, (c) and let a circle be drawn around the diameter ΛN, (d) and let
a sphere be imagined (e) and let ΛKNM be its great circle, (f) and let
NΛ be cut at P, so that it is: as ΘΠ to ΠB, NP to PΛ, (g) and let the sur-
Eut. 348   face <=of the sphere KΛMN> be cut by a plane <passing> through
the <point> P, right to ΛN, (h) and let ΛM be joined; (1) therefore
the segments of the circles on the lines KM, AΓ are similar;[155] (2) so
that the segments of the sphere are similar as well. (3) And since it is:
as ΘB to BΠ, so NΛ to ΛP ((4) for the <things shown> according
to division, too);[156] (5) But also: as ΠB to BΓ, so PΛ to ΛM,[157] (6)
therefore also: as ΘB to NΛ, BΓ to ΛM.[158] (7) And it was also: as ΘB
to ΛN, BΓ to EZ; (8) therefore EZ is equal to ΛM;[159] (9) so that the

---

[154] Refers only to the part of the construction that is applicable at this stage, i.e. that
on the two already given spheres. The third sphere with its internal structure will be
constructed more explicitly in the following.

[155] See Eutocius. The argument is essentially the same as that already seen from the
analysis, Step 6.

[156] A reference to *Elements* V.18 through which (from Step f) Step 3 derives.

[157] The same argument seen several times above, based on the similarity of the triangles which results from the similarity of the segments.

[158] *Elements* V.22, 16.   [159] *Elements* V.9.

circle whose radius is EZ, also, is equal to <the circle whose radius is equal to> ΛM.[160] (10) And the circle having EZ as the radius is equal to the surface of the segment ΔEZ, (11) while the circle, whose radius is equal to ΛM, is equal to the surface of the segment KΛM; (12) for this is proved in the first <book>.[161] (13) Therefore the surface of the segment KΛM is equal, too <to the surface of the segment of the sphere ΔEZ>. (14) and KΛM is similar to ABΓ.

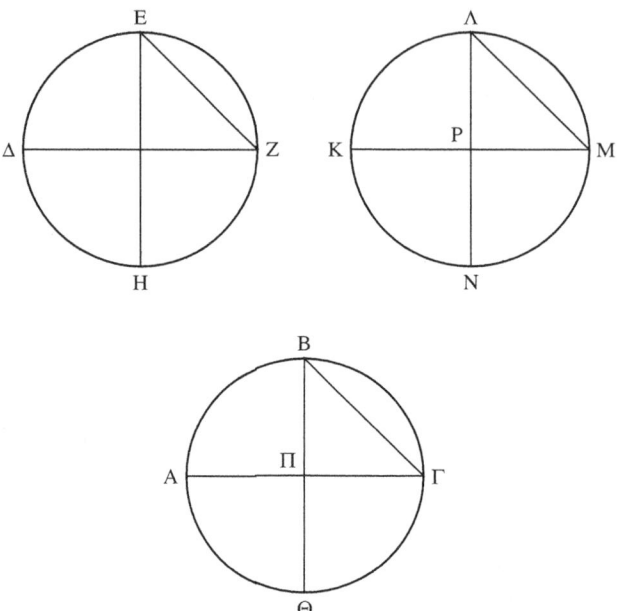

II.6 Second diagram Here codex C breaks off for the final time. Still, I use the diagram I suppose codex C would have had, assuming once again it would have been identical with that of the analysis. The diagram of codex A is not identical with that of the analysis, however, as this time the horizontal lines pass through the centers. (They have been changed by codex G to pass above the centers.)

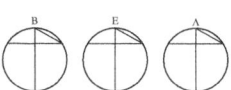

TEXTUAL COMMENTS

Heiberg brackets Steps 3 and 9 in the analysis, with some justification, since both are relatively simple backwards-looking justifications. Why he did not bracket Step 4 of the synthesis I do not know. The remarkable thing is not so much the deviant terminology, as the deviant position: this is a reference to a proportion-theory manipulation, placed as a backwards-looking justification. References to proportion-theory manipulations have a formulaic position, forwards-looking ("dividedly," "alternately . . ."). Scholia, of course, tend to be backwards-looking; this may be one. Step 12, another backwards-looking justification, is certainly a scholion, referring as it does to the "first" book (Archimedes would refer, if at all, to the "previous" letter).

Rejecting too little, Heiberg might also have inserted too much. I refer to Step 9 of the analysis which, as standing in the manuscripts, merely implies

---

[160] *Elements* XII.2.   [161] *SC* I.42–3.

the very long omitted phrase "<the circle whose radius is equal to>." Heiberg inserts it into the text itself (following Coner). In other words, he sees this as a lacuna in the text – a result of textual corruption. This is quite possible, but one should also note the possibility of a genuine abbreviation in the original. Note a similar case, that of the words "<to the surface of the segment of the sphere ΔEZ>" in Step 13: either the scribes, or Archimedes, were very impatient. I suspect we see here the tendency, already apparent in the preceding proposition, to move into some private shorthand, instead of explicit argument.

GENERAL COMMENTS

The relation between analysis and synthesis

As this is a relatively simple case, it may be used as a convenient instance for the general question of the relation between analysis and synthesis. In particular, it is often considered that the synthesis is somehow an "inversion" of the analysis. It is therefore worth seeing in detail how much more complex the relation is. In the following, I go through the steps of the synthesis (aS, . . . , 1S . . .) and try to correlate them to the steps of the analysis (aA, . . . , 1A, . . .).

aS: a partial mirroring of cA, dA, eA, fA, gA.

bS: related to 12A(which last however is only implicitly phrased).

cS: related to dA, but not as its direct mirror, since the objects have different meanings in the two steps.

dS: related to aA, in a roundabout way (similar to the previous relation cS/dA).

eS: indirectly related, once again, to dA (the meaning of the object narrowing down on that of dA).

fS: this step stipulates the ratio of 6A, *inverted*. 7A manipulates the ratio of 6A by *inverting and compounding* it. Thus fS is directly related to neither 6A nor 7A, but is somewhere in between the two.

gS: a partial mirroring of cA, dA, eA, fA, gA, nearly wholly complementing aS.

hS: a partial mirroring of gA, now wholly complementing aS.

1S: indirectly related to the analysis: from the derivation 5A→6A, one may infer fS→1S.

2S: implicit in aA→5A.

3S: 7A, inverted (not in the directional sense of the analysis/synthesis relation, but in the technical sense of inversion as a proportion-theory operation, i.e. *Elements* V.7 Cor.).

4S: partially recovers the argument of 7A (naturally, inverted, now in the analysis/synthesis sense).

5S: similar in nature to the argument of 8A, though referring to different objects.

6S: implicit in 10–11A.

7S: again invoking bS, which is implicit in 12A.

8S: mirrors 4A.

9S: mirrors 2A.

10S, 11S: the two taken together repeat the thrust of the argument of
3A (which, however, is generally phrased – and which may well be
interpolated).

12S: this has no mirroring in the analysis, but then the step is almost
certainly interpolated.

13S: mirrors 1A.

14S: mirrors 5A.

The relation is obviously very intricate indeed, even in this simple proposition.
We see the following principles at work:

First, the analysis and the synthesis are allowed to "bundle" their claims and
results differently. Generally speaking, since the synthesis follows the analysis,
it is allowed to use it to "bundle" its statements more tightly, e.g. in Step aS,
which bundles together several statements from the analysis, explicitly referring
backwards to it.

Second, as a matter of logic, what gets "mirrored" are not specific steps of
the argument, but the overall understanding of the geometrical situation: so
that the relation is neither one-to-one nor direct. The most striking case for this
is fS, which "mirrors" a step *between* 6A and 7A.

Third, while this proposition is simple, it does, naturally, have structure, it
is articulated: in this case, the main articulation is the duality, of *similarity* and
equality. The composition of the two components differs between the analysis
and the synthesis, and is not correlated by a simple mirroring.

In general, then, we see that the synthesis is not the analysis mirrored. It is
essentially a standard forwards-looking argument. It is conducted, indeed, *in
view* of a certain understanding of the geometrical situation, an understanding
gathered during the process of the analysis. There is, however, no question of
the analysis in any way *determining* the synthesis.

/7/

From the given sphere, to cut, by a plane, a segment, so that the segment
has the given ratio to the cone having the base the same as the segment,
and an equal height.

Let there be the given sphere, whose great circle is ABΓΔ, and its
diameter BΔ; so it is required to cut the sphere by a plane <passing>
through AΓ, in such a way that the segment of the sphere ABΓ will
have to the cone ABΓ a ratio the same as the given.

(a) Let it come to be (b) and let E be center of the sphere, (c) and
<let it be:> as EΔZ taken together to ΔZ, so HZ to ZB; (1) therefore
the cone AΓH is equal to the segment ABΓ;[162] (2) therefore the ratio of

---

[162] *SC* II.2.

the cone AHΓ to the cone ABΓ is given as well; (3) therefore the ratio of HZ to ZB is given.[163] (4) And as HZ to ZB, EΔZ taken together to

Eut. 349

ΔZ; (5) therefore the ratio of EΔZ taken together to ΔZ is given; [(6) so that also <the ratio> of EΔ to ΔZ;[164] (7) therefore ΔZ is given as

Eut. 350

well]; (8) so that AΓ <is given> as well.[165] (9) And since EΔZ taken together has to ΔZ a greater ratio than EΔB taken together to ΔB,[166] (10) and EΔB taken together is three times EΔ,[167] (11) and BΔ is twice EΔ, (12) therefore EΔZ taken together has to ΔZ a greater ratio than that which three has to two. (13) And the ratio of EΔZ taken together to ZΔ is the same as the given; (14) therefore the ratio given for the construction must be greater than that which three has to two.

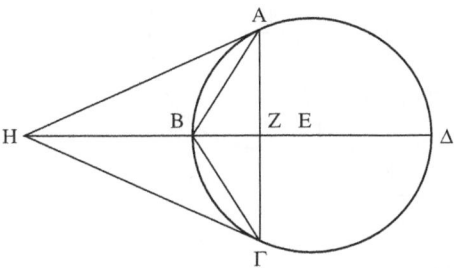

II.7

So the problem shall be constructed like this: (a) let there be the given sphere, whose great circle is ABΓΔ, and diameter BΔ, and center E, (b) and let the given ratio <be> the <ratio> of ΘK to KΛ, greater than that which three has to two. (1) but it is: as three to two, EΔB taken together to ΔB. (2) Therefore ΘK has to KΛ a greater ratio than that which EΔB taken together has to ΔB, as well; (3) therefore, dividedly: ΘΛ has to ΛK a greater ratio than EΔ to ΔB.[168] (c) And let it be made: as ΘΛ to ΛK, so EΔ to ΔZ,[169] (d) and let AZΓ be drawn through Z in right <angles> to BΔ, (e) and let a plane be drawn through ΓA, right

---

[163]   *Elements* XII.14.   [164]   *Elements* V.17.

[165]   See Eutocius. Essentially (though Eutocius does not spell this out) the argument is the same as that seen already in his earlier comments, to Proposition 6. Besides using results from the *Data*, the argument offered by Eutocius relies upon *Elements* VI.8 Cor.

Here the analysis proper comes to an end, and the following steps are dedicated to establishing a limit on the conditions of solvability (*diorismos*). Notice that Archimedes did not separate the *diorismos* in any formal way from the analysis. In this case at least, he saw the study of the limits on the condition of solvability as an integral part of the analysis itself.

[166]   See Eutocius. The argument is based upon *Elements* V.8, 18.

[167]   The closest rendering of the Greek would have been "thrice" rather than "three times." See also general comments.

[168]   An extension to inequalities of *Elements* V.17.

[169]   Defining the point Z.

to BΔ; I say that the segment of the sphere [from] ABΓ has to the cone ABΓ the same ratio as ΘK to KΛ.

(f) For let it be made: as EΔZ taken together to ΔZ, so HZ to ZB;[170] (4) therefore the cone ΓHA is equal to the segment of the sphere ABΓ.[171] (5) And since it is: as ΘK to KΛ, so EΔZ taken together to ΔZ,[172] (6) that is HZ to ZB, (7) that is the cone AHΓ to the cone ABΓ,[173] (7) and the cone AHΓ is equal to the segment of the sphere, (8) therefore as the segment ABΓ to the cone ABΓ, so ΘK to KΛ.

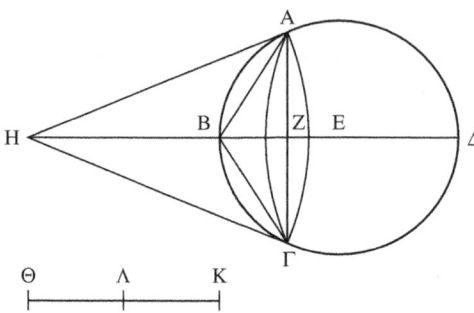

II.7 Second diagram Codex D has ΘΛ greater than ΛK (Heiberg chose the opposite).

TEXTUAL COMMENTS

The situation with Steps 6–7 of the analysis is as follows. Eutocius' commentary explains at length the argument these two steps contain. As printed by Heiberg, however, Eutocius' lemma for this comment consists of Step 5, immediately followed by Step 8. It is natural to assume, then, with Heiberg, that a late scribe has padded Archimedes' text by adding in a summary of Eutocius' argument. However, Eutocius' manuscripts do not have Steps 5 *and* 8 as the lemma for Eutocius' comment, but Step 5 alone: Step 8 was added *by Heiberg* to Eutocius' lemma. Perhaps something is indeed missing from Eutocius' text, as suggested by Heiberg. If so, the missing piece of text may be either Step 8 alone, as suggested by Heiberg (who is then probably right in excising Steps 6–7 from Archimedes' text), or it may be the whole of Steps 6–8. (Then Steps 6–7 are Archimedean, and Eutocius' comment is meant not to supply a completely missing argument, but to expand a brief one.) Or finally, we may even keep the manuscripts' reading, and assume that Eutocius' lemma had Step 5 alone, with an implicit "etc.:" possibly, Eutocius' way of referring to the whole of Steps 5–8?

In the definition of goal following Step e of the synthesis, Heiberg brackets the word "from" in the expression "The segment of the sphere [from] ABΓ." In this of course he may be right, but let us notice the variability of expressions as used in the text. For instance, in Step 7 of the synthesis, the very same object is called "The segment of the sphere," a deviant expression again in the

---

[170] Defining the point H.   [171] *SC* II.2.
[172] Step c, *Elements* V.18.   [173] *Elements* XII.14.

sense that the diagrammatic letters are missing. Heiberg, following the *Editio Princeps*, appended here those letters, "ABΓ" – which do not occur in the manuscripts, and which are not required at this stage. Finally, the last step of the proposition refers to the same object in yet another way, "The segment ABΓ" (the qualification "of the sphere" dropped), to derive the elegant expression: "As the segment ABΓ to the cone ABΓ." In short: it is not that the original language is arbitrary. Phrases are indeed repeated, often with great precision. But they are also to some extent varied. Against the background of a generally regimented language, such variations may be meaningful, aesthetically and sometimes mathematically.

GENERAL COMMENTS

The nature of numbers in standard Greek geometrical practice

The limit on the conditions of solubility (Steps 9–14 of the analysis) raises the issue of the relation between ratios and fractions. In general on this subject, see Fowler (1999), chapter 7. The main result shown by Fowler is that the concept of ratio – even when it is between two numbers – never approaches the modern concept of a fraction. Ratios, simply, are not arithmetized. They are mysterious relations, holding directly between objects, irreducible to numerical expressions. Hence the interesting expression of Step 12: "a greater ratio than that which three has to two," rather than, for instance, *"a greater ratio than three to two" or, let alone, *"a greater ratio than three over two."

Similarly, numbers are not introduced as entities into the geometrical world. Instead of speaking of "the line EΔ multiplied by three," Archimedes speaks of "thrice EΔ" (see footnote to Step 10 in the analysis). The geometrical world may have arithmetical properties, and it is possible to speak of the object that contains EΔ three times. But this does not bring the object *three*, itself, into the geometrical world, as an independent constituent object in the relations in which geometrical objects stand.

This may serve as an introduction, setting out the clear demarcations of numbers and ratios in standard Greek mathematical practice. These demarcations come under serious strain in what follows, as Archimedes' mathematics begins to take off.

/ 8 /

If a sphere is cut by a plane (not through the center), the greater segment has to the smaller: a smaller ratio than duplicate that which the surface of the greater segment has to the surface of the smaller, yet greater than half as large again <the same ratio of the surfaces>.[174]

---

[174] "Duplicate" and "half as large again" signify here what we would call "square" and "3/2 exponent," respectively. The concept of duplicate ratio is standard Greek practice,

Let there be a sphere and in it a great circle, ΑΒΓΔ, and a diameter, ΒΔ, and let it be cut through ΑΓ by a plane right to the circle ΑΒΓΔ, and let ΑΒΓ be a greater segment of the sphere; I say that the segment ΑΒΓ has to the ΑΔΓ a smaller than duplicate ratio than the surface of the greater segment to the surface of the smaller segment, yet greater than half as large again <the same ratio of the surfaces>.

(a) For let the <lines> ΒΑΔ be joined, (b) and let Ε be center, (c) and let it be made: as ΕΔΖ taken together to ΔΖ, ΘΖ to ΖΒ, (d) while as ΕΒΖ taken together to ΒΖ, so ΗΖ to ΖΔ, (e) and let cones be imagined, having the circle around the diameter ΑΓ as base, and the points Θ, Η as vertices; (1) so the cone ΑΘΓ will be equal to the segment of the sphere ΑΒΓ,[175] (2) and the <cone> ΑΓΗ to the ΑΔΓ,[176] (3) and it is: as the <square> on ΒΑ to the <square> on ΑΔ, so the surface of the segment ΑΒΓ to the surface of the segment ΑΔΓ; (4) for this has been demonstrated already.[177]

[It is to be proved that the greater segment of the sphere has to the smaller a smaller ratio than duplicate the surface of the greater segment to the surface of the smaller segment.] I say that the cone ΑΘΓ, too, has to the <cone> ΑΗΓ ((5) that is ΖΘ to ΖΗ)[178] a smaller ratio than duplicate that which the <square> on ΒΑ has to the <square> on ΑΔ ((6) that is ΒΖ to ΖΔ).[179]

(7) And since it is: as ΕΔΖ taken together to ΔΖ, so ΘΖ to ΖΒ [(8) and as ΕΒΖ taken together to ΒΖ, so ΖΗ to ΖΔ] (9) it shall also be: as ΒΖ to ΖΔ, ΘΒ to ΒΕ; (10) for ΒΕ is equal to ΔΕ[180] [(11) for this has been proved by the above]. (12) Again, since it is: as ΕΒΖ taken together to ΒΖ, ΗΖ to ΖΔ, (f) let ΒΚ be equal to ΒΕ; (13) for it is clear that ΘΒ is greater than ΒΕ,[181] (14) since ΒΖ, too, <is greater> than ΖΔ; (15) and it shall be: as ΚΖ to ΖΒ, ΗΖ to ΖΔ.[182] (16) And as ΖΒ to ΖΔ, ΘΒ was proved to be to ΒΕ,[183] (17) and ΒΕ is equal to ΚΒ; (18)

Eut. 350

---

but the "half as large again ratio" may well be an Archimedean invention, made specifically for this proposition. See general comments.

[175] *SC* II.2.   [176] *SC* II.2.

[177] From *SC* I.42–3 (with *Elements* XII.2).   [178] *Elements* XII.14.

[179] See Eutocius. The argument is based on *Elements* VI.8, with some help from *Elements* VI. 20 Cor., V.22.

[180] Both are radii in the sphere. From Step 7, ΕΔΖ:ΔΖ::ΘΖ:ΖΒ, it is possible to get, through *Elements* V.17, an implicit (7′) ΕΔ:ΔΖ::ΘΒ:ΒΖ, and then through Step 10 (ΒΕ=ΔΕ), another implicit (7″) ΒΕ:ΔΖ::ΘΒ:ΒΖ, which, with *Elements* V.7 Cor., 16, yields Step 9. Step 11 is probably an interpolation (see also textual comments).

[181] And therefore – so Step 13 implies – we know that the point Κ falls between the points Β, Θ.

[182] Step f. We also get implicitly (through *Elements* V.16), (15′) ΚΖ:ΗΖ::ΖΒ:ΖΔ. This proportion, although implicit, is used several times in this proposition.

[183] So with the implicit proportion (15′) ΚΖ:ΗΖ::ΖΒ:ΖΔ, we can now get a new implicit proportion: (16′) ΚΖ:ΗΖ::ΘΒ:ΒΕ, which is used in the following.

Eut. 350   therefore as ΘB to BK, so KZ to ZH.[184] (19) And since ΘZ has to ZK a
            smaller ratio than ΘB to BK,[185] (20) and as ΘB to BK, KZ was proved
            to be to ZH, (21) therefore ΘZ has to ZK a smaller ratio than KZ to
Eut. 351   ZH; (22) therefore the <rectangle contained> by ΘZH is smaller than
Eut. 352   the <square> on ZK.[186] (23) Therefore the <rectangle contained> by
            ΘZH has to the <square> on ZH ([that is ZΘ to ZH])[187] (24) a smaller
            ratio than that which the <square> on KZ has to the <square> on
            ZH[188] [(25) and the <square> on KZ has to the <square> on ZH
            a ratio duplicate than KZ to ZH];[189] (26) therefore ΘZ has to ZH a
            smaller ratio than duplicate that which KZ has to ZH [(27) KZ has to
            ZH a smaller ratio than duplicate that which BZ has to ZΔ];[190] We have
Eut. 352   looked for this. (28) And since BE is equal to EΔ,[191] (29) therefore
            the <rectangle contained> by BZΔ is smaller than the <rectangle
Eut. 353   contained> by BEΔ;[192] (30) therefore ZB has to BE a smaller ratio than
            EΔ to ΔZ,[193] (31) that is ΘB to BZ,[194] (32) therefore the <square> on
            ZB is smaller than the <rectangle contained> by ΘBE,[195] (33) that is
            than the <rectangle contained> by ΘBK. (g) Let the <square> on BN
Eut. 353   be equal to the <rectangle contained> by ΘBK;[196] (34) therefore it is:
Eut. 354   as ΘB to BK, the <square> on ΘN to the <square> on NK.[197] (35) But
            the <square> on ΘZ has to the <square> on ZK a greater ratio than the
            <square> on ΘN to the <square> on NK[198] [(36) therefore also the
            <square> on ΘZ has to the <square> on ZK a greater ratio than ΘB
Eut. 354   to BK, (37) that is ΘB to BE, (38) that is KZ to ZH]; (39) therefore ΘZ
            has to ZH a greater ratio than half as large again the <ratio> of KZ to
            ZH [(40) for this is <proved> at the end].[199] (41) And it is: as ΘZ to
            ZH, the cone AΘΓ to the cone AHΓ,[200] (42) that is the segment ABΓ

---

[184] By a substitution of BE/KB (warranted by Step 17) on the implicit proportion
(16′) KZ:HZ::ΘB:BE (as well as an inversion of order, as always taken as a notational
equivalence).

[185] See Eutocius, whose argument relies on *Elements* V.8, 18 extended to inequalities.

[186] An extension to inequalities of *Elements* VI.17; see Eutocius.

[187] *Elements* VI.1.   [188] *Elements* V.8.   [189] *Elements* VI.20 Cor. 2.

[190] An irrelevant, syntactically infelicitous sentence. See textual comments.

[191] Radii to the sphere.

[192] See Eutocius. The argument relies upon *Elements* II.5.

[193] See Eutocius. The argument is an extension to inequalities of *Elements* VI.16.

[194] This is the implicit result (7′) EΔ:ΔZ::ΘB:BZ mentioned in n. 180 above.

[195] An extension to inequalities of *Elements* VI.17.

[196] Defining the point N.

[197] See Eutocius. The argument is based on *Elements* V.16, 18, VI.17, 20 Cor. 2.

[198] See Eutocius. The argument is based on *Elements* VI.22, as well as an extension
to inequalities of *Elements* V.18.

[199] See Eutocius, who manages to derive this result (as a general statement in proportion theory), from definitions alone: *Elements* V. Def. 10, 11. Also see textual comments.

[200] *Elements* XII.14.

to the segment $A\Delta\Gamma$,[201] (43) while as KZ to ZH, BZ to $Z\Delta$,[202] (44) that is the <square> on BA to the <square> on $A\Delta$, (45) that is the surface of the segment $AB\Gamma$ to the surface of the segment $A\Delta\Gamma$;[203] (46) so that the greater segment has to the smaller a ratio smaller than duplicate that which the surface of the greater segment has to the surface of the smaller segment, yet greater than half as large again.

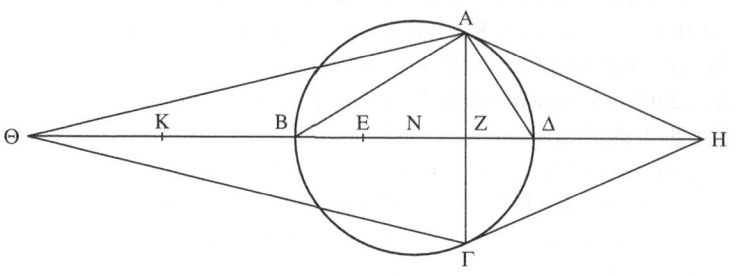

TEXTUAL COMMENTS

Step 27 cannot be right. Archimedes has derived in Step 26 the proportion $\Theta Z{:}ZH<(KZ{:}ZH)^2$. Then Step 27, as it stands in the manuscripts, asserts that $KZ{:}ZH<(BZ{:}Z\Delta)^2$, and then the text goes on to make the second-order comment: "We have looked for this," which must be read as a reference to the secondary definition of goal (as transformed in Steps 5–6): $Z\Theta{:}ZH<(BZ{:}Z\Delta)^2$.

Clearly Step 27 then is not what we have actually looked for. Even in combination with Step 26 it does not yield the definition of goal. Worse: Step 27 is not quite right. The implicit Step 15′ has secured another result with these ratios, namely, KZ:HZ::ZB:ZΔ from which one could argue (though this would be alien to Archimedes' geometry) that since KZ:HZ>1, we can in fact derive Step 27: but why should we try to derive such a result? In short, Step 27 has no bearing on the definition of goal, and it results from 15′ only by accident, as it were. Worse still, the syntax of Step 27 as it stands in the manuscripts is deviant, what is known as an "asyndeton:" the sentence does not have a connecting particle ("and," "but," "therefore" . . .).

There are therefore two possibilities. Either Step 27 is a late and not very intelligent scholion, or it is a textually corrupt version of what Archimedes himself had written.

Let us pursue the second option. Now, the words following Step 27 do call for something different from Step 26 alone. I therefore suggest (words to be inserted are inside pointed brackets): "(27′) <and> KZ to ZH <was proved to be the same as BZ to ZΔ. (27″) Therefore ΘZ has to ZH> a smaller ratio than duplicate that which BZ has to ZΔ."

II.8
Codex A has omitted lines BA, AΔ (reinstated by codex B, followed by Heiberg). The lines are very clearly "intended" to be in the diagram. If indeed we believe the diagrams do ultimately derive from Archimedes himself, then possibly we may have here an authorial slip, but a loss in the transmission process (likely with codex A itself) is much more likely.

---

[201] Steps c, d, *SC* II.2.   [202] This is Step 15′ (see n. 182 above).
[203] *SC* I.42–3.

This of course is merely a suggestion (which I imagine I would have criticized had it been offered by Heiberg – we editors are like this). Note that this suggestion has Archimedes refer to 15′ (in 27′), as if it was explicitly stated. This is comparable to the way in which 7′ is used in Step 31; indeed, 15′ itself is used in 43. This type of practice – referring to implicit results as if they were explicit – has some antecedents so far in the book, and will blossom in the next two proofs: see general comments to the alternative proof of 8.

In general this is a difficult proposition, in content and in text. Heiberg has been very critical – perhaps with some justice. Step 11 shows a lack of understanding of the real basis for Step 9, and therefore must be an interpolation. Probably the same is true of the mathematically redundant Step 8 (since one reason it might be there is as another feeble attempt to support 9). Step 40 is probably a reference to Eutocius by a later scholiast – unless here is another piece of appended material lost from our textual tradition (like the lost proof referred to in Proposition 4).

On the other hand, Heiberg may have been hyper-critical elsewhere. The first formulation of the secondary definition of goal (immediately following Step 4) is indeed stylistically redundant: but this is bad style, not bad mathematics, and no one has promised us that Archimedes, besides being a great mathematician, was also a great stylist. As for Steps 23, 25 and 36–8, they are all bracketed for the poor reason that they do not agree with Heiberg's text of Eutocius.

GENERAL COMMENTS

"Exponent" ratios

The concepts of "duplicate ratio," "half as large again ratio" call for some explanation.

First, we should resist the temptation to see such concepts as a metaphorical transfer from multiplication to exponents. The reason we should avoid this is that the concepts "double," etc., were not as deeply entrenched in the practice of multiplication as they are for a contemporary mathematician. In Greek mathematics, multiplication was less frequent than ratio-manipulation. The word "duplicate," *diplasios* or *diplasion*, in itself signifies no more than a certain doubling, which can equally well accommodate a duplicate ratio – going twice the same ratio – or a duplicate magnitude – having twice the same magnitude. It will only be much later, when the everyday practices of calculation have become a significant background to mathematics (i.e. in the Arabic and Latin Middle Ages), that it shall become proper to speak of a transfer of concepts from calculation to ratio-manipulation.

Euclid speaks freely of "duplicate ratios" (e.g. *Elements* VI.19), and there indeed the sense is clear. The expression "half as large again ratio" is not attested outside this Archimedean passage, and is much more difficult to make sense of. Yet notice the casual way in which the concept is introduced by Archimedes. If he had indeed been the one to have introduced it, then this is a sign of how much of the conceptual work was left by Greek mathematicians for their readers.

To make this effort, imagine then (following Dijksterhuis (1956) 210) a geometrical progression:

$a_1, a_2, a_3, \ldots$

We may say that $a_1:a_3$ is duplicate $a_1:a_2$ and the sense of "duplicate" is now obvious for us: it really is traversing *double the distance in the progression*. So now let us introduce "half as large again" the distance. All we need to do is to insert $a_{2.5}$, a *geometrical* mean between a_2, a_3. Then there is again a clear sense in which $a_1:a_{2.5}$ is half as large again $a_1:a_2$. Or, if we do not wish to introduce "fractions," we may understand "half as large again" as the ratio which three has to two. Then let us append another term to the progression:

a_1, a_2, a_3, a_4

And now it shall be clear that $a_1:a_4$ is half as large again $a_1:a_3$ (three steps in the progression to two). It is this understanding of half as large again that Eutocius uses in his explication of Step 39.

Steps f–13: the language and the diagram

The sequence of Steps f–13 is instructive: "(f) let BK be equal to BE; (13) for it is clear that ΘB is greater than BE."

What is the sense of "for" in Step 13? This translates *gar*, a very important Greek particle that, outside of mathematics, has a very wide semantic range. In mathematics, however, "A *gar* B" almost always means nearly the same as "B, therefore A:" it is inverted derivation, with an added sense that the B clause in "A *gar* B" somehow helps you, the reader, really understand how A could have been asserted. The main claim of "A *gar* B" is almost formal, then: B$\to$A. This strains our understanding of the text, since no such relation seems to hold between Steps 13 and f.

To make sense of this, we need to bear in mind the nature of the use of language, as well as the diagram. First, to the extent that Greek mathematics is still felt to be written in a (subset of) natural Greek – not in some artificial language – we should expect particles to possess, potentially, a wider set of meanings. Now, *gar* may mean, for instance, something like "indeed," "in truth:" not to make any claim that the preceding statement is true, but to focus closely on the claim that its statement is legitimate. And we may say that the utterance of Step f is in some sense valid or legitimate, thanks to the fact mentioned in Step 13. To see this, note that Step f is more than the string of words "let BK be equal to BE." Part of the information is encapsulated not in the text, but in the diagram. The *content* of Step f is partly in the diagram – and this is the content that K falls between the points B, Θ, *as seen in the diagram*. Step 13, finally, legitimates the implication that the line equal to BE falls as seen in the diagram.

Archimedes takes off

This is no ordinary Greek mathematics. One feature I have mentioned already in the textual notes is the willingness to take implicit assertions as starting-points for later arguments. Another feature is the quasi-analytic nature of the proof: that is, Archimedes states in the course of the proof an interim goal (the secondary definition of goal), whose achievement then would be equivalent to the proof itself. This is like the analysis of a theorem (in the analysis and synthesis structure): the result is obtained by showing its equivalence to some other, obtainable result.

Both the taking of implicit results as bases for further argument, and the analysis of a theorem, make it necessary to read the proof in a radically new way. This can no longer be seen as an attempt to persuade a reader. Instead, we see the reasoning used by Archimedes – and we are led to believe that we see it in its very process of being articulated. This is more of an invitation inside Archimedes' study, to come and peer over his shoulder. Is this a deliberate decision, a form adopted by Archimedes? Or is this text indeed in a less final form – are we really allowed to see a raw, unedited Archimedes?

I shall return to discuss such questions in discussing the following, startlingly raw proof.

/In another way/

(a) Let there be a sphere, in it a great circle ABΓΔ, (b) and diameter AΓ, (c) and center E, (d) and let it be cut, through BΔ, by a plane right to the <line> AΓ; I say that the greater segment ΔAB has to the smaller, BΓΔ, a ratio smaller than duplicate that which the surface of the segment ABΔ has to the surface of the segment BΓΔ, yet greater <ratio> than half as large again <as the same>.

(e) For let AB, BΓ be joined; (1) and the ratio of the surface to the surface is the <ratio> of the circle whose radius is AB to the circle whose radius is BΓ,[204] (2) that is the <ratio> of AΘ to ΘΓ.[205] (f) Let Eut. 357   each of AZ, ΓH be set equal to the radius of the circle.[206] (3) And the ratio of the segment BAΔ to the BΓΔ is combined of: that which the segment BAΔ has to the cone whose base is the circle around the diameter BΔ while <its> vertex is the point A; and the same cone to the cone having the same base, and the point Γ as vertex; and the said Eut. 357   cone to the segment BΓΔ.[207] (4) But the ratio of the segment BAΔ to

---

[204] *SC* I.42–3.

[205] The argument for this is based on *Elements* XII.2, VI.8 Cor. 2, and is given by Eutocius as the first comment on the proposition itself.

[206] I.e. the radius of the great circle.

[207] A composition of *three* ratios (and not two, as is the usual practice). Effectively, Archimedes is noting that the ratio of A:D is the same as the ratio composed of A:B, B:C and C:D; the beauty of the operation is that one may choose B and C in a completely free way. In this case, between the segments BAΔ (=A), BΓΔ, (=D), we introduce: the cone BAΔ (=B), and the cone BΓΔ (=C). Also, see Eutocius' comment.

Eut. 358   the cone BAΔ is the <ratio> of HΘ to ΘΓ,[208] (5) while the <ratio>
Eut. 358   of the cone to the cone is the <ratio> of AΘ to ΘΓ,[209] (6) and the
             <ratio> of the cone BΓΔ to the segment BΓΔ is the <ratio> of AΘ
Eut. 358   to ΘZ;[210] (7) and the <ratio> combined of the <ratio> of HΘ to ΘΓ
             and AΘ to ΘΓ is <the ratio> of the <rectangle contained> by HΘA
Eut. 358   to the <square> on ΘΓ,[211] (8) and the <ratio> of the <rectangle
             contained> by HΘ, ΘA to the <square> on ΘΓ, together with the
             <ratio> of AΘ to ΘZ, is the <ratio> of the <rectangle contained>
Eut. 358   by HΘ, ΘA, on ΘA, to the <square> on ΘΓ, on ΘZ,[212] (9) and the
             <ratio> of the <rectangle contained> by HΘA, on ΘA, is the <ratio>
Eut. 358   of the <square> on ΘA, on ΘH;[213] (10) therefore that the <square>
             on ΘA, on ΘH, has to the <square> on ΓΘ, on ΘZ, a smaller ratio than
             the duplicate <ratio> of AΘ to ΘΓ[214] [(11) duplicate the <ratio> of
             AΘ to ΘΓ is: the <ratio> of the <square> on AΘ to the <square> on
Eut. 359   ΘΓ]. (12) Therefore the <square> on AΘ, on ΘH, has to the <square>
             on ΘΓ, on ΘZ, a smaller ratio than the <square> on AΘ, on ΘH, to
Eut. 359   the <square> on ΓΘ, on ΘH.[215] (13) Therefore that the <square>
Eut. 359   on ΓΘ, on ZΘ, is greater than the <square> on ΓΘ, on ΘH.[216] (14)
             Therefore that ΘZ is greater than ΘH.[217]

---

[208] *SC* II.2 Cor., Step f (see Eutocius' comment). In terms of the preceding note, this is A:B::HΘ:ΘΓ.

[209] *Elements* XII.14. In terms of n. 207, this is B:C::AΘ:ΘΓ.

[210] By the same argument as Step 4 above (*SC* II.2 Cor., Step f). In terms of n. 207, this is C:D::AΘ:ΘZ. We therefore have from Steps 3–6 that the ratio of the original segments is composed of the three ratios HΘ:ΘΓ, AΘ:ΘΓ, AΘ:ΘZ. Archimedes now moves on to compose the three ratios together, in Steps 7–8.

[211] *Elements* VI.23 (see Eutocius' comment).

[212] See Eutocius' lemma (preceding his commentary to this proposition). The expression "area on line" means something like "the solid figure defined by the area as basis and the line as height" or, alternatively, "the area multiplied by the line." Archimedes' text allows both readings.

[213] What it is a ratio *to*, is left unstated. This is, indeed, irrelevant to the main claim – which is a mere re-accounting of the very same object, switching lines between "base" and "height." Perhaps Archimedes feels so uneasy about this strange object that he prefers not to name it in isolation, and to refer to it instead within a proportion statement – be it even a truncated, meaningless one.

[214] The meaning of Step 10 is: "therefore <(part of) the *demonstrandum* will be obtained by proving> that the <square> on ΘA on ΘH has to the <square> on ΓΘ on ΘZ a smaller <ratio> than the duplicate <ratio> of AΘ to ΘΓ." See general comments for this startlingly original stylistic innovation.

[215] The claim is that the formulation of Step 12 is equivalent to that of Step 10, and could equally serve as the *demonstrandum*. See Eutocius for the specific argument (briefly, an extension of *Elements* VI.1), and see general comments for the structure of argumentation.

[216] *Elements* V.10.

[217] An extension of *Elements* VI.1. Note that the first branch of the proof now comes to a halt. It has been shown that the *demonstrandum* is equivalent to the inequality of

Eut. 360   So I claim, as well, that the greater segment has to the smaller a ratio greater than half as large again the ratio of the surface. (15) But the <ratio> of the segments was proved to be the same as that which the <square> on AΘ, on ΘH, has to the <square> on ΘΓ, on ΘZ, (16) while the <ratio> of the cube on AB to the cube on BΓ is half as large

Eut. 360   again the ratio of the surface.[218] (17) So I claim that the <square> on AΘ, on ΘH, has to the <square> on ΓΘ, on ΘZ, a greater ratio than [the cube on AB to the cube on BΓ, (18) that is] the cube on AΘ to the cube on ΘB,[219] (19) that is the <square> on AΘ to the

Eut. 360   <square> on BΘ, and AΘ to ΘB.[220] (20) But the <ratio> of the <square> on AΘ to the <square> on ΘB, taking in the <ratio> of AΘ to ΘB, is the <ratio> of the <square> on AΘ to the <rectangle

Eut. 361   contained> by ΓΘB,[221] (21) and the <ratio> of the <square> on AΘ to the <rectangle contained> by BΘΓ is the <square> on AΘ, on ΘH,

Eut. 361   to the <rectangle contained> by BΘΓ, on ΘH;[222] (22) so I claim that, [therefore], the <square> on AΘ, on ΘH, has to the <square> on ΓΘ, on ΘZ, a greater ratio than [the <square> on AΘ to the <rectangle contained> by BΘΓ, (23) that is] the <square> on AΘ, on ΘH, to the <rectangle contained> by BΘΓ, on ΘH.

Eut. 361   (24) Now, it is to be proved that the <square> on ΘΓ, on ΘZ, is smaller than the <rectangle contained> by BΘΓ, on HΘ.[223] (25) Which is the same as proving that the <square> on ΓΘ has to the <rectangle contained> by BΘΓ a smaller ratio than HΘ to ΘZ[224]

Step 14. This in turn is obviously true (by the assumption of the definition of goal, AΘ is greater than ΘΓ, and Step f constructs ZA=ΓH). Eutocius goes on to offer a "synthetic" proof, based on this Archimedean "analysis."

[218] Structure of the original Greek: "Of the ratio of the surface, the <ratio> of the cube on AB to the cube on BΓ is half as large again." The claim seems to be based on an understanding that the ratio of cubes with given sides, is half as large again as the squares on the same sides. This indeed could have been Archimedes' *definition* of a ratio "half as large again;" see also Eutocius.

[219] *Elements* VI.8, 4, and an extension to solids of *Elements* VI.22.

[220] "And" here stands for a composition of ratios. (Compare the standard practice, to allow "and" to represent addition: composition of ratios, as was suggested by Saito (1986), may be, for the Greeks, more an "addition" than a "multiplication:" it is a form of putting things together.)

[221] In this proof virtual skyscrapers of solids and their ratios project into the air from the meager diagram. Now Archimedes juggles a few of those skyscrapers – lets them all go simultaneously – and as he catches them again – lo and behold! – the ratio of solids has been transformed into a ratio of areas, squarely set upon the ground of the diagram (on which ground new skyscrapers are immediately built in the next step). See Eutocius for this extraordinary feat (based upon *Elements* VI.8, 1). The main intuition is that ratios may be cancelled out through the operation of composition of ratios: the ratios A:B, B:C, cancel each other out and leave only A:C).

[222] An extension of *Elements* VI.1.   [223] *Elements* V.8.

[224] An extension of *Elements* VI.16.

[(26) therefore it is required to prove that HΘ has to ΘZ a greater ratio than ΓΘ to ΘB].[225]

Eut. 362

(g) Let EK be drawn from E at right <angles> to EΓ (h) and, from B, let a perpendicular, <namely> BΛ, be drawn on it <=on EK>; (27) it remains for us to prove that HΘ has to ΘZ a greater ratio than ΓΘ to

Eut. 362

ΘB. (28) And ΘZ is equal to AΘ, KE taken together;[226] (29) therefore it is required to prove that HΘ has to ΘA, KE taken together a greater ratio than ΓΘ to ΘB; (30) and therefore, when ΓΘ is subtracted from ΘH, and EΛ from KE ((31) <=EΛ> being equal to BΘ),[227] (32) it shall be required that it be proved that the remaining, ΓH, has to the remaining, AΘ, KΛ taken together, a greater ratio than ΓΘ to ΘB,[228]

Eut. 362

(33) that is ΘB to ΘA,[229] (34) that is ΛE to ΘA,[230] (35) and alternately, that KE has to EΛ a greater ratio than KΛ, ΘA taken together to ΘA,[231] (36) and dividedly: KΛ has to ΛE a greater ratio than KΛ to ΘA.[232] (37) That ΛE is smaller than ΘA.[233]

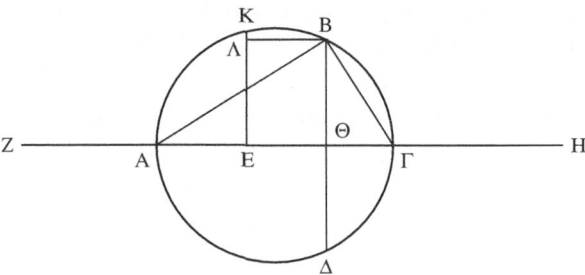

TEXTUAL COMMENTS

As shall be noted in the general comments, this proposition is very unlike the Euclidean standard. This makes it, I believe, more likely to be genuine. Arguably, no one but the author would dare introduce such a radical, massive interpolation.

This does not mean that everything here is by Archimedes. Eutocius commented extensively on this proposition, in the process quoting most of it. This means that there is more scope than usual for mismatch between manuscript

II.8 Second diagram Codex A had an understandable difficulty with the fine structure of K/Λ. the result was as in the thumbnail: the two points conflated and permuted.   Codex B alone corrected the permutation. Codices BD have detached the two points by moving the point E to the center of the circle; Codex G alone has the structure represented in the main diagram.   Codex G omits Δ, line BΓ.

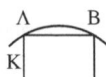

---

225   *Elements* VI.1.

226   Step f. That KE is a radius is understood on the basis of the diagram.

227   Steps d, g, h; *Elements* I.34.

228   An extension to inequalities of *Elements* V.17.

229   *Elements* VI.8. Cor.   230   Steps d, g, h, *Elements* I.34.

231   *Elements* V.16.   232   *Elements* V.17.

233   In other words, this is the final *demonstrandum*, obviously attainable (ΛE is less than a radius, ΘA is more than a radius).

authority for Archimedes, and manuscript authority for Eutocius. Heiberg (1910) lists those discrepancies in a note (217, n.3), and then relies mainly on the manuscripts for Archimedes. This decision I follow.

In some cases, the text must be corrected to restore mathematical sense (for this Heiberg relies, again rightly, on Eutocius). An especially interesting episode is the introduction of the *epi* locution in phrases such as "the rectangle *epi* the line" (which I translate as "the rectangle, on the line"). Some (mathematically educated?) scribe did not believe his eyes, and had changed three times *epi* (Steps 8–9) into *pros*, resulting with nonsense such as: "(8) . . . the <ratio> of the <rectangle contained> by HΘ, ΘA to ΘA to the <square> on ΘΓ on ΘZ."

From Step 10 onwards, this scribe returned to copy *epi* as *epi*. The scribe was also confused by the first two occurrences of "cube" (Step 16): he rendered those cubes, κύβος/κύβον as "circles:" κύκλος/κύκλον. From Step 17 onwards, cubes were copied as cubes. Briefly then, while the proposition is textually corrupt, these corruptions are usually so transparent that Heiberg's corrections are very safe. I follow them without notice.

There is one interesting complication. In the sentence preceding Step 15, my translation has: "So I claim, as well, that the greater segment has to the smaller a ratio greater than half as large again the ratio of the surface." The manuscripts have, however: "So I say why the greater segment has to the smaller a ratio greater than half as large again the ratio of the surface." In terms of the Greek, the manuscripts have διότι where I translate δὴ ὅτι (following corrections in the manuscripts and Heiberg's text). In Steps 27 and 29 there is a very similar situation, the manuscripts having διότι which Heiberg corrects into δεῖ ὅτι. In these cases, however, he has at least the textual authority of Eutocius to follow. One could choose to keep the manuscripts' reading, and to develop a whole interpretation of Greek mathematical proof on its basis: it would be tempting to argue that the analysis is perceived as looking for the *reasons* for the truth of the claim. But it must be admitted that our text is in such a bad shape that it is better to stick, tentatively, to Heiberg's emendations.

Steps 11, 17, 22 and 26 are, in part or in whole, bracketed by Heiberg. The reason for bracketing them is that they look like elementary explanations, which may be reflections of Eutocius' commentary. I am doubtful, because it seems to me that whoever would be capable of reading such a proposition in an intelligent way, would hardly need to insert elementary clarifications at all. But, of course, it is impossible to prove that they are genuine, either.

GENERAL COMMENTS

Structural properties of the proof

In overall structure, this proposition is comparable to propositions such as *SC* I.23: the construction starts directly with lettered objects, with no general enunciation. This, while not standard in the Euclidean sense, is an *Archimedean* form, then: deviant, but in a recognized way.

A new formal departure occurs in Step 10, and is pursued through-out the proposition. In a footnote, I filled in Step 10 to say: "Therefore <(part of) the *demonstrandum* will be obtained by proving> that the <square> on ΘA on ΘH has to the <square> on ΓΘ on ΘZ a smaller ratio than the duplicate <ratio> of AΘ to ΘΓ." The underlying logical structure is this: First, Steps 3–9 have implicitly established that the ratio of the segments, "segment ΔAB to segment BΓΔ," is the same as the ratio "the <square> on ΘA on ΘH to the <square> on ΓΘ on ΘZ." Add to this that Steps 1–2 have implicitly established that the ratio of the surfaces, "surface ΔAB to surface BΓΔ," is the same as the ratio "AΘ to ΘΓ." Now, part of the *demonstrandum* was that the ratio of the segments is less than duplicate the ratio of the surfaces. So this can now be reformulated as in Step 10: "therefore that the <square> on ΘA, on ΘH, has to the <square> on ΓΘ, on ΘZ, a smaller ratio than the duplicate <ratio> of AΘ to ΘΓ."

The thrust of Steps 1–10 taken together is, in a sense, meta-geometrical. Their primary intention is not to show that some geometrical proposition, say P, is true. Their primary intention is to show that if some geometrical proposition, say P, is true, then so is another geometrical proposition, say Q. I repeat: Steps 1–10 make an assertion of the type P—>Q, rather than the type P. In this they resemble the analysis stage of geometrical problems. This does not show how to construct the required object, but shows, instead, that the required construction may be effected given certain prerequisites. And, similar to a geometrical analysis, the proposition goes on, from Step 12 onwards, as if it *assumed* the *demonstrandum*, showing its equivalent formulations – until finally such an equivalent formulation is found, which is obviously true (Step 14), where this part of the proposition comes to a halt. A similar, more explicit structure is followed in the remaining part of the proposition. In short: the proposition before us is best seen (as Eutocius already implied) as the analysis of a geometrical *theorem*.

One result of the "analytic" nature of the proof is that one is continuously asked to bear in mind not only the immediately proven result, but also the possible implications it carried with it. Besides the assertion which is proved, there is always the shadow of other, implied assertions. One feature of this proof is that, even outside the basic analysis structure, such shadow assertions are treated as real ones: to imply an assertion is the same as to prove it. Thus, for instance, Step 15 asserts that: ". . . the <ratio> of the segments was proved to be the same as that which the <square> on AΘ, on ΘH has to the <square> on ΘΓ on ΘZ" – a claim which was never stated earlier, and which can be obtained only as the implicit result of Steps 3–8 – no less than six steps are required to derive this result!

Extensions of plane geometry to solid figures

There is a set of interconnected practices original to this proposition:

First, a ratio is described as "half as large again" another ratio.

Second, in Step 3, a ratio is decomposed into three component ratios (and not two, as elsewhere in Greek mathematics).

Third, the expression "{area} *epi* {line}" is used, meaning something like "the multiplication of an area by a line."

These practices have the following in common:

First, all are extensions of established practices. The "half as large again ratio" is an extension of "duplicate ratio" and "triplicate ratio." The composition into three ratios is an extension of a composition into two ratios. The "{area} *epi* {line}" is an extension of the practice in calculations (where it is customary to speak of a "{number} *epi* {number}."

Second, all derive from a need to extend Greek mathematics into dealing with solids in unprecedented ways. Greek mathematics has as its main tool the application of proportion to geometry. This, however, happens mainly at the level of plane geometry. Book VI of the *Elements*, which provides the tools for this, deals only with plane geometry. Thus Archimedes must be innovative here.

Finally, all these innovations, by the very act of innovation, underline the arbitrariness of the borderline which they have crossed. Archimedes may not intend anything beyond an extension into solids, but it is open for the readers to conclude that further extensions are legitimate. The reader can now see the possibility of fractional ratios, many term compositions of ratios,[234] and the arithmetization of geometry.

/9/

The hemisphere is greater than spherical segments contained by an equal surface.

Let there be a great circle in a sphere, <namely> ABΓΔ, and its diameter AΓ, and another sphere, whose great circle is EZHΘ, and its diameter EH. And let them be cut by a plane: one sphere through the center, the other not through the center. And let the cutting planes be right to the diameters AΓ, EH, and let them cut <the great circles> at the lines[235] ΔB, ZΘ.

(1) So the segment of the sphere at the circumference ZEΘ is a hemisphere [(2) and, among the other sections[236] at the circumference

---

[234] Note that the combination of "fractional ratios" and "many term compositions of ratios" would allow the concept of a "positive rational power."

[235] Instead of the standard Greek abbreviation for "straight line," which is "straight <line>" (but which, to accommodate modern practice, I translate by "line"), we have here a very rare "<straight> line" so that, for once, my translation is literally correct.

[236] "Sections," here, mean the same as "segments." Perhaps the word is seen as needed to have both "segments" in the strict sense, as well as the hemisphere. (Possibly, that is, the limiting case of the hemisphere may not be considered a "segment" at all: notice the enunciation, that does not refer to the hemisphere being greater than "other" segments, but simply its being greater than segments – as if it was not one itself.)

BAΔ:[237] in the first figure (next to which is the sign ↻) it is greater than a hemisphere, in the other, it is smaller than a hemisphere], <the surfaces being> equal <to one another; I say>[238] that the hemisphere at the circumference ZEΘ is greater than the segment at the circumference BAΔ.[239]

(3) For since the surfaces of the said segments are equal, (4) it is obvious that BA is equal to the line EZ [(5) for the surface of each segment has been proved to be equal to a circle, whose radius is equal to the line drawn from the vertex of the segment on the circumference of the circle, which is the base of the segment.[240] (6) And since the circumference BAΔ in the other figure (next to which is the sign ↻), is greater

Eut. 364than half a circle] (7) it is clear that BA is, in square, smaller than twice AK,[241] (8) and greater than twice the radius.[242] (a) And also, let ΓΞ be equal to the radius of ABΔ, (b) and let MA have to AK that ratio which ΓΞ has to ΓK, (c) and let there be a cone on the circle around the diameter BΔ, having the point M as vertex; (9) so this <cone> is equal

Eut. 364to the segment of the sphere at the circumference BAΔ.[243] (d) Also, let EN be equal to EΛ, (e) and let there be a cone on the circle around the diameter ΘZ, having the point N as vertex; (10) so this <cone>, as

Eut. 365well, is equal to the hemisphere at the circumference ΘEZ.[244] (11) So, the <rectangle> contained by APΓ is greater than the <rectangle>

Eut. 365contained by AKΓ ((12) for the reason that it <=rect. APΓ> has the smaller side greater than the smaller side of the other<=AKΓ>),[245]

Eut. 365(13) and the <square> on AP is equal to the <rectangle> contained

Eut. 365by AK, ΓΞ; (14) for it is half the <square> on AB;[246] (15) Now, both

---

[237] Note that there are *two* such segments in the diagram.

[238] There is a lacuna in the text, which Heiberg filled in, following the *editio princeps*, in a slightly different way from the one printed here ("let them be equal . . ."). In general, for the lacunose nature of this proposition, see textual comments.

[239] Notice that from now on the argument refers only to the figure marked by ↻ (the "greater than hemisphere" case).

[240] *SC* I.42–3. One needs *Elements* XII.2 to derive Step 4 from Step 5.

[241] Anachronistically: $BA^2 < 2AK^2$.

[242] Anachronistically: $BA^2 > 2(radius)^2$. For both Steps 7 and 8 see Eutocius, whose argument relies on *Elements* I.47, II.12.

[243] Steps a–c, *Elements* V.18; and then *SC* II.2.

[244] See Eutocius, who uses *Elements* XII.10 and *SC* I.34 Cor. (the core result of *SC* I).

[245] See Eutocius, whose argument ultimately relies on *Elements* II.5. Where P is, we are supposed at this stage to know on the basis of the diagram (see textual and general comments).

[246] Step 14 is where the point P is *defined* (see textual and general comments). For the derivation of Step 13 from Step 14 see Eutocius, who uses Step b, *Elements* III.31, VI.8, 4.

taken together are greater than both taken together, as well [(16) therefore the <rectangle> contained by ΓAP is greater than the <rectangle
contained> by ΞKA].[247] (17) And the <rectangle contained> by MKΓ
is equal to the <rectangle contained> by ΞKA;[248] (18) so that ΓA has
to KΓ a greater ratio than MK to AP.[249] (19) But the ratio which AΓ
has to ΓK, is that which the <square> on AB has to the <square> on
BK.[250] (20) So it is clear that the half of the <square> on AB – which
is equal to the <square on ZΛ[251] – (21) has>[252] to the <square> on
BK a greater ratio than MK to twice AP, (22) which <=twice AP> is
equal to ΛN;[253] (23) therefore the circle around the diameter ZΘ, too,
has to the circle around the diameter BΔ a greater ratio than MK to
NΛ.[254] (24) So that the cone having the circle around the diameter ZΘ
as base, and the point N as vertex, is greater than the cone having the
circle around the diameter BΔ as base, and the point M as vertex;[255]
(25) so it is clear that the hemisphere at the circumference EZΘ, too,
is greater than the segment at the circumference BAΔ.

Eut. 366
Eut. 366
Eut. 366

Eut. 367

Eut. 367

Eut. 367

II.9
The arrangement of
codex D is somewhat
changed to that of the
thumbnail. All three
circles are also made
equal.   Codices E4
have the two left
figures better aligned,
one on top of the other;
for *lectio difficilior*
reasons I prefer, with
little certainty, the less
well-aligned GH.
The sign 𝖉 is omitted in
Codex A, while the
letter M is introduced
at the right end of the
line NH. It has also had
Z instead of Ξ, in the
top-left figure.
Codex 4 permutes Z/Θ.

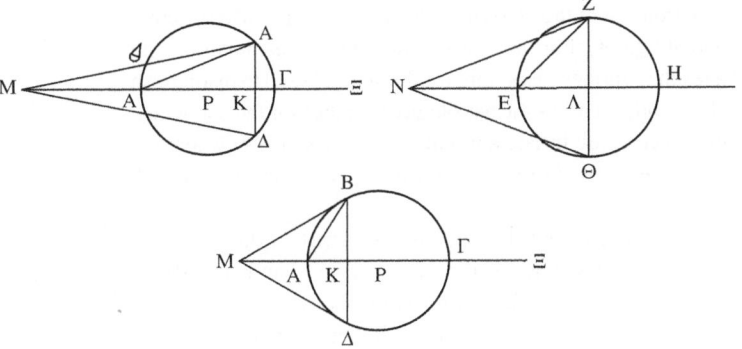

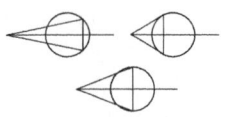

[247] See Eutocius, who uses *Elements* II.1, 3.

[248] The original structure is: ". . . to the <rectangle contained> by ΞKA, is equal
the <rectangle contained> by MKΓ." See Eutocius, who uses Step b, *Elements* V.18,
VI.16. The implicit result of Steps 16–17 is (17′) (rect. ΓAP) > (rect. MKΓ). This is
used by the following step (see comments).

[249] From 17′ (see n. 248), with an extension to inequalities of *Elements* VI.16.

[250] See Eutocius, who uses *Elements* III.31, VI.4, 8 Cor., 20 Cor. 2. The implicit result
of Steps 18–19 is (19′) (sq. AB):(sq. BK) > MK:AP. This is used in Steps 20–1.

[251] *Elements* I.47.

[252] There is a lacuna in the text, which Heiberg filled in, following the *editio princeps*,
in a different way from the one printed here (he has AP instead of my ZΛ). See textual
comments.

[253] For an explanation of the derivation of Step 22 from Steps 4 and 12, see Eutocius.
The argument uses Step d, as well as *Elements* I.47. The implicit result of Steps 20–2 is
(22′) (sq. ZΛ):(sq. BK) > MK:ΛN. This is used in the following step.

[254] See Eutocius, who uses *Elements* XII.2, V.8.

[255] See Eutocius, who uses *Elements* XII.14, 15.

TEXTUAL AND GENERAL COMMENTS

In this proposition, the text cannot be separated from the mathematics, and I run the textual and general comments together. The main question is the textual status and meaning of the "double figure:" the presence of two figures for the cases ("greater than a hemisphere" and "smaller than hemisphere"), labeled by the same letters and distinguished by the sign ☉ next to one of the figures (the "greater than hemisphere"). I first offer a preliminary discussion of this double figure. Then I move on to the general textual and mathematical nature of the proposition. Then I return to the double figure and offer a number of possible hypotheses.

The double figure, a preliminary discussion

Heiberg argued that only one of the two cases was drawn by Archimedes (the "greater than hemisphere"). From this it would follow that the sign ☉ was introduced by whoever added the second figure and that the references in the text to ☉ must all be late interpolations.

Heiberg's case is very strong. Such a meta-sign as ☉ is natural coming from a commentator, not from an author (Archimedes could simply use different letters for the different figures). Further, it is natural for a commentator to note the extension of a result to further cases. Finally, the symbol used to distinguish the figures (☉) is best understood as the astrological symbol for the sun which may have been introduced only in late antiquity (see Jones (1999) where the symbol is used in six papyri, five of which may be dated, all from the fourth century or later).[256]

There are two complications. First, the argument as stated here does not seem to hold for both cases. It holds only for the case "greater than hemisphere" (the one marked by ☉). So Archimedes' proof is invalid: he does not prove what he sets out to prove. Another complication is that Eutocius does refer to the sign ☉ as if it were part of the text of Archimedes. Thus there are two possibilities. One is that this reference in Eutocius' text to the sign ☉ was added by the same interpolator who added the same sign into Archimedes' text. The other possibility is that the sign was there before Eutocius had written his commentary. Heiberg opted for the first option, that the mention of ☉ in Eutocius' text was a late interpolation. If so, this is the only interpolation of its kind. (Clearly there are late interpolations that take Eutocius for granted, but these are not interpolations into *Eutocius'* text).

Heiberg's main argument ran like this. The text of the Archimedean manuscripts for the beginning of Step 7 is "it is clear that . . ." (referring backwards to Step 6), but Eutocius' quotation is "*And* it is clear that," which

---

[256] Datable are papyri 4142, 4168, 4272 (all fourth century), 4274, 4275 (early sixth century). Papyrus 4171 is a tiny fragment and thus can not be dated with certainty, but it is clearly from Ptolemy's Handy Tables and thus of course not earlier than the second century. The evidence for the presence of ☉ rather than some other sign is from Eutocius' text alone. (Codex A either omits or changes it for Archimedes.) The *Lecto difficilior* is, however, powerfully in its support.

does not seem to refer backwards to Step 6 but to be a local fresh start (because of the connector "*and*"). Heiberg therefore argued that Step 6 was an interpolation later than Eutocius, and with it goes the reference to δ in the manuscript of Archimedes which Eutocius seems to quote. This, however, rests on Heiberg's misleading assumption, that Eutocius' lemmas are quotations in the modern sense. The stronger probability is therefore that Eutocius' text of Archimedes contained the extra figure, the sign δ, and the steps referring to it. If this is an interpolation, then, it has to antedate Eutocius: this much is almost certain. This is a useful starting-point, because it leads to the following conclusion: that Eutocius thought that the Archimedes' proof, *as it is preserved in our manuscripts*, was valid (while, of course, standing in need of Eutocius' own many elucidations). This does not show anything conclusively: neither Archimedes nor Eutocius are infallible. But it is a useful starting-point when studying the mathematics of the argument. The argument, as it stands, should be right.

Another preliminary point is that Archimedes understood well that the cases of the "greater than hemisphere" and "smaller than hemisphere" may call for separate proofs. *SC* I.42–3 are two propositions offering just this set of two cases. It is also clear that the "smaller than hemisphere" case was not a more complex case, beyond the reach of Archimedes' mathematical techniques. So the decision to focus on just one case becomes problematic. I shall later on try to see if the general validity of this proof can be shown on the basis of the proof as given. But first, we must widen our understanding of the textual position.

Textual and mathematical problems other than the double figure

This proposition is characterized in general by a strange combination of the over-explicit and the under-explicit. It is over-explicit in that it is, linguistically, less elliptic than usual. Formulaic expressions are spelled out, with nouns actually inserted instead of just being implied by their article. Thus we have, in Step 11, "the <rectangle> contained by" (instead of the more common, in such advanced contexts, 'the <rectangle contained> by"). Similarly, in Steps c and e, a point is explicitly called a "point;" in Step 4, a line is explicitly called a "line." Altogether, I count in this proposition at least seven such "redundancies." Another, less clearly "redundant" case worth mentioning is the formula "the segment of the sphere at the circumference ABC." This is used elsewhere – in *SC* II.6, Step a – but there, Archimedes often uses various abbreviations for this cumbersome formula, whereas here the same formula is repeated eight times.[257]

---

[257] There is another redundancy yet, at a purely grammatical level. In the Greek formula for ratio "as the <line> AB to the <line> CD" it is customary to drop the second article (since the preposition *pros* already settles the case of the following lettered object, and there is no need for an article). In this text, the article is kept five out of six times. In three places, Heiberg brackets this article, following Eutocius, but clearly Eutocius can not be used as a textual guide on such details.

On the other hand, the text is elliptic to an unprecedented extent at the level of logic. Some crucial steps of the argument are left out.

First, the end of Step 2 and the beginning of the definition of the goal are printed by me as: "<the surfaces being> equal <to one another; I say> that the hemisphere . . ." that is, the manuscripts have: "equal that the hemisphere . . ." (which is a simplification: my English here departs a little from the Greek, which, however, is clearly false). Heiberg, following the *editio princeps*, printed (I translate his Greek): "And let the surfaces of the segments be equal; now I say, that the hemisphere." He might be right, in which case what we see is textual corruption leading to the loss of a whole sentence. Or, alternatively, we have here a very non-standard expression. At any rate, the central fact of the diagram – that the segments have equal surfaces – was delegated to a late moment of the construction and, perhaps, was left completely tacit.

Second, as I mention in a note to Step 12, the identity of the point P is not set out explicitly in the text. It is customary in Greek mathematics to have the qualitative position of a point decided by the diagram. However P must be defined quantitatively, as the point satisfying $(sq. AB)^2 = 2*(sq. AP)$, or (equivalently) $AP = E\Lambda$. It is hard to think of many other cases in Greek mathematics where such quantitative relations are left to be seen through the diagram.[258] Heiberg, following the *editio princeps*, considered this a lacuna in the text, and supplied a step of the construction following Step 8, "So let BA be twice AP in square." (This is a plausible emendation of the text, especially since an homoeoteleuton might explain the lacuna: in Heiberg's text as printed, these are the words δυνάμει ἔστω in 224.6 and 224.7).

Third, as I explain in notes to Steps 17, 19 and 22, the argument moves onwards in these three cases not on the basis of assertions made in the text, but on the basis of implicit logical extensions of the text. This in itself is not exceptional: we have seen similar implicit arguments before, especially in the preceding, alternative proof to Proposition 8. However, in the context of this textually difficult proposition, it is more tempting to ascribe such gaps in the argument to gaps in the text. In one case (that following Steps 16–17) Heiberg, following the *editio princeps*, prints a step which is not in the manuscript, supplying the supposed lacuna.

Finally, there is no question that the text of Step 20 must be lacunose. There are four missing words (which might be explained by the short homoeoteleuton ἀπό: Heiberg 226.20). My emendation differs from Heiberg slightly, but some emendation is necessary.

What we see therefore is that, on the level of linguistic expression, the text is redundant while, on the level of mathematical argument, the text is lacunose. And this returns us to the main difficulty of the proposition. There

---

[258] Of course it is impossible to see the complex relation $(sq. AB)^2 = 2*(sq. AP)$ in any diagram, but conceivably $AP = E\Lambda$ could be visualized within certain diagrammatic practices. But not in Greek mathematics (where quantitative relations are not shown in diagrams), and not with these particular diagrams (the three figures are not set out to allow such an inspection).

is a redundancy at the level of the signs used: a double figure, with a special extra sign next to one of the figures. And there is also a logical lacuna: the text proves for only one case, while two are required. I shall now discuss this logical lacuna, before offering a conclusion.

How can the proof be seen to apply to both cases?

When we say that the proof does not apply directly to both cases, what we mean is that some assertions in the text do not apply to both cases. This is true only of a single sequence of assertions, Steps 6–8: "... (6) And since the circumference BAΔ in the other figure (next to which is the sign ♂), is greater than half a circle] (7) it is clear that BA is, in square, smaller than twice AK, (8) and greater than twice the radius."

Heiberg thought Step 6 had been interpolated. This leaves Archimedes in the uncomfortable position of stating Steps 7–8 although they are in fact *wrong*: given the way in which B, A and K were defined (assuming the absence of the ♂ layer of the text), assertions do not apply, unless one adds a gratuitous assumption (that we deal only with the "greater than hemisphere" case). However, assuming that Heiberg was wrong and that something like Step 6 was written by Archimedes, this still does not get Archimedes off the hook completely. The problem now is that Archimedes makes an assertion which is true for one case, without noting that, in the other case, the assertion no longer holds. Since the argument later relies upon Steps 7–8, this means that the argument is only valid for one case. So there are two options: either Steps 7–8 are false because they are based on a gratuitous assumption, or the whole proof is false because it is based on Steps 7–8, known to hold for only a single case.

To study the second option, we must see *how* the proof is dependent upon Steps 7–8. These steps are used, implicitly, in Steps 11–12. As pointed out by Dijksterhuis in his analysis of the theorem (Dijksterhuis (1956) 219–21), the "nucleus to the proof" is this implicit argument leading up to Step 11. Once Step 11 is obtained, we need no longer look back, and from Step 11 onwards, the proof never relies on the identity of the case and on Steps 7–8. "(11) So, the <rectangle> contained by APΓ is greater than the <rectangle> contained by AKΓ (12) for the reason that it <=rect. APΓ> has the smaller side greater than the smaller side of the other<=AKΓ>."

This can be seen along the following line of reasoning:

(A) it is understood that P is defined by $2*(AP)^2=(AB)^2$ (see preceding discussion).

(B) Steps 7–8 together provide $2*r^2<(BA)^2<2*(AK)^2$ (I use "r" to refer to the radius of the circle ABΓΔ, whose diameter is AΓ. In other words, r is half AΓ).

(C) Using A, we can substitute $2*(AP)^2$ for $(AB)^2$ in the expression (B), yielding: $2*r^2<2*(AP)^2<2*(AK)^2$, or $r^2<(AP)^2<(AK)^2$, or $r<AP<AK$.

(D) In geometrical terms, (C) amounts to stating that the line AP is longer than the radius, but shorter than the line AK. That is, starting from the point A along the line AΓ, we first reach the center (the shortest line of the three – the radius), we then reach the point P (the middle line – AP), finally reaching the point K (the longest line – AK). Or to put it differently: P is between K and the center.

(E) Step 11 claims that (rect. AΡΓ) > (rect. AKΓ), and this is a direct result of (D) above, together with a simple application of *Elements* II.5. The general rule can be stated like this: *if a line* (in this case, AΓ) *is cut at two points* (in this case, K and P), *and rectangles are constructed on the segments produced by the cuts* (in this case, AΡΓ and AKΓ), *then the rectangle produced by the cut nearer the center will be greater* (in this case, AΡΓ>AKΓ).

To sum up so far: Step 11 and with it the entire proposition are true because P falls between K and the center, and this in turn is based upon the inequalities stated at Steps 7–8 which, in turn, rely upon this being the case "greater than hemisphere."

This then seems to be a gross mistake (always, assuming that the proposition is intended to apply for both cases). However, as suggested already, we have one real clue to the proof, and this is that Eutocius thought that it was a valid one. I therefore quote now from his comment on Steps 7–8:

"For, a <line> being joined from B to the center, with the resulting angle <subtended> by BA being obtuse, the <square> on AB is greater than the <squares> on the <lines> containing the obtuse <angle>,[259] which are equal[260] . . ." (in what follows, Eutocius deduces easily Steps 7–8, i.e. $2*r^2 < (BA)^2 < 2*(AK)^2$). ". . . And these hold in the case of the figure, on which is the sign ϐ, while in the other figure the opposite may be said correctly."

In other words, Eutocius says that in the "smaller than hemisphere" case he could make the same comment, simply inverting everything. This is true. In the "smaller than hemisphere" case the angle will be *acute* instead of *obtuse* and so the square on AB would be *smaller* than the sum of the squares,[261] and so on, until we finally reach the opposite inequality: $2*r^2 < (BA)^2 > 2*(AK)^2$.

This is not only a true remark, but also a pertinent remark, showing an understanding of the function of Steps 7–8. For from the inequality $2*r^2 > (BA)^2 > 2*(AK)^2$ we may, along the lines of (C) above, have the derivation: $2*r^2 > 2*(AP)^2 > 2*(AK)^2$, or $r^2 > (AP)^2 > (AK)^2$, or finally r>AP>AK, and so, once again, P is between K and the center, *only on the other side*. We switch sides but the crucial feature for the proof remains the same. Thus Eutocius could have convinced himself that the proof was generally valid.

It remains to ask, how did *Archimedes* convince himself of the general truth of Step 11. After all, from the point of view of his own text, the proof seemed to rely on Steps 7–8, that is on one case only. However, the Archimedean text has a further clue for the argument underlying Step 11. And this is Step 12: ". . . (12) for the reason that it <=rect. AΡΓ> has the smaller side greater than the smaller side of the other<=AKΓ>."

---

[259] *Elements* II.12 (which Eutocius does not quote explicitly). Calling the center X, we have $(AB)^2 > (AX)^2 + (XB)^2$.

[260] Both are radii.

[261] *Elements* II.13 instead of II.12 – but remember that Eutocius did not quote II.12 explicitly, so the "opposite" claim is still literally true. We do not need to make any other change in the form of the argument, besides inverting the terms.

This is the explicit argument for Step 11: if you have four lines on isoperimetric rectangles, a, b, c, d, so that a>b>c>d, it will also be true that b*c>a*d.[262] In this case we have AK>AP>PΓ>KΓ, from which AP*PΓ>AK*KΓ. Thus Step 11 derives from Step 12. But Archimedes does not show us how the general rule of Step 12 holds in the specific case of Step 11, and sends us to look ourselves to prove the inequality between the specific lines. This we can do through Steps 7–8 – but only indirectly. Archimedes sends us to prove a complicated result.

Yet, finally, what *does* Archimedes send us to prove? What is the *demonstrandum* implied by Steps 11–12 taken together?

We are now very near the end of the first volume of this translation, so this is an appropriate moment to offer a libation for the divine genius of Archimedes. Dear reader, Archimedes does not send us in Step 12 to prove that AK>AP>PΓ>KΓ. He sends us to prove *that either* AK>AP>PΓ>KΓ *or* AK<AP<PΓ<KΓ. His text is clear:

". . . (12) for the reason that it has the smaller side greater than the smaller side of the other."

The smaller side – *whichever that may be*. In the "greater than hemisphere" case, the smaller side happens to be KΓ; in the "smaller than hemisphere" case, the smaller side happens to be AK,[263] but Step 12 is deliberately indeterminate about the choice of cases. Thus the crucial moment of the proof – Step 11 – is proved generally, for it is proved through Step 12 which does not assume, as stated, one case or the other.

CONCLUSION

The above is, I believe, a nice story, which leaves a number of questions unanswered. Eutocius clearly understood the basic structure of the general proof, but he was not explicit about it in the way I have been above. He may have felt that this imparity between the cases is, after all, a strange aspect of the proof, better left unmentioned. And in fact this is a strange feature of the proof as it is stated in the manuscripts. If we have Archimedes' text as he has written it, then we can say that Archimedes, while certainly not mistaken himself about the generality of his proof, still produced a very misleading argument – which he may well have done, even intentionally. It is clear that towards the end of this treatise, he moved on to unprecedented heights of complexity and originality: the alternative proof to 8, this proposition, and finally the (since lost) lemma to *SC* II.4. This is not for the outsider, for the trespasser; traps may be laid.

This is all assuming that Archimedes wrote the proposition roughly as we have it. This I find implausible. We saw that the proposition has more textual difficulties than usual, and that those difficulties form a pattern: redundancy at the level of form, lacunosity at the level of content. This might be explained

---

[262] In this case we can derive a>b from c>d, since a+d=b+c=(the diameter of the circle).

[263] As can be seen from Eutocius' comment on Steps 7–8.

in many ways: for instance, that the text could have been *restored* (this does come at the end of the roll; a following piece of text was even lost from the main tradition!). A damaged text, repaired by a reader with modest skills in mathematics, should have the textual pattern described above. In the micro-level of presentation, such a scribe will tend to be less elliptic than the practicing mathematician. Such a scribe will use fuller forms throughout, when filling in gaps in the papyrus. On the other hand, such a scribe will be unable to reconstruct the more complex arguments. We may even imagine that such a scribe, with his limited (though not non-existent) mathematical skills, might reconstruct a damaged diagram, supplying the double figure and some notes related to it. Another obvious possibility suggests itself: that Steps 7–8 were not alone in Archimedes' original. Perhaps he went on to state explicitly the correlate to them in the "smaller than hemisphere" case, and this, again, is a lacuna in our text.

To sum up, there is a range of possibilities. While it is probable that some "restoration," whatever that may have been, was done on this proposition, its extent is unclear. It may have included the double diagram and especially the sign δ, but these may also conceivably have been inserted by Archimedes himself (I do not think this is likely). In the latter case, this is another example of the way in which Archimedes breaks away, towards the end of this treatise, from Greek mathematical conventions. And while it is clear that much is missing in the text, it is possible that the text never had a correlate of 7–8 for the "smaller than hemisphere" case (I think this is likely). If so, this would be another example of the way in which Archimedes plays with his readers, a game that is almost malicious: proving, in fact, much more than he seems to do; perhaps inviting criticisms he knows he may refute at ease. What an achievement, in the Greek game of proofs and refutations!

EUTOCIUS' COMMENTARY TO *ON THE SPHERE AND THE CYLINDER* I

TO BOOK 1

As I found that no one before us had written down a proper treatise on the books of Archimedes *on Sphere and Cylinder*,[1] and seeing that this has not been overlooked because of the ease of the propositions (for they require, as you know, precise attention as well as intelligent insight), I desired, as best I could, to set out clearly those things in it which are difficult to understand; and I was more led to do this by the fact that no one had yet taken up this project, than I was deterred by the difficulty; as I was also reasoning in the Socratic manner that, with god's support, most probably we shall reach the end of my efforts. And third, I thought that, even if, through my youth, something will strike out of tune, this will be made right by your scientific comprehension of philosophy in general, and especially of mathematics; and so I dedicate it to you, Ammonius, the best of philosophers.[2] It would be fitting that you help my effort. And if the book seems to you slight, then do not allow it to go from yourself to anyone else, but if it has not strayed completely off the mark, make your view upon it clear for, if it comes

---

[1] Following the requirements of his genre (see Mansfeld [1998]), Eutocius begins with a survey of the literature – non-existent, in this case.

[2] A Neo-platonist philosopher, a pupil of Proclus at Athens (A.D. 412–485), he taught at Alexandria and especially wrote commentaries on Aristotle. He also wrote a (lost) commentary on the slight arithmetical treatise of Nicomachus. To write commentaries on Platonist/Aristotelian treatises, with excursions into elementary mathematics, was the standard among Neo-platonists. Eutocius, evidently a Neo-platonist himself in some sense, deviates from this pattern, in that he goes into much more complicated mathematics.

to be established by your own judgment, I shall try to explicate some other of the Archimedean treatises.

To the definitions

After stating the theorems that he is about to set out, he follows the custom of all geometers in their exposition and tries to clarify at the start of the work those expressions, which he himself used in his own fashion, and the terms of the hypotheses and the hypotheses themselves; and he says first that "there are in a plane some curved lines, which are either all on the same side as the straight lines joining their limits or have nothing on the other side."[3] The assertion will be clear if we realize which lines he calls "curved lines in a plane." Now, it should be known that he calls "curved lines" not simply the circular or the conical or those which have continuity without breaking, but he terms "curved" any line in a plane, without qualification, which is other than straight; any single line in a plane, compounded in whatever way, so that even if it is composed of straight lines . . .[4]

Arch. 35

---

[3] Here is a crucial textual question that I will discuss once and for all. In his quotations from Archimedes, Eutocius does not exactly follow Archimedes' text. How to account for this? I shall focus on this definition. The differences between Eutocius' text and Archimedes' text are: 1. Eutocius, in quoting, changes the syntactic structure of the original, turning the original independent clause into a dependent clause (a difference more marked in the original Greek). 2. Eutocius has "curved lines" for Archimedes' "limited curved lines." 3. Eutocius has "all" for Archimedes "wholly." 4. A few minor differences, of no mathematical significance: Eutocius has αἵτινες where Archimedes has αἱ; the ἤτοι in Archimedes becomes ἤ in Eutocius; the word order in the phrase "the straight lines joining their limits" differs slightly between the two versions.

Point 1 shows that Eutocius is willing to change the original text while incorporating it into his own discursive prose. This is no direct quotation. Point 2 shows that, here at least, Eutocius did not use an earlier, better preserved Archimedean text. For while the word "limited" could have been lost, perhaps, in some copying of the text, it could not have been *added*. This is not a standard epithet; it would not just be added by a copier. Hence either Eutocius did not set out to copy his text *verbatim*, or he has a text that is, at this particular point, inferior to ours. To sum up, the working assumption employed in this translation is that Eutocius' work is a mathematical, not a textual commentary, and that Eutocius is not interested in *verbatim* quotations. Naturally, from time to time, differences between Eutocius' and Archimedes' formulations may be the result of textual corruption in either. My suggestion is not that textual corruption does not happen, but that it is not the main reason for differences between Archimedes' and Eutocius' texts.

[4] There is a large lacuna in the text here (some early scribes report that a whole page was missing from their source). Eutocius explicated at length Archimedes' use of "curved lines," and went on to discuss the expression "on the same side" and its application for the definition of "concave in the same direction." Only the diagram for

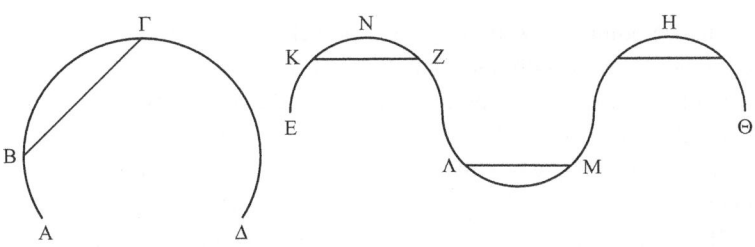

In Def. I
DH had the EKNZ arc
greater than a
semicircle, the ΛM arc
smaller than a
semicircle (H further
had the HΘ arc greater
than a semicircle).
Perhaps they should
have been followed.
Codex D has been
somewhat
miscalculated, so that
the arc HΘ reached the
end of the page
prematurely, the letters
HΘ consequently
pushed rightwards. He
further had BΓ less
tilted to the left – as if
to compensate, it then
had KZ slightly tilted
to the left.   A has
been omitted on codex
A (reinstated on BD).
Since the diagram
accompanies a lost
piece of text, we cannot
say for certain that BD
were indeed right.

... to ABΓΔ. But since, as was already said above, he calls "curved lines" not only the circumferences of curved figures, but also those which are composed of straight lines, and it is among these that lines concave in the same direction are identified, it is possible to take, on some line, concave in the same direction, two chance points, so that the line joined to them will fall on neither side of the line, but will

Arch. 35

coincide with the line itself. Therefore he says that he calls "concave in the same direction, a line in which the straight lines drawn between any two points whatever, either all fall on the same side of the line, or some fall on the same side, and some on the line itself, but none on the other side." And the same can be understood for surfaces as well.

Then in the following he gives the names "solid sector" and "solid rhombos," clearly explaining the concepts for these names.

Following this he sees fit to make some postulates, useful to him for the following proofs, which, while agreeing in themselves with perception, are no less capable of being proved, too, from the common notions and the results in the Elements.[5]

Arch. 36

The first of the postulates is the following: "of all the lines which have the same limits, the smallest is the straight."

For let there be in a plane some limited line, AB, and some other line, AΓB, having the same limits, A, B. So he says that it is given to him[6] that AB is smaller than AΓB. Now, I say that this was postulated while being *true*.

(at least part of) this discussion is preserved, and it appears that, among other things, Eutocius had pointed out that the lines KZ, MΛ are not on the same side of ENHΘ.

[5] Apparently, Eutocius suggests that Archimedes merely chose to put as postulate what, in principle, could have been *proved*. This reflects Eutocius' understanding of the nature of "postulates" (that they are not absolute unprovables but are merely assertions which are, locally to a given treatise, left without proof). It is likely (but not necessary) that Eutocius misunderstood Archimedes' intention and that Archimedes had a stronger grasp of the logical possibilities, seeing that the postulates are independent of other standard assumptions.

[6] "Given to him" = "he gets this for free." In other words: there is no need to offer a special argument, as the conclusion follows from a postulate. To postulate – to "require" – is to ask for something to be given, in this special sense.

(a) For let a chance point be taken on AΓB, <namely> Γ, (b) and
let AΓ, ΓB be joined; (1) so it is obvious that AΓ, ΓB are greater than
AB.[7] (c) Once again, let other chance points be taken on the line AΓB,
<namely> Δ, E, (d) and let AΔ, ΔΓ, ΓE, EB be joined. (2) So here,
too, similarly, it is clear that the two <lines> AΔ, ΔΓ are greater than
AΓ; (3) and the two <lines> ΓE, EB <are greater> than ΓB. (4) So
that AΔ, ΔΓ, ΓE, EB are greater by much than AB. (5) So similarly,
if by taking other points between those already taken, we join straight
lines to those lines which were taken before, we shall find that those
lines are even greater than AB, (6) and doing this continuously we shall
find the closer straight lines to the line AΒΓ to be even greater. (7) So
that it is evident from this that the line itself <=AΒΓ> is greater than
AB, (8) since it is possible to take a line, by joining straight lines on
every point of the line itself <=AB>, which is composed of straight
lines and is alike to the line itself, (9) and which is proved, through the
same arguments, to be greater than AB; for, in the proofs of agreed
things, there is nothing absurd in adding such conceptions as well.[8]

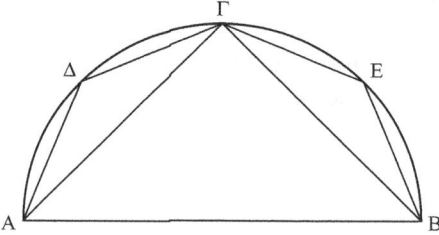

In Def. II
Codex D inserts a
semi-diameter going
down from Γ.

Arch. 36   Following this he says that he postulates also that "among those lines
which have the same limits, those are unequal which are concave in the
same direction" (in the way mentioned above).[9] But being concave in
the same direction alone does not suffice for their being unequal, but
also "when either: one is wholly contained by the other, or a part is
contained, and a part it has common" (and the container is then greater
than the contained).

For let there be imagined – so that this, too, will be made manifest –
two lines in a plane, AΒΓΔEZ and AHΘZ, having the same limits A, Z
and concave in the same direction, and yet again, that AHΘZ is wholly
contained by the line AΒΓΔEZ and <by> the straight line having the
same limits as themselves. Now I claim that: the lines set forth are
unequal; and the container is greater.

---

[7] *Elements* I.20.

[8] Eutocius says that his proof is invalid as such, but may be used since the truth of the
conclusion is not in doubt. It is unclear whether he was hoping for a more valid proof
when he asserted, initially, that the postulate can be proved.

[9] Must be a reference to Eutocius' explication of "concave in the same direction"
which we lost in the lacuna.

(a) For let BΘ, ΓZ, ΔZ be joined. (1) Now since, if a line ΘA is imagined joined, AH, HΘ will then be constructed internally on one of the sides of ABΘ – (2) therefore AH, HΘ are smaller than AB, BΘ.[10] (3) Let ΘZ be added <as> common; (4) therefore AH, HΘ, ΘZ are smaller than AB, BΘ, ΘZ. (5) But BΘ, ΘZ are smaller than ΒΓZ ((6) for, again, they are constructed internally on one <side> of ΒΓZ); (7) therefore AB, ΒΓ, ΓZ are greater by much than AH, HΘ, ΘZ. (8) But ΓΔ, ΔZ are greater than ΓZ; (9) and ΔE, EZ are <greater> than ΔZ; (10) therefore ΑΒΓ, ΔΕΖ are much yet greater than ΑΗΘΖ.

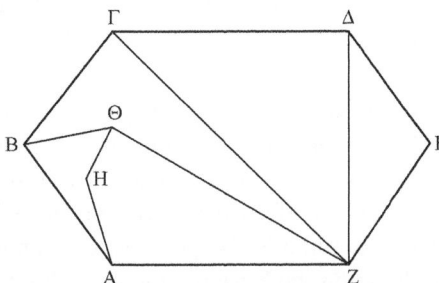

For the sake of clarity, let other lines, too, be hypothesized (similarly to those mentioned above) such as ΑΒΓΔΕ, ΑΖΗΘΚΕ. I say that the container is greater.

(a) For let AZ, HΘ be imagined to be produced to Λ. (1) Now again, since the two ZΛ, ΛH are greater than ZH, (2) let AZ, HΘ be added <as> common; (3) therefore ΑΛ, ΛΘ are greater than AZ, HZ, HΘ. (4) But ΑΛ, ΛΘ are smaller than ΑΒΘ. (5) Therefore ΑΒΘ are by much greater than ΑΖΗΘ. (6) Let ΘΚ be added <as> common; (7) therefore ΑΒΘΚ are greater than ΑΖΗΘΚ. (8) But ΒΘΚ are smaller than ΒΓΚ. (9) Therefore ΑΒΓΚ are by much greater than ΑΖΗΘΚ. (10) Let ΚΕ be added <as> common; (11) therefore ΑΒΓΚΕ are greater than ΑΖΗΘΚΕ. (12) But ΓΚΕ are smaller than ΓΔΕ; (13) therefore ΑΒΓΔΕ are greater by much than ΑΖΗΘΚΕ.

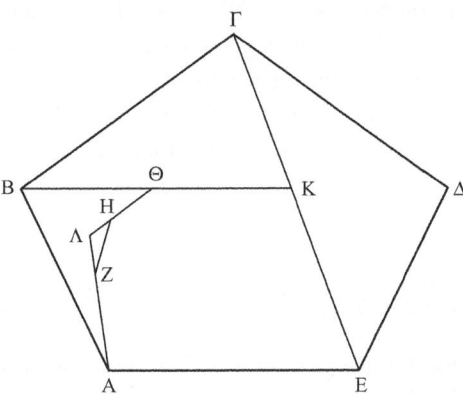

In Def. III
Codex D has the line AH perfectly vertical, and the line BΘ perfectly horizontal. It also (not unrelated) has AZ greater than ΓΔ. This is one of the very rare cases where both codex B, as well as Heiberg's edition, are virtually identical to codex A.

In Def. IV
Codex B has BΘΚ as two separate lines, Θ higher than both B and K (the two in the same height). Codex D has E higher than A; G has A higher than E. Codex D has omitted K.

---

[10] *Elements* I.21. This and *Elements* I.20 are the only tools of this argument.

And if they be circumferences,[11] either the container or the contained or even both, the same can be understood. For taking on them <=on the given curved lines> continuous points, and with straight lines being joined to these <points>, then there shall be taken lines composed of straight lines – to which applies the proof above – as the <lines> composed of straight lines come to be alike the lines set out originally – through its being considered, too, that every line has its existence in a continuity of points.[12]

A further point: he did right, in that he did not characterize the inequality of the lines just by their being concave in the same direction; instead, he added that the one must also be contained by the other and by a line having the same limits:

For if this is not so, nor would the inequality of the lines always hold true, as can be perceived in the attached diagrams. For the line ABΓΔ, and AEZΔ, having the same limits, are also concave in the same direction, and it is not clear which of the two is greater; indeed, it is possible that they are equal, too. But it is also possible to imagine each as concave in the same direction, both having the same limits, but set in a position opposite to each other, as each of the said lines is to AHΘKΔ[13] – for in this case, too, their equality or inequality is not clear. Therefore he set forth well, "that the one must be wholly contained by the other and by a line having the same limits, or that some will be contained, and some it will have <as> common," as in AHΘKΔ and AΛMNΞΔ; for in these some is contained, but some is common, namely AΛ, MN.

And it was quite of necessity that this, too, was added in for the sake of the judgment of inequality: that it is necessary that the lines have the same limits; for if this is not so then, even if they may be contained one by the other, neither will they all be unequal, but in some cases equal, or the contained may even be greater. So that this shall be made clear, let two lines be imagined in a plane, ABΓ, containing an obtuse angle, that at B, and let a chance point be taken on BΓ, <namely> Δ, and let AΔ, AΓ be joined. (1) Now since AΔ is greater than AB, (a) let ΔE be set equal to AB, (b) and let AE be bisected at Z, (c) and let ZΓ be joined. (2) Now since the two AZΓ are greater than AΓ,[14] (3) and AZ is equal to ZE, (4) therefore EZΓ, too, are greater than AΓ. (5) Let

---

[11] "Circumferences" here mean any curved lines.

[12] Eutocius repeats the argument used in his proof of his Postulate 1, and once again his unease is palpable.

[13] Eutocius rightly perceives that two lines may each be "concave in the same direction," the "same direction" being different in each case.

[14] *Elements* I.20.

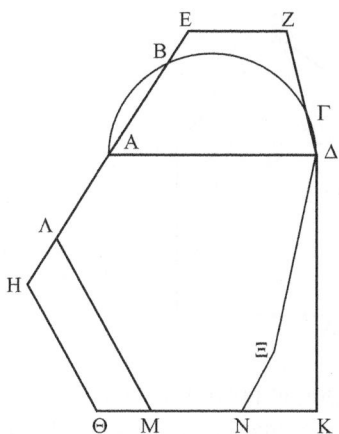

In Def. V
Codex D has the arc
greater than a
semicircle, so that the
line ΓΔ is more clearly
inside it.   Codex G
has the lines EA, AH
separate (AH tilted
upwards at H).

AB, ΔE be added <as> common; (6) therefore ΔZΓ are greater than
BAΓ. (7) So that with one line, BAΓ, imagined concave to the same
direction, and another, ΔZΓ, contained by the other and not having the
same limits; it was proved not only that the container is not greater, but
also that it is smaller.

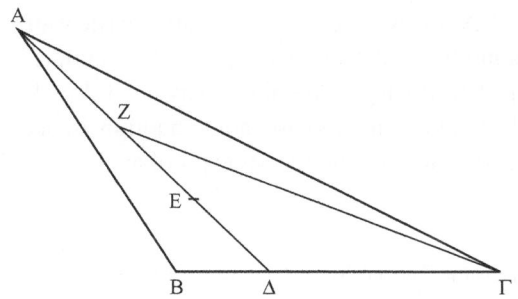

In Def. VI
Codex D has the angle
at B nearly right.
Codex 4 has the three
segments AZ, ZE, EΔ
nearly equal.

And it is possible to see the same thing in lines composed of several
straight lines. (a) For let there be imagined in a plane two straight lines,
ABΓ, (b) and a chance point, Δ, and AΔ, joined. (c) So again, let ΔE
be set equal to AB, (d) and let EA be bisected by Z, (e) and let AH be
drawn at right <angles> to AΔ; (f) and let ZH be joined; (g) and let
ZΘ be set equal to AH, (h) and again let ΘH be bisected at K, (i) and
let HΛ be drawn at right <angles> to ZH, (j) and let KΛ be joined; (k)
and again <let> KM <be> equal to HΛ, (l) and let MΛ be bisected
by N, (m) and again let ΛΓ be drawn at right <angles> to KΛ, (n) and
let NΓ be joined. (1) Then it is obvious, through what has been proved
above, (2) that ΔZ is greater than AB, (3) ZK than AH, (4) KN than
HΛ, (5) and NΓ than ΛΓ; (6) so that the whole line ΔZKNΓ is greater
than BAHΛΓ.

Therefore he did well in adding, for unequal <lines>, that they have
the same limits.

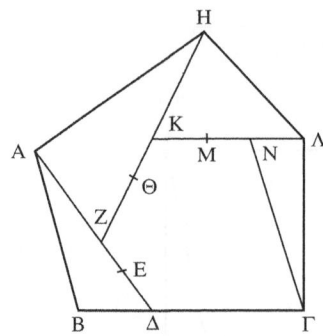

In Def. VII
Codices DE, followed
by Heiberg, have the
angle at Γ obtuse.
Codex D further
slightly tilts the whole
figure
counterclockwise.
Codex E has Δ instead
of Λ, omits E.

It is possible, with some thought, to prove the same things for surfaces as well, concerning all that was mentioned above, if the surfaces
taken have the limits in planes.[15]

To Theorem 2

Arch. 43   "And ΓA being added onto itself will exceed Δ." Clearly, if AB is
either a superparticular of Δ, or even some chance superpartient of
it.[16] But if AB is either a multiple of Δ or a superparticular-multiple,
then subtracting BΓ (equal to Δ) from AB, the remainder ΓA will
exceed Δ, so that it will be required, in this case, not to multiply it, but
to set out AΘ right away, equal to AΓ, and the same proof applies.

---

[15] It is interesting that, apparently, Eutocius had nothing to say on "Archimedes'
axiom."

[16] The terminology used here is contained in Nicomachus' *Arithmetic*, on which
Ammonius, Eutocius' addresses, wrote a commentary. The use of the terminology may
therefore be understood as a gesture of respect towards Ammonius and his tradition, and
has little to do with Archimedes.

Nicomachus classifies integer ratios into five classes, in ascending complexity: multiples (of the form n:1, "twice" or more), superparticular (of the form $\frac{(n+1)}{n}$, e.g. 3:2.
Note that all superparticulars are smaller than "twice"), superpartiens (etymologically,
ratios which are greater than unity by a certain number of parts of unity, but less than
by unity itself, i.e. they are still less than "twice." Effectively, superpartiens ratios are
all integer ratios bigger than unity, smaller than twice, which are not superparticulars),
superparticular-multiple (instead of the form $\frac{(n+1)}{n}$, as superparticulars are, these are of
the form $\frac{(mn+1)}{n}$, e.g. 5:2), and superpartiens-multiple (effectively, all remaining integer
ratios greater than "twice").

This classification ill befits Eutocius' purposes. He required a distinction between
those ratios that are not smaller than twice, and those that are (if AB:Δ is not smaller
than twice, ΓA must be multiplied to exceed Δ; otherwise, it exceeds it straight away).
Nicomachus' system is too fine-grained, and, concentrating as it does on integer ratios,
it is inappropriate to the geometrical ratio of this proposition.

Arch. 44   "And compoundly, ZE has to ZH a smaller ratio than AB to BΓ." For
it should be proved as follows that if a first has to a second a smaller
ratio than a third to a fourth, then, compoundly, too, the same ratio
follows:[17]

Let there be four magnitudes AB, BΓ, ΔE, EZ, and let AB have to
BΓ a greater ratio than ΔE to EZ. I say that compoundly, too, AΓ has
to ΓB a greater ratio than ΔZ to ZE.

(a) For let it come to be: as ΓB to BA, so ZE to ZΘ. (1) Therefore
inversely: as AB to BΓ, so ΘZ to ZE.[18] (2) But AB has to BΓ a greater
ratio than ΔE to EZ; (3) therefore ΘZ, too, has to ZE a greater ratio
than ΔE to EZ. (4) Therefore ZΘ is greater than EΔ[19] (5) and the
whole ΘE <is greater> than ΔZ, (6) and through this ΘE has to EZ
a greater ratio than ΔZ to ZE.[20] (7) But, as ΘE to EZ, AΓ to ΓB,
(8) through the compounding;[21] (9) therefore also AΓ has to ΓB a
greater ratio than ΔZ to EZ.

But then, let AΓ have to ΓB a greater ratio than ΔZ to ZE. I say that
dividedly, too, AB has to BΓ a greater ratio than ΔE to EZ.[22]

(1) For again, similarly, if we make: as BΓ to ΓA, so ZE to EΘ,
EΘ will be greater than ΔZ.[23] (2) And subtracting EZ <as> common,
(3) ΘZ will be greater than ΔE, (4) and through this ΘZ will have to
ZE, that is AB to BΓ[24] ((5) through the division[25]) (6) a greater ratio
than ΔE to EZ.[26]

And it is clear through similar <arguments> that, when AB has
to BΓ a smaller ratio than ΔE to EZ, both compoundly and, again,
dividedly, the same reasoning shall hold.

From the same, the argument for the conversion is made clear, too.
For let AΓ have to BΓ a greater ratio than ΔZ to ZE. I say that conversely, too, ΓA has to AB a smaller ratio than ZΔ to ΔE.[27]

---

[17] Eutocius sets out to prove the extension into proportion-inequalities of *Elements*
V.18. In modern terms, he proves: from a:b>c:d, derive (a+b):b>(c+d):d.

[18] *Elements* V.7 Cor.   [19] *Elements* V.10.

[20] *Elements* V.8.   [21] *Elements* V.18.

[22] This is the extension into proportion-inequalities of *Elements* V.17 (the converse
of V.18). In modern terms, this is: from (a+b):b>(c+d):d, derive a:b>c:d.

[23] A repetition of Steps a, 1–4 above. Start from the construction here, BΓ:ΓA::
ZE:EΘ, invert it with *Elements* V.7 Cor. to get ΓA:BΓ::EΘ:ZE, substitute EΘ:ZE
for ΓA:BΓ in the formulation given in the setting-out (AΓ:ΓB>ΔZ:ZE) and you get
EΘ:ZE>ΔZ:ZE, hence through *Elements* V.10 the claim of this step.

[24] The effective assertion of Step 4 is ΘZ:ZE::AB:BΓ.

[25] *Elements* V.17.   [26] *Elements* V.8.

[27] Extension to proportion-inequalities of *Elements* V.19 Cor., in modern terms: from
a:b>c:d derive a:(a-b)<c:(c-d). (Note that this is far less intuitive than the preceding
results.)

(1) For since AΓ has to ΓB a greater ratio than ΔZ to ZE, (2) dividedly, too, AB has to BΓ a greater ratio than ΔE to EZ,[28] (3) inversely, BΓ has to BA a smaller ratio than ZE to EΔ,[29] (4) and compoundly: ΓA has to AB a smaller ratio than ΔZ to ΔE.[30]

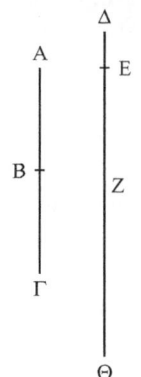

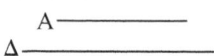

In I.2
The figure rotates in codex D as in the thumbnail.   Codex G, followed by Heiberg, has AB>BΓ; Codex D has BΓ>AB. Codices DG have EZ>ZΘ.

To 3

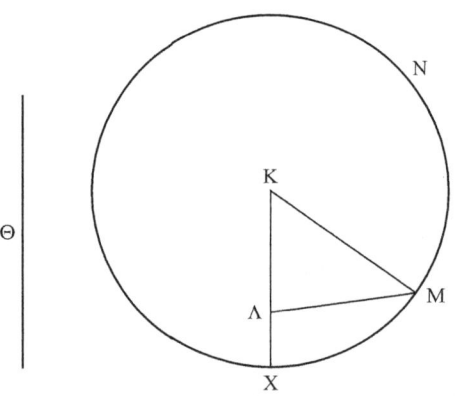

In I.3
Codex D has the line Θ to the right of the main circle.   Codices BD, followed by Heiberg, in a sense correctly, have the angle at Λ right. Codex G has the circle tilted slightly clockwise, and so (rather more) does B. Possibly, there was some such tilt in codex A itself.

Arch. 47   "And let KM be drawn down from K, equal to Θ." For this is possible,[31] with KΛ being produced as to X, and setting KX equal to Θ, and having

---

[28] As proved immediately above.

[29] Assuming an extension into proportion-inequalities of *Elements* V.7 Cor., in modern terms: from a:b>c:d derive b:a<d:c. Eutocius does not prove this extremely intuitive result. (Nor does he prove the extension of *Elements* V.16, the "alternately" operation, into proportion inequalities: in modern terms, (a:b>c:d)→(a:c>b:d). Although this extension is not required right now, it is often used by Archimedes.)

[30] As proved immediately above.

[31] The first words of Eutocius' commentary are "for this is possible." These are also the words of the (interpolated?) Step 1 in Archimedes' own proposition. According to

a circle drawn with a center K and a radius KX, namely the <circle> XMN; for KM will be equal to KX, that is to Θ.

Arch. 47   "Therefore NΓ is a side of an equilateral and even-sided polygon." For, since the single right <angle> stands on a fourth part <of a circle>, and the cutting from the right <angle> was made according to an even division, it is clear that the circumference of the fourth part will also be divided into equal circumferences, even-times-even in number; so that <it follows>, too, that the line subtending one of the circumferences is a side of an equilateral and even-sided polygon.

Arch. 47   "So that ΟΠ, too, is a side of the equilateral polygon." For if, after we have made the <angle contained> by ΠΗΔ equal to the angle <contained> by ΞΗΝ, we join <a line> from Π to Δ and produce it as far as ΗΘ, which together with ΗΔ contains an angle equal to the <angle contained> by ΠΗΔ, then ΠΘ will be equal to ΠΟ, and <it will be> a tangent to the circle. (1) For since ΞΗ is equal to ΗΔ,[32] (2) and ΗΠ is common, (3) and they contain equal angles, (4) therefore the base is equal,[33] too – (5) ΞΠ to ΠΔ – (6) and the angle <contained> by ΠΞΗ, too, which is right,[34] (7) <is equal> to ΠΔΗ; (8) so that ΔΠ is a tangent.[35] (9) Now, since the angles at Δ are right, (10) and also, the <angles contained> by ΠΗΔ, ΔΗΘ are equal, (11) and the <line> next to the equal <angles>, ΔΗ, is common: (12) ΠΔ is equal, too, to ΘΔ.[36] (13) But ΞΠ was shown equal to ΠΔ; (14) therefore ΘΠ, too, is equal to ΠΟ, (15) and to all similar <lines that are> similarly tangent. (16) So that ΘΠ is a side of an equilateral and even-sided polygon circumscribed around the circle.

That it <= the polygon> is also similar to the inscribed, is immediately clear. (1) For, ΟΗ being equal to ΗΠ, (2) and ΓΗ to ΗΝ, (3) therefore ΟΠ is parallel to ΓΝ.[37] (4) Through the same, ΠΘ, too, <is parallel> to ΝΚ. (5) So that the <angle contained> by ΓΝΚ is equal to the <angle contained> by ΟΠΘ.[38] (6) And through this, the circumscribed is similar to the inscribed.

Heiberg's theory, this Step 1 in Archimedes' text was indeed an interpolation, a brief pointer added by a late reader to refer to the contents of Eutocius' commentary. An alternative theory is that Archimedes' Step 1 is genuine (or at least that it existed in Eutocius' own text), and should be considered as part of Eutocius' *quotation* of the Archimedean text.

[32] *Elements* I. Def.15.   [33] *Elements* I.4.   [34] *Elements* III.18.
[35] The claim is based on *Elements* III.18.   [36] *Elements* I.26.
[37] *Elements* VI.2.   [38] *Elements* I.29.

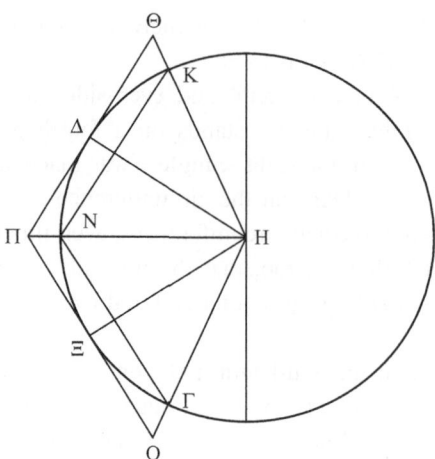

In I.3 Second diagram
Codices E4 have the
figure tilted slightly
clockwise, and so
(rather more) does D.
Codex A had the letter
K positioned at the top
of the vertical diameter
(corrected by codices
BD, and a later hand in
G).   O is omitted on
codex H.

Arch. 48   "Therefore MK has to KΛ a greater ratio than ΓH to HT." (1) For,
the angle at K being greater than the <angle contained> by ΓHT,[39] (2)
if we set up the <angle contained> by ΛKP (P imagined between Λ,
M),[40] equal to ΓHT, the triangle ΛKP is similar to ΓHT,[41] (3) and it is:
as PK to KΛ, so ΓH to HT;[42] (4) so that, MK has to KΛ, too, a greater
ratio than ΓH to HT.[43]

To 6

Arch. 55   "So, through this, the circumscribed is smaller than the <circle and
area> taken together." (1) For since the circumscribed has to the inscribed a smaller ratio than the <circle and area> taken together to the
circle, (2) much more, therefore, the circumscribed has to the circle a
smaller ratio than the <circle and area> taken together to the circle;[44]

---

[39]   Based on Step g of Archimedes' proposition. We have resumed Archimedes' diagram (K, T are not present in Eutocius' diagram) . . .

[40]   . . . And therefore interventions inside the diagram now take the form of imagination
instead of actual drawing. Eutocius does not draw his own diagram, but expects the reader
to look at Archimedes' text with its Archimedean diagram.

[41]   The angle at Λ is right through Step c, while the angle at T is right through *Elements*
III.3. Then through *Elements* I.32 the triangles are similar.

[42]   *Elements* VI.4.

[43]   That MK>PK can be shown through *Elements* I.32, I.19 (though this is probably
obvious to Eutocius, based on the diagram).

[44]   That the circle is greater than the inscribed is asserted by Archimedes in the passage
following the postulates.

(3) so that the circumscribed is smaller than the <circle and area> taken together.[45]

And taking away the circle <as> common, the remaining <segments> that are left are smaller than the area B.

To 8

Arch. 60   "Therefore the <lines> joined from the vertex to A, B, Γ are perpendiculars on them <= the sides of the base triangle>." For let the cone be imagined apart, and let H be its vertex, and <the> center of its base Θ, and let ΘA be joined from Θ to A – and HA from H. I say that HA is a perpendicular on ΔE.

(1) For since HΘ is perpendicular to the plane of the circle, (2) <so are> also all the planes through it <= through HΘ>;[46] (3) so that the triangle HΘA, too, is right to the base. (4) And ΔE was drawn in one of the planes at right <angles> to the common section of the planes, ΘA; (5) therefore ΔE is at right <angles> to the plane HΘA;[47] (6) so that <it is at right angles> to HA, too.[48] (7) And similarly, the <lines> joined from the vertex to Γ, B, too, will be proved to be perpendiculars on ΔZ, EZ.

It should be understood that, in the preceding, it was rightly added that the inscribed pyramid must in all cases have its base equilateral; for otherwise the <lines> from the vertex to the sides of the base could not have been equal; but in the <proposition> before us he did not add that the base is equilateral, because the same may follow, no matter which kind it <= the pyramid> is.

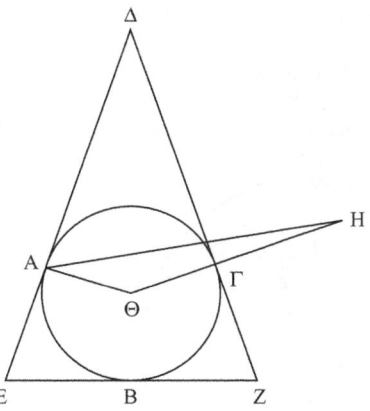

In 8
Codex D has the triangle ΔEZ nearly equilateral.

---

[45] *Elements* V.10.   [46] *Elements* XI.18.
[47] *Elements* XI. Def. 3.   [48] *Elements* XI. Def. 3.

To 9

Arch. 64   "Therefore the triangles ABΔ, BΔΓ are greater than the triangle
AΔΓ." (1) For since there is a solid angle, the <angle> at Δ, (2)
the <angles contained> by AΔB, BΔΓ are greater than the <angle
contained> by AΔΓ,[49] (3) and, if we join <a line> from the vertex
to the bisection of the base,[50] as ΔE (which is then perpendicular on
AΓ),[51] (4) the <angle contained> by AΔB will be greater than the
<angle contained> by AΔE.[52] (a) Now let the <angle contained> by
AΔZ be set up equal to the <angle contained> by AΔB, (b) and, setting ΔZ equal to ΔΓ, (c) let AZ be joined. (5) Now since two <sides>
are equal to two <sides>, (6) but also angle to angle, (7) the triangle
ABΔ, too, is equal to the triangle AΔZ,[53] (8) which is greater than
the <triangle> AΔE;[54] (9) therefore the triangle ABΔ, too, is greater

---

[49] *Elements* XI. 20.

[50] As pointed out by Heiberg, "base" here is the *line* AΓ – the base of the triangle
AΔΓ – and not (as the word means in Archimedes) the *triangle* ABΓ, the base of the
pyramid.

[51] AΔ is equal to ΓΔ (isosceles cone), ΔE is common and AE was hypothesized
equal to ΓE, hence through *Elements* I.8 the triangles are congruent and the two angles
at E are equal and right.

[52] Each are halves: AΔB is half the sum AΔB, BΔΓ, and AΔE is half AΔΓ and so
Step 4 derives from Step 2.

[53] *Elements* I.4.

[54] This statement is not necessarily true: it seems that Eutocius takes a feature of the
particular diagram he has drawn, and assumes it must hold in all cases. I do not refer to
the fact that AΔZ *appears*, in the diagram, to contain AΔE and therefore appears to be
greater. Of course, the diagram represents a three-dimensional structure, and the appearance of containment is meaningless. In all probability, Eutocius would not commit such
a trivial mistake. However, I think he might have reasoned like this. If we call the point on
AΓ, directly "below" the point Z, by the name X (by "below" I mean in the surface of the
page where the diagram is drawn, as in figure to this note), then it is indeed true, about

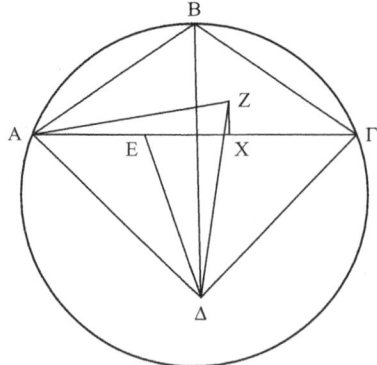

this configuration, that triangle AZΔ is greater than triangle AXΔ (in the configuration of the diagram, we can show AZ>AX, ΔZ>ΔX), which in turn is greater

than the <triangle> AΔE. (10) And similarly, the <triangle> ΔBΓ, too, <is greater> than the <triangle> ΔEΓ; (11) therefore the two <triangles> AΔB, ΔBΓ are greater than the <triangle> AΔΓ.

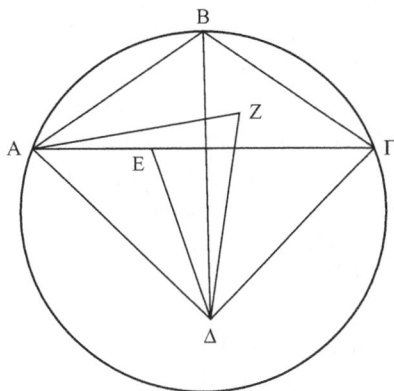

In 9
Codex D has Δ at the center of the circle.

To 10

Arch. 68 "For let HZ be drawn, tangent to the circle, also being parallel to AΓ, the circumference ABΓ being bisected at B." For it will be proved that the <line> drawn in this way is parallel to AΓ, (a) by joining ΘA, ΘΔ, ΘΓ from the center, Θ. (1) For since AΔ is equal to ΔΓ, (2) and ΔΘ <is> common, (3) two <sides> are equal to two. (4) But the base, too, AΘ, <is equal> to the base, ΘΓ;[55] (5) therefore angle is equal to angle, too.[56] (6) But the angles <contained> by HBΔ, ΔBZ are right, as well; (7) for ΘB has been drawn from the center to the touching-point;[57] (8) so that the remaining <angle contained> by ΔHB, too, is equal to the <angle contained> by ΔZB.[58] (9) And through this, HΔ is equal to ΔZ;[59] (10) so that ZH is parallel to AΓ.[60]

Arch. 69   "So, circumscribing polygons around the segments (the circum-ferences of the remaining <segments> being similarly bisected, and tangents being drawn), we will leave some segments smaller than the

---

than triangle AEΔ by simple containment. Unfortunately, this relation is true only of this particular configuration. X is not necessarily between Γ and E. I suspect Eutocius was misled, as it were, by the very sophistication required to "see" that AZΔ>AEΔ: proud of his acute perception, he failed to perceive beyond the particular case, the result being a very rare case for the Greeks: a mistake taken to be a mathematical argument.

[55] *Elements* I. Def. 15.   [56] *Elements* I.8.

[57] *Elements* III.18.   [58] *Elements* I.32.

[59] *Elements* I.6: ΔH is equal to ΔZ, and then through Step 1 the claim is seen to be true.

[60] The line HZ cuts equal parts from the equal lines ΔA, ΔΓ, i.e. it cuts them proportionally, so through *Elements* VI.2 it is parallel to the base.

area Θ." In the case of inscribed <polygons> it has been proved in the
Elements that the triangles inscribed inside the segments are greater
than half their respective segments,[61] and through this it was possible,
bisecting the circumferences and joining lines, to have as remainders
some segments smaller than the given area;[62] but in the case of circumscribing this is no longer proved in the *Elements*.

Now since he says this in the <proposition> under discussion (and
the same can be deduced from the sixth theorem), that it is to be proved
that the tangent takes away a triangle greater than half its respective
remaining , for instance (as in the same diagram[63]), that
the triangle HΔZ is greater than half the <area> contained by AΔ,
ΔΓ and by the circumference ABΓ:

(a) For, the same <lines> joined, (1) since the <angle contained> by
ΔBZ is right, (2) ΔZ is greater than BZ.[64] (3) But ZB is equal to ZΓ ((4)
for each of them is a tangent);[65] (5) therefore also, ΔZ is greater than
ZΓ. (6) So that the triangle ΔBZ is greater than the triangle BZΓ ((7)
for they are under the same height);[66] (8) therefore it <=the triangle
ΔBZ> is greater by much than the remaining BZΓ. (9)
So through the same, ΔBH is greater than BHA, as well; (10) therefore
the whole ΔZH is greater than half the remaining AΔΓ.

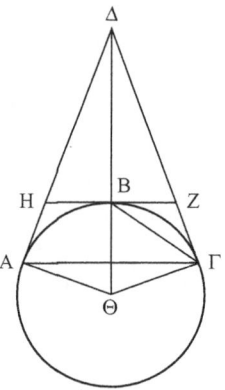

To 13

In 10
Codex E has Θ beneath
the centre of the circle.
The line BΓ is removed
in codices BDG; codex
H introduces the line
AB.

Arch. 83   "So let a circumscribed <polygon> be imagined inside the circle B,
and an inscribed, and a circumscribed <rectilinear figure> around the
circle A, similar to the <polygon> circumscribed around B." Now,

---

[61] Proved as an interim result, in *Elements* XII.2.   [62] *Elements* X.1.
[63] I.e. the diagram of the preceding comment.   [64] *Elements* I.18.
[65] Can be deduced from *Elements* III.36.   [66] *Elements* VI.1.

it is clear how to inscribe, inside a given circle, a polygon similar to the <polygon> inscribed in another <circle>, and this has also been said by Pappus in the Commentary to the *Elements*;[67] but we no longer have this similarly said: <how> to circumscribe around a given circle a polygon similar to <another polygon> circumscribed around another circle; so this should be said now:

For let a <polygon> similar to the <polygon> inscribed inside the circle B be inscribed, and around the same <circle> A <let a polygon be circumscribed> similar to the <polygon inscribed> inside it <=the circle A>, as in the third theorem; and it will also be similar to the <polygon> circumscribed around B.

Arch. 84   "And since the rectilinear <figures> circumscribed around the circles A, B are similar, they will have the same ratio, which the radii <have> in square." The same is proved in the *Elements* for inscribed <polygons>,[68] but not for the circumscribed; and it will be proved like this:

(a) For let the circumscribed and inscribed rectilinear <figures> and the joined radii KE, KM, $\Lambda\Theta$, ΛN be imagined on their own; (1) so it is obvious that KE, $\Lambda\Theta$ are radii of the circles around the circumscribed polygons, (2) and are to each other, in square, as the circumscribed polygons.[69] (3) And since the <angles contained> by KEM, $\Lambda\Theta N$ are halves of the angles in the polygons,[70] (4) the polygons being similar, (5) it is also clear that they themselves <=the angles> are equal. (6) But, also, the <angles> at M, N are right;[71] (7) therefore the triangles KEM, $\Lambda\Theta N$ are equiangular,[72] (8) and it shall be: as KE to $\Lambda\Theta$, so KM to ΛN;[73] (9) so that the <squares> on them, too. (10) But as the <square> on KE to the <square> on $\Theta\Lambda$, so the circumscribed to each other;[74] (11) and therefore, as the <square> on KM to the <square> on ΛN, so the circumscribed to each other.

Arch. 85   "Therefore the triangle $TK\Delta$ has to the rectilinear <figure> around the circle B the same ratio which the triangle $KT\Delta$ <has to> the triangle

---

[67] The first mention of a mathematician other than Archimedes and the only reference by Eutocius to this commentary to Euclid. Pappus, hard to pigeon-hole (commentator? Mathematician?), lived in Alexandria in the fourth century A.D. He is known to us chiefly through a work – mostly extant – titled the *Collection*. As the title suggests, this is a miscellany with some parts more resembling a commentary on pieces of early mathematics, some parts resembling original, creative mathematics. Whatever Pappus has written as formal commentary to Euclid, it has not survived in the Greek manuscript tradition (a commentary to Book X of the *Elements* is extant in Arabic).

[68] *Elements* XII.1.   [69] *Elements* XII.1.

[70] This is proved in the course of the third comment to Proposition 3 above.

[71] *Elements* III.18.   [72] *Elements* I.32.

[73] *Elements* VI.4.   [74] *Elements* VI.20 Cor.

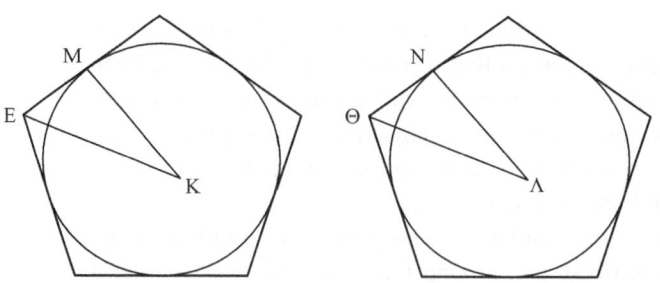

ZPΛ." (1) For since the rectilinear <figures> around the circles A, B are to each other as the radii in square, (2) that is TΔ to H in square, (3) that is TΔ to PZ in length, (4) that is as the triangle KTΔ to the <triangle> ZPΛ, (5) but the <triangle> KTΔ is equal to the <figure> circumscribed around the circle A, (6) therefore it is: as the <triangle> KTΔ to the <figure> circumscribed around the circle B, so the same triangle KTΔ to the triangle ZPΛ.

Arch. 85   "Therefore, alternately: the prism has to the cylinder a smaller ratio than the <figure> inscribed inside the circle B to the circle B; which is absurd." (1) If we make: as the surface of the prism to the surface of the cylinder, so the <figure> inscribed inside the circle B to some other <figure>, it <=the figure inscribed inside the circle B> will be <in the said ratio> to a <figure> smaller than the circle B;[75] (2) to which <=to the hypothetical, smaller figure inside the circle> the inscribed <figure> has a greater ratio than <it has> to the circle,[76] (3) that is the surface of the prism has to the surface of the cylinder a greater ratio than the inscribed <figure> to the circle; (4) but it was proved to have a smaller <ratio>, too (5) which is absurd.

In 13
Codex G has the figure upside down, and the points E, Θ consequently lower; codex D has two internal pentagons, and also has the figure upside down, with the points E, Θ consequently higher (see thumbnails). (The arrangement between the pentagons is changed in codex B, but the overall structure of each is kept as in the figure). Codices BDG have K, Λ as centers.

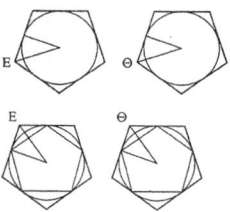

To 14

Arch. 93   "But Γ has to Δ a greater ratio than the polygon inscribed in the circle A to the surface of the pyramid inscribed inside the cone[″]. For the radius of the circle has to the side of the cone a greater ratio than the perpendicular drawn from the center on one side of the polygon to

---

[75] I.e., the "some other <figure>" will be smaller than the circle B. The substantive argument may be put like this. The prism is greater than the cylinder, hence the ratio mentioned here is that of the greater to the smaller. Thus, it is also the ratio of the inscribed to something *smaller* than the inscribed; and smaller-than-the-inscribed must also be smaller than the circle (from the passage following the postulates).

[76] *Elements* V.8.

the perpendicular drawn on the side of the polygon from the vertex of the cone<″>.[77] (a) For let the diagram specified in the text be imagined on its own, (b) and a polygon, ZΘK, <imagined> inscribed inside the circle A, (c) and let a perpendicular AH be drawn from the center of the circle A on one side of the polygon, <namely on> ΘK; (1) so it is obvious that the <rectangle contained> by the perimeter of the polygon and <by> AH is twice the polygon.[78] (d) So let the vertex of the cone, the point Λ, be imagined as well, (e) and ΛH <imagined> joined from Λ to H – (2) which is then a perpendicular on ΘK, (3) as was proved in the comment to Theorem 8. (4) Now since the inscribed polygon is equilateral, (5) and, also, the cone is isosceles, (6) the perpendiculars drawn from Λ on each of the sides of the polygon are equal to ΛH; (7) for each of them is, in square, the <squares> on the axis, and on the <line> equal to AH.[79] (8) And, through this, the <rectangle contained> by the perimeter of the polygon and <by> ΛH is twice the surface of the pyramid; (9) for the <rectangle contained> by each side and <by> the perpendicular drawn on it from the vertex (<a perpendicular which is> equal to ΛH) (10) is twice its respective triangle;[80] (11) so that it is: as AH to HΛ, the polygon to the surface of the pyramid (the perimeter of the polygon taken as a common height).[81] (12) So, HN being drawn parallel to MΛ, it shall be: as AM to MΛ, AH to HN.[82] (13) But AH has to HN a greater ratio than to HΛ; (14) for ΛH is greater than HN;[83] (15) therefore also: AM has to MΛ (that is Γ to Δ) (16) a greater ratio than AH to HΛ ((17) that is the polygon to the surface of the pyramid).

---

[77] A very interesting textual issue. The lemma, as marked in the manuscripts by marginal sigla, ends with what I mark as ["]. Heiberg thought that this is where the lemma ended in fact. But our manuscripts for Archimedes go on with another passage ("for the radius ... vertex of the cone"), practically identical to what Heiberg takes to be Eutocius' first paragraph of commentary. Therefore Heiberg goes on to square-bracket that passage in the Archimedean text (clearly he thinks someone copied it from Eutocius into the main text of Archimedes). Heiberg's hypothesis is quite possible. However, it is not necessary, unless one goes for Heiberg's ruthless eradication of backwards-looking justifications from the Archimedean text. Otherwise, then, it is simpler to end the lemma where I do (marked by <″>). In this case, of course, the square brackets in the Archimedean text ought to be removed.

[78] This is made obvious by dividing the polygon into triangles whose bases total as the perimeter of the polygon, and whose heights are the radius of the circle; then *Elements* I.41.

[79] *Elements* I.47. (Since they are all equal to a constant sum, they are also equal to each other.)

[80] *Elements* I.41.   [81] *Elements* VI.1.

[82] *Elements* VI.2, 3.   [83] *Elements* I.32, 19.

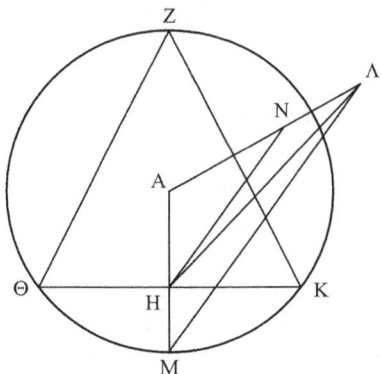

To 16

In 14
Codex D has the line
HN aligned so that the
point N is on the line
ZK.   Codex E has
Δ instead of A.
Codex H has the lines
"NH", "ΛH" start from
a point above H.

Arch. 98   "And since the <rectangle contained> by BH, HA is equal to: the
<rectangle contained> by BΔZ and the <rectangle contained> by
AΔ and <by> ΔZ, AH taken together, through ΔZ's being parallel
to AH." (1) For since ΔZ is parallel to AH, (2) it is: as BA to AH,
BΔ to ΔZ;[84] (3) and through this, the <rectangle contained> by the
extremes BA, ΔZ is equal to the <rectangle contained> by the means
BΔ, AH.[85] (4) But the <rectangle contained> by BA, ΔZ is equal to
the <rectangle contained> by BΔ, ΔZ and the <rectangle contained>
by AΔ, ΔZ, (5) through the first theorem of the second book of the
Elements; (6) therefore the <rectangle contained> by BΔ, AH, too,
is equal to the <rectangle contained> by BΔ, ΔZ and the <rectangle
contained> by AΔ, ΔZ. (7) Let the <rectangle contained> by ΔA, AH
be added <as> common; (8) therefore the <rectangle contained> by
BΔ, AH together with the <rectangle contained> by ΔA, AH (which is
the <rectangle contained> by BA, AH),[86] (9) is equal to the <rectangle
contained> by BΔ, ΔZ and the <rectangle contained> by AΔ, ΔZ
and also the <rectangle contained> by AΔ, AH.

To 23

Arch. 118   "And let the number of the sides of the polygon be measured by four."
He wants that the sides of the polygon be measured by four because
it will be of use to by him, in the <propositions> following this one,
that (with the circle moving around the diameter AΓ) all the sides

---

[84] *Elements* VI.2, 4.   [85] *Elements* VI.16.   [86] *Elements* II.1.

are carried along conical surfaces. For if the sides of the polygon are
not measured by four, then it is possible – even if it is an even-sided
<polygon> – that not all <sides> are carried along conical surfa-
ces – which can be understood in the case of the sides of the hexagon;
for two opposite parallel sides of it are in fact carried along a cylindrical
surface. Which, as was said, is not of use to him in the following.

To 30

Arch. 136   "But KΘ is equal to the diameter of the circle ABΓΔ." For if we join
<a line> from X to the point at which KZ touches the circle ABΓΔ
(imagined as the <point> M),[87] and similarly <we join> XK: (1)
since XK is equal to XZ, (2) and, also, the <angles> at M are right,[88]
(3) KM will then be equal to MZ, as well.[89] (4) But then, ZX is equal
to XΘ, as well; (5) therefore XM is parallel to KΘ[90] (6) and through
this it will be: as ΘZ to ZX so KΘ to XM. (7) But ΘZ is double XZ;
(8) therefore KΘ <is> also double XM, (9) which is the radius of the
circle ABΓΔ.

To 32

Arch. 141   "But the diameter of the circle M, also, has to the diameter of the
<circle> N a ratio which EΛ has to AK." (a) For if HΛ, ΓK are joined,
(1) the angles at K, Λ are then right,[91] (2) and, AK being parallel to ΛE,
(3) the triangle HΛE is then equiangular with the triangle ΓKA, (4) and
through this, it is: as HΛ to ΛE, so is ΓK to KA.[92] (5) But as HΛ to ΛE,

---

[87] The "imagined" probably shows that the letter M is not in Eutocius' copy of
Archimedes. Hence the letter M in the Archimedean text "for <it is> twice XM which
is a radius of the circle ABΓΔ" is very probably post-Eutocian: a reader may have in-
serted Eutocius' letter M into the diagram, together with a brief sentence indicating the
Eutocian argument. This is an example of a relatively clear interpolation into the text of
Archimedes.

[88] *Elements* III.18.

[89] This seems to invoke a specific congruence theorem, from the equality of "two
sides and a right angle" (the shared line XM is tacitly assumed as a premise for the
argument). This theorem is not in Euclid's *Elements*. Eutocius had probably relied on
the *Elements* roughly as we know them. I do not know how to account for this argument,
then.

[90] *Elements* VI.2.   [91] *Elements* III.31.   [92] *Elements* VI.4.

so all the <lines> joining the angles of the circumscribed <polygon>, to the diameter of the circle around the circumscribed <polygon>,[93] (6) while as ΓK to KΛ, so all the <lines> joining the angles of the inscribed <polygon>, to the diameter of the circle ABΓΔ; (7) therefore as all the <lines> joining the angles of the circumscribed <polygon>, to the diameter of the circle around it <=the circumscribed polygon>, so all the <lines> joining the angles of the inscribed <polygon>, to the diameter of the circle ABΓΔ.[94] (8) But as the diameter to the side, so the diameter to the side, (9) since also, as HE to EΛ, so ΓΛ to AK; (10) therefore through the equality, too:[95] as all the <lines> joining <the angles>, to EΛ, so all the <lines> joining <the angles>, to AK. (11) But as all <the lines> to the side EΛ, so the <rectangle contained> by all <the lines> and by EΛ – that is the <square> on the radius of M[96] (12) to the <square> on EΛ ((13) taking EΛ as a common height);[97] (14) while as all <the lines>, to AK, so the <rectangle contained> by all <the lines> and <by> AK – that is the <square> on the radius of N[98] – (15) to the <square> on AK ((16) again, taking AK as a common height);[99] (17) therefore it is: as the <square> on the radius of M to the <square> on EΛ, so the <square> on the radius of N to the <square> on AK.[100] (18) And therefore as the radius itself,[101] of M, to EΛ, so the radius of N to AK. (19) Alternately: as the radius of M to the radius of N, so EΛ to AK,[102] (20) twice the antecedents, as well:[103] as the diameter of M to the diameter of N, <so> EΛ to AK.

---

[93] *SC* I.21.   [94] *Elements* V.11.   [95] *Elements* V.22.   [96] *SC* I. 29.

[97] *Elements* VI.1. The structure of Steps 11–13 is somewhat involved. Step 11 effectively asserts the result of *SC* I.29, rect. (Lines, EΛ) = sq. (radius of M). Steps 11–12, without bringing to bear this equality, assert that lines:EΛ::rect. (lines, EΛ):sq. (EΛ). (This claim, as explained by Step 13, is an obvious truth (stated in *Elements* VI.1): all we do is to add the height EΛ to both sides of the ratio lines: EΛ and the proportion therefore must hold.) With the addition of the effective claim of Step 11 we get the effective claim of Steps 11–12, which we may call Step 12*, lines:EΛ::sq. (radius of M):sq. (EΛ). It is this Step 12* that Eutocius uses in what follows.

[98] *SC* I.24.

[99] Analogously to Steps 11–13, Eutocius in Steps 14–16 effectively derives the implicit result, which we may call 15*, lines:AK::sq. (radius of N):sq. (AK).

[100] From Step 10 we have lines:EΛ::lines:AK. With Steps 12* and 15*, lines:EΛ::sq. (radius of M):sq. (EΛ) and lines:AK::sq. (radius of N):sq. (AK), we now get the conclusion of Step 17, sq. (radius of M):sq. (EΛ)::sq. (radius of N):sq. (AK).

[101] "Itself:" so far we have mentioned the *square on the radius*, and now we mention the *radius* itself: we reach Step 18 by cutting off one dimension of the proportion obtained in Step 17.

[102] *Elements* V.16.   [103] *Elements* V.4.

To 34

Arch. 148 "And I, Θ taken, so that they exceed each other, K <exceeding> I, and I <exceeding> Θ, and Θ <exceeding> H, <all> by an equal <difference>." The proposition is, given two lines, to find two mean proportionals in an arithmetical proportion, which is the same as "exceeding each other by an equal <difference>." And this is to be done like this: let the two given lines be AB, ΓK (unequal), and, (a) taking away from AB a <line> equal to ΓK, (b) let the remainder AΔ be cut into three <equal parts> at E, Z,[104] (c) and let H be set equal to EB, (d) and <let> Θ <be set> equal to ZB. (1) So there shall be Θ, H, producing the proposition.

Now I say that AB has to ΓK a greater ratio than triplicate the <ratio> which AB has to H.

(a) For let it come to be: as AB to H, so H to some other <line> Λ. (1) And since AB exceeds H by that part of itself, by which part of itself H, too, exceeds Λ,[105] (2) and the same part of AB is greater than the <same> part of H, (3) therefore AB exceeds H by a greater <difference> than H <exceeds> Λ. (4) But AB exceeds H and H exceeds Θ by the same <difference>; (5) so that H exceeds Θ by a greater <difference> than H <exceeds> Λ; (6) so that Λ is greater than Θ. (b) So if, again, we make: as H to Λ, so Λ to M, (7) it <=M> will be greater by much than ΓK. (8) And since four lines AB, H, Λ, M are in continuous proportion, (9) AB has to M a ratio triplicate of AB to H;[106] (10) so that AB has to ΓK a greater ratio than triplicate <of the ratio it has> to H.

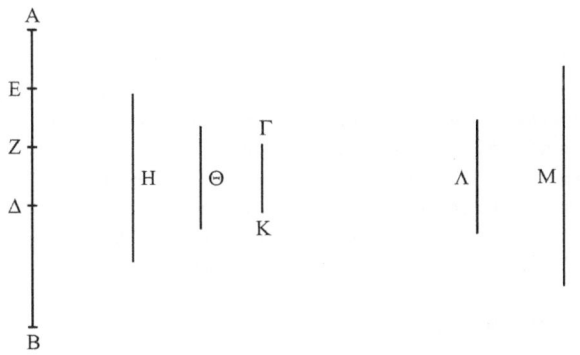

In 34
Codex D has ZΔ
roughly equal to ΔB,
both being rather
greater than AE, EZ.

---

[104] *Elements* VI.9.

[105] We have AB:H::H:Λ. Assume the ratio is "three times" (36:12::12:4). AB exceeds H by two thirds of itself (36 exceeds 12 by 24), and so does H of Λ (12 exceeds 4 by 8).

[106] *Elements* V. Def. 11.

To 37

Arch. 158   "But the <rectangle contained> by EΘ and <by> EZ, ΓΔ, KA has been proved equal to the <rectangle contained> by EΛ, KΘ." (1) For it was proved in the twenty-second theorem that EZ, ΓΔ, KA have to ΘK the same ratio which ΛE <has> to EΘ (2) so that the <rectangle contained> by the extremes is equal to the <rectangle contained> by the means.[107]

Arch. 158   "And the <rectangle> contained by EΛ, KΘ is smaller than the <square> on ΘA." (1) For <it is smaller> than the <rectangle contained> by ΛΘ, ΘK, too,[108] (2) which is equal to the <square> on ΘA – (a) as is clear with the <line> AΛ joined (4) and, through this,[109] with the resulting triangle ΘAK being similar to ΘAΛ[110] (4) for it will be: as ΛΘ to ΘA, so AΘ to ΘK,[111] (5) and the <rectangle contained> by the extremes is equal to the <square> on the mean.[112]

To 39

Arch. 163   "So it will have the same center as the circle ABΓ." (a) For if lines are drawn from Δ to Θ, E, Λ, (1) they shall be equal, (2) through <the fact that> the lines joined from Δ to the touching-points are perpendicular on the tangents, (3) as well as <the fact that> these tangents are bisected at the touching-point.[113]

Arch. 164   "And when this is <the case>, the surface is then greater than the surface." (1) For since MZ is carried along a conical surface, (2) it will

---

[107]   *Elements* VI.16.

[108]   *Elements* III.15. (This clause "For <it is> also <smaller> than the <rectangle contained> by ΛΘ, ΘK" is reflected, abbreviated further, in the Archimedean Step 4: "For <it is> also <smaller> than the <rectangle contained by> ΛΘ, KΘ." The absence of "by" in Archimedes' Step 4 – the only difference between the two texts – turns a common formulaic ellipsis, "the by XY," into a meaningless expression, "the XY." Thus there is a likelihood that the Archimedean Step 4 is a hyper-abbreviated scholion based on Eutocius.)

[109]   The "through this" probably refers to the construction of Step a.

[110]   *Elements* VI.8; III.31.   [111]   *Elements* VI.4.   [112]   *Elements* VI.17.

[113]   Eutocius envisages an argument along the following lines: draw the lines mentioned in Step a and the perpendiculars mentioned in Step 2; now you have two triangles around each of the perpendiculars; through Steps 2, 3 (and then *Elements* I.4), they are congruent, hence the remaining sides (which are the radii of the greater circle) are equal, so there are three equal lines from a point inside the greater circle to its circumference, and then through *Elements* III.9 the center of the smaller circle is also that of the greater circle.

be carried along a surface of a truncated[114] cone which is equal to a circle, whose radius has a mean ratio between ZM, and the half of: ZH and MN taken together.[115] (3) So, similarly, the surface of the truncated cone resulting from MA is equal, too, to a circle, whose radius has a mean ratio between MA, and the half of: AB and MN taken together. (4) And ZM is greater than MA, (5) while ZH <is greater> than AB; (6) therefore the mean, too, is greater than the mean; (7) so that the surface, too, <is greater> than the surface. (8) Therefore the <surface resulting> from ZM, NH is greater than the surface <resulting> from MA, NB.

To 40

Arch. 167 "Therefore the surface of the figure KZΛ is greater than the circle etc." How the assertion is to be inferred seems very unclear, but it is clear if you say it like this: (1) since the circle N is equal to the surface of the figure, (2) and the radius of N is, in square, the <rectangle contained> by MΘ, ZH, (3) but the <rectangle contained> by MΘ, ZH is greater than the <rectangle contained> by ΓΔ, ΔΞ ((4) for MΘ has been proved equal to ΓΔ, (5) and ZH greater than ΔΞ), (6) therefore the circle N is greater than the circle whose radius is, in square, the <rectangle contained> by ΓΔ, ΔΞ. (7) But the <rectangle contained> by ΓΔ, ΔΞ is equal to the <square> on ΔΑ;[116] (8) therefore the circle N, that is the surface of the circumscribed <figure> (9) is greater than the circle whose radius is equal to ΔΑ.

To 41

Arch. 171 "But the said areas are to each other, as the <square> on the side EK to the <square> on the side ΑΛ." (a) For if ΔΛK is joined, (1) EK being parallel to ΑΛ, (2) it is: as EΔ to ΔΑ, EK to ΑΛ.[117] (3) But as EΔ to ΔΑ, EZ to ΑΓ;[118] (4) and therefore as EK to ΑΛ, EZ to ΑΓ, (5) and the half of EZ to the half of ΑΓ.[119] (6) So similarly it will be proved for all the <lines> joining the angles of the polygon, as well, that they have the same ratio to each other, which EK <has> to ΑΛ. (7) Therefore also as one to one, so all to all;[120] (8) therefore as EK to ΑΛ, so all the

---

[114] A technical term invented by Eutocius: note how supple the language is (Archimedes speaks of "the surface of the cone between the parallel planes at . . .").

[115] *SC* I.16.   [116] *Elements* VI.8 Cor.   [117] *Elements* VI.2.

[118] *Elements* VI.2.   [119] *Elements* V.15.   [120] *Elements* V.12.

<lines> joining the angles of the circumscribed <polygon> together
with the half of the base of the greater segment, to all the <lines> joining <the angles of the inscribed polygon> together with the half of the
base of the smaller segment. (9) So that also: as the <square> on EK to
the <square> on AΛ, so the <rectangle contained> by EK and <by>
all <the lines joining the angles of the circumscribed figure> to the
<rectangle contained> by AΛ and <by> all <the lines joining the an-
gles of the inscribed figure>. (10) For the similar rectilinear <figures>
are in a duplicate ratio of the homologous sides,[121] (11) and duplicate
the <ratio> of EK to AΛ <is> the <ratio> of the <square> on EK to
the <square> on AΛ (12) while of: the <lines>
joining the <angles> of the greater , to the <lines> joining the <angles> of the smaller – is the <ratio> of: the
<rectangle contained> by EK and by all <the lines joining the angles
of the greater segment>, to the <rectangle contained> by AΛ and by
all <the lines joining the angles of the smaller segment>; (13) for they
<=the rectangles> are similar, too, (14) through <the fact that> their
sides being proportional.

Arch. 172   "And it is: as EK to the radius of the smaller sphere, so AΛ to the
perpendicular drawn from the center on AΛ." (a) For if we join a line
from the center to the touching-point, (1) the joined <line> will be a
perpendicular on both EK, AΛ,[122] (2) and it will be: as EΔ to ΔA, that
is EK to AΛ,[123] (3) the <line> joined from the center to the touching-
point, that is the radius of the smaller sphere (4) to the perpendicular
<drawn> from the center on AΛ.

Arch. 172   "And it was proved that as EK to AΛ, so the radius of the circle M to
the radius of the circle N." (1) Since it has been proved that it is: as the
polygon to the polygon, so the circle M to the circle N,[124] (2) that is the
<square> on the radius of M to the <square> on the radius of N.[125]

To 42

Arch. 175   "For each of the ratios is duplicate the <ratio>, which the side of the circumscribed polygon has to the <side> of the inscribed <polygon>."
(1) For it was proved[126] in the preceding <proposition> that it is: as the

---

[121] *Elements* VI.20.   [122] *Elements* III.3, 18.   [123] *Elements* VI.2, or VI.4.

[124] Refers to Step 4 in the proof of *SC* I.41 – a step bracketed earlier by Heiberg
(but here he commented, somewhat surprisingly to my mind, that "this, which Eutocius
claims to have been proved, is in fact unproved").

[125] *Elements* XII.2.

[126] Another interesting use of the verb "to prove." The reference is to Step 10 of
Proposition 41, where Archimedes does not give a proof. Instead, Archimedes, in turn,

radius of the circle equal to the surface of the circumscribed <figure>, to the radius of the circle equal to the surface of the inscribed <figure>, so the side of the circumscribed polygon, to the side of the inscribed <polygon>. (2) And the circles are to each other in a duplicate ratio of the radii; (3) therefore[127] the surface, too, has a ratio duplicate that of the side to the side.

To 44

Arch. 180   "Therefore the circumscribed solid has to the inscribed a smaller ratio than the solid sector to the cone Θ." (1) For if the circumscribed solid has to the inscribed a smaller ratio than triplicate of the <ratio>, which Δ has to Z, (2) and Δ has to E <a ratio> greater than triplicate <of the same>, (3) therefore the circumscribed has to the inscribed a smaller ratio than Δ to E. (4) But Δ has to E <a smaller ratio> than the sector to the cone; (5) therefore the circumscribed, too, has to the inscribed a smaller ratio than the sector to the cone.

[The commentary of Eutocius of Ascalon to the first <book> of Archimedes' *On the Sphere and Cylinder*; the edition being collated by the Milesian mechanical author, Isidore, our teacher].[128]

---

states there that this "was proved." Either Eutocius is effectively quoting Proposition 41, or he is using "prove" in the sense of "to make a warranted assertion."

[127] This "therefore" is simply the wrong connector. It should have been "and also," and then one could have a concluding result, "therefore both are in a duplicate ratio of the side to the side." In a hurry, Eutocius collapsed the two into one.

[128] The Isidore mentioned here must be the mathematician and architect, known from elsewhere as a somewhat younger associate of Anthemius (another mathematician and architect, and the dedicatee of Eutocius' commentary on Apollonius). The notice, of course, need not be Eutocius' own. There are similar remarks at the end of Eutocius' commentaries to the second book and the *Dimension of Circle*. The nature of such remarks was studied in Cameron (1990), while a posthumous article by Knorr (unpublished ms. from 1991) argues extensively against many of Cameron's positions in his article. In particular, Cameron suggests we may remove the square brackets surrounding this notice and read it as part of Eutocius' own text.

What the "collation" in question could have meant, poses the most difficult question. It could be anything ranging from mere proofreading to a major re-edition – either of Eutocius' text or of Archimedes' text itself. My own private guess would go like this. Since the Archimedes text as we have it (the archetype of manuscripts A and C) already seems to contain interpolations based on Eutocius, while at the same time this archetype seems to make good mathematical sense (judging from the relatively small number of obvious mistakes in it), some "edition" of the Archimedes text was probably prepared in early Byzantine times, later than Eutocius' own work. I therefore guess, together with Knorr, that the notice here derives from this edition of Archimedes' text made soon after Eutocius' own work.

EUTOCIUS' COMMENTARY TO *ON THE SPHERE AND THE CYLINDER* II

Now that the proofs of the theorems in the first book are clearly discussed by us, the next thing is the same kind of study with the theorems of the second book.

First he says in the 1st theorem:

Arch. 188

"Let a cylinder be taken, half as large again as the given cone or cylinder." This can be done in two ways, either keeping in both the same base, or the same height.[1] And to make what I said clearer, let a cone or a cylinder be imagined, whose base is the circle A,[2] and its height AΓ, and let the requirement be to find a cylinder half as large again as it.

(a) Let the cylinder AΓ be laid down, (b) and let the height of the cylinder, AΓ,[3] be produced, (c) and let ΓΔ be set out <as> half AΓ; (1) therefore AΔ is half as large again as AΓ. (d) So if we imagine a cylinder having, <as> base, the circle A, and, <as> height, the line AΔ, (2) it shall be half as large again as the <cylinder> set forth, AΓ; (3) for the cones and cylinders which are on the same base are to each other as the height.[4]

(e) But if AΓ is a cone, (f) bisecting AΓ,[5] as at E, (g) if, again, a cylinder is imagined having, <as> base, the circle A, and, <as>

---

[1] There are infinitely many other combinations, of course, as Eutocius will note much later: his comment is not meant to be logically precise, but to indicate the relevant mathematical issues.

[2] Eutocius learns from Archimedes to refer to a circle via its central letter. This is how ancient mathematical style is transmitted: by texts imitating texts.

[3] This time AΓ designates "height," not "cylinder:" no ambiguity, as the Greek article (unlike the English article) distinguishes between the two.

[4] *Elements* XII.14.

[5] This time AΓ is a line, not a cone; again, this is made clear through the articles.

height – AE, (4) it will be half as large again as the cone AΓ; (5) for the cylinder having, <as> base, the circle A, and, <as> height, the line AΓ, is three times the cone AΓ,[6] (6) and twice the cylinder AE; (7) so that it is clear that the cylinder AE, in turn, is half as large again as the cone AΓ.

So in this way the problem will be done keeping the same base in both the given <cylinder>, and the one taken. But it is also possible to do the same with the base coming to be different, the axis remaining the same.

For let there be again a cone or cylinder, whose base is the circle ZH, and <its> height the line ΘK. Let it be required to find a cylinder half as large again as this, having a height equal to ΘK. (a) Let a square, ZΛ, be set up on the diameter of the circle ZH, (b) and, producing ZH, let HM be set out <as> its half, (c) and let the parallelogram ZN be filled; (1) therefore the <parallelogram> ZN is half as large again as the <square> ZΛ, (2) and MZ <is half as large again> as ZH. (d) So let a square equal to the parallelogram ZN be constructed,[7] namely <the square> ΞΠ, (e) and let a circle be drawn around one of its sides, <namely> ΞO, as diameter. (3) So the <circle> ΞO shall be half as large again as the <circle> ZH; (4) for circles are to each other as the squares on their diameters.[8] (f) And if a cylinder is imagined, again, having, <as> base, the circle ΞO, and a height equal to ΘK, (5) it shall be half as large again as the cylinder whose base is the circle ZH, and <its> height the <line> ΘK.[9]

(g) And if it is a cone, (h) similarly, doing the same,[10] and constructing a square such as ΞΠ, equal to the third part of the parallelogram ZN, (i) and drawing a circle around its side ΞO, (j) we imagine a cylinder on it, having, <as> height, the <line> ΘK; (5) we shall have it half as large again as the cone put forth. (6) For since the parallelogram ZN is three times the square ΞΠ, (7) and <it is> half as large again as ZΛ, (8) the <square> ZΛ shall be twice the <square> ΞΠ, (9) and through this the circle, too, shall be twice the circle (10) and the cylinder <twice> the cylinder.[11] (11) But the cylinder having, <as> base, the circle ZH, and, <as> height, the <line> ΘK, is three times the cone <set up> around the same base and the same height;[12] (12) so that the cylinder having, <as> base, the circle ΞO, and a height equal to ΘK, is in turn half as large again as the cone put forth.

---

[6] *Elements* XII.10.   [7] *Elements* II.14.

[8] *Elements* XII.2.   [9] *Elements* XII.11.

[10] Refers to Steps (b–d) in this argument (not to (e–g), (4–7) in the preceding argument).

[11] *Elements* XII.11.   [12] *Elements* XII.10.

And if it is required that neither the axis nor the base shall be the same, the problem, again, will be made in two ways; for the obtained cylinder will have either its base equal to a given <base>, or its axis <equal to a given axis>. For first let the base be given, e.g. the circle ΞO, and let it be required to find, on the base ΞO, a cylinder half as large again as the given cone or cylinder. (a) Let a cylinder be taken (as said above), half as large again as the given cone or cylinder, having the same base as that set forth <=in the given>, <namely the cylinder> ΦY, (b) and let it be made: as the <square> on ΞO to the <square> on TY, so the height of ΦY to PΣ. (1) Therefore the cylinder on the base ΞO, having, <as> height, the <line> PΣ, is equal to the <cylinder> ΦY; (2) for the bases are reciprocal to the heights;[13] (3) and the task is then made.

And if it is not the base being given, but the axis, then, obtaining ΦY by the same principle, the things mentioned in the proposition will come to be.

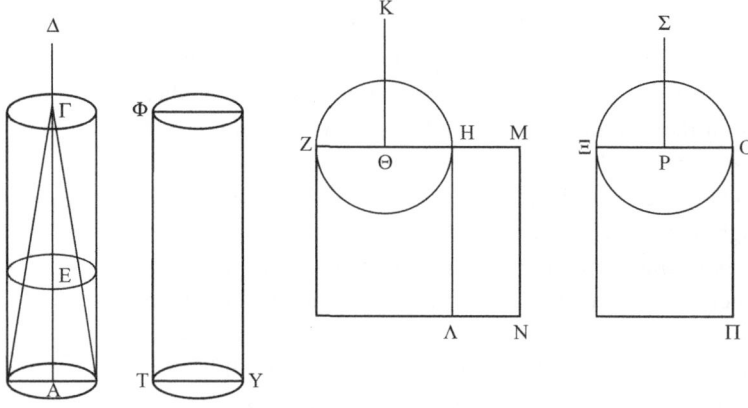

To the synthesis of the 1st

This being taken,[14] now that he has advanced through analysis the <terms> of the problem – the analysis terminating <by stating> that it is required, given two <lines>, to find two mean proportionals in continuous proportion – he says in the synthesis: "let them be found," the finding of which, however, we have not found at all proved by him,

Arch. 189

---

[13] *Elements* XII.15.

[14] Referring to the construction just provided by Eutocius. Eutocius' own self-reference is not an accident: the text suddenly becomes more discursive. We move from commentary to a mini-treatise, as it were, "On the Finding of Two Mean Proportionals."

In II.1
It appears that the following may have happened. Codex A had the diagram, to begin with, at the top of the right-sided page on the opening (i.e. the top of a verso side of a leaf). The text itself, however, ended at the left-sided page in the preceding opening (i.e. the bottom of the recto side of the same leaf). The scribe of codex A thus decided to copy the diagram twice, once at the bottom of the recto, again at the top of the verso. Precisely this structure of two consecutive, identical diagrams is preserved in codices E4. Codex D has the first diagram at the bottom of the recto, and a space for the second diagram, at the top of the verso. Codex H, which does not follow

but we have come across writings by many famous men that offered this very problem (of which, we have refused to accept the writing of Eudoxus of Cnidus, since he says in the introduction that he has found it through curved lines, while in the proof, in addition to not using curved lines, he finds a discrete proportion and uses it as if it were continuous,[15] which is absurd to conceive, I do not say for Eudoxus, but for those who are even moderately engaged in geometry). Anyway, so that the thought of those men who have reached us will become well known, the method of finding of each of them will be written here, too.[16]

As Plato[17]

Given two lines, to find two mean proportionals in continuous proportion.

Let the two given lines, whose two mean proportionals it is required to find, be ABΓ, at right <angles> to each other. (a) Let them be produced along a line towards Δ, E,[18] (b) and let a right angle be constructed,[19] the <angle contained> by ZHΘ, (c) and in one side, e.g. ZH, let a ruler, KΛ, be moved, being in some groove in ZH, in such a way that it shall, itself <=KΛ>, remain throughout parallel to HΘ. (d) And this will be, if another small ruler be imagined, too, fitting with ΘH, parallel to ZH: e.g. ΘM; (e) for, the upward surfaces[20] of ZH, ΘM being grooved in axe-shaped grooves (f) and knobs being made,

(cont.) so closely the original layout of codex A, has the two diagrams consecutive *on the same page*. I edit here the first of the two diagrams; the second is largely identical, with the exception that Φ was omitted in codex A, and M is omitted in codex H. Codex D adds further circles to the rectangles: see thumbnail. Codices DH have genuine circles, instead of almond shapes, at Φ, TY; codex D has them also at Γ, A. Codex G has all base lines on the same height; D has all on the same height except for TY which is slightly higher; H has A at the same height as Π, both higher than ΛN, in turn higher than TY; B has the figures arranged vertically, rather than horizontally. Perhaps the original arrangement cannot be reconstructed. The basic proportions, however, are remarkably constant between the codices. Codex E has X (?) instead of Λ.

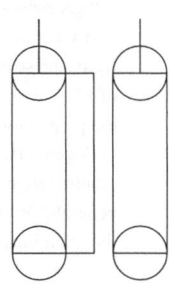

---

[15] That is, instead of a:b::b:c::c:d, all the pseudo-Eudoxus text had was a:b::c:d.

[16] In paraphrase: "although strictly speaking I merely write a commentary on Archimedes, here I have come across many interesting things that are less well known and, to make them better known, I copy them into my new text." It is interesting that Eutocius' bet came true: his own text, because of its attachment to Archimedes, survived, whereas his sources mostly disappeared.

[17] It is very unlikely that Plato the philosopher produced this solution (if a mathematical work by Plato had circulated in antiquity, we would have heard much more of it). The solution is either mis-ascribed, or – much less likely – it should be ascribed to some unknown Plato. In general, there are many question marks surrounding the attributions made by this text of Eutocius: Knorr (1989) is likely to remain for a long time the fundamental guide to the question. In the following I shall no more than mention in passing some of these difficulties.

[18] For the time being, Δ, E are understood to be as "distant as we like." Later the same points come to have more specific determination.

[19] The word – *kataskeuasthō* – is not part of normal geometrical discourse, and already foreshadows the mechanical nature of the following discussion. Notice also that we have now transferred to a new figure.

[20] "Upward surfaces:" notice that the contraption is seen from above (otherwise, of course, there is nothing to hold KΛ from falling).

fitting KΛ to the said grooves, (1) the movement of the <knobs>[21] KΛ shall always be parallel to HΘ. (g) Now, these being constructed, let one chance side of the angle be set out, HΘ, touching the <point> Γ,[22] (h) and let the angle and the ruler KΛ be moved to such a position where the point H shall be on the line BΔ, the side HΘ touching the <point> Γ,[23] (i) while the ruler KΛ should touch the line BE on the <point> K, and on the remaining side[24] <it should touch> the <point> A,[25] (j) so that it shall be, as in the diagram: the right angle <=of the machine, namely ΘHK> has <its> position as the <angle contained> by ΓΔE, (k) and the ruler KΛ has <its> position as EA has;[26] (2) for, these being made, the <task> set forth will be <done>. (3) For the <angles> at Δ, E being right, (4) as ΓB to BΔ, ΔB to BE and EB to BA.[27]

---

[21] The manuscripts – not Heiberg's edition – have a plural article, which I interpret as referring to the knobs.

[22] Imagine that what we do is to put the contraption on a page containing the geometrical diagram. So we are asked to put the machine in such a way, that the side KΛ touches the point Γ. This leaves much room for maneuver; soon we will fix the position in greater detail.

[23] The freedom for positioning the machine has been greatly reduced: H, one of the points of HΘ, must be on the line BΔ, while some other point of HΘ must pass through Γ. This leaves a one-dimensional freedom only: once we decide on the point on BΔ where HΘ stands, the position of the machine is given. Each choice defines a different angle ΓΔB. (Notice also that it is taken for granted that HΘ is not shorter than BΓ.)

[24] "The remaining side" means somewhere on the ruler KΛ, away from K and towards Λ, though not necessarily at the point Λ itself.

[25] The point K must be on BE, while some point of the ruler KΛ must be on the point A. Once again, a one-dimensional freedom is left (there are infinitely many points on the line BE that allow the condition). Each choice of point on BE, once again, defines a different angle AEB. Thus the conditions of Steps h and i are parallel. They are also inter-dependent: AE, ΓΔ being parallel, each choice of point on BΔ also determines a choice on BE. Of those infinitely many choices, the closer we make Δ to B, the more obtuse angle ΓΔE becomes, and the further we make Δ from B, the more acute angle ΓΔE becomes. Thus, by continuity, there is a point where the angle ΓΔE is right, and this unique point is the one demanded by the conditions of the problem – none of the above being made explicit.

[26] Now – and only now – Δ and E have become specific points.

[27] Note also that the lines AE, ΔΓ are parallel, and also note the right angles at B (all guaranteed by the construction). Through these, the similarity of all triangles can be easily shown (*Elements* I.29 suffices for the similarity of ABE, ΓBΔ. Since Δ, E are right, and so are the sums BΓΔ+BΔΓ, BAE+BEA (given *Elements* I.32), the similarity of ΓEΔ with the remaining two triangles is secured as well). *Elements* VI.4 then yields the proportion.

A general observation on the solution: it uses many expressions belonging to the semantic range of "e.g., such as, a chance". This can hardly be for the sake of signaling generalizability. Rather, the hypothetical nature of the construction is stressed. Further, the main idea of the construction is to fix a machine on a *diagram*. So the impression is

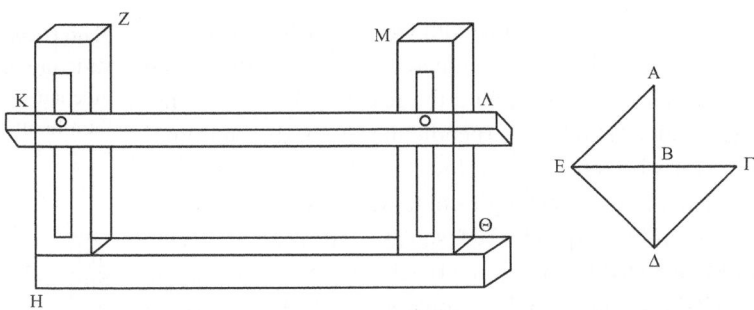

As Hero in the *Mechanical Introduction* and in the *Construction of Missile-Throwing Machines*[28]

Let the two given lines, whose two mean proportionals it is required to find, be AB, BΓ. (a) Let them <=the two given lines> be set out, so that they contain a right angle, that at B, (b) and let the parallelogram BΔ be filled, and let AΓ, BΔ be joined [(1) So it is obvious, that they <=AΓ, BΔ> are equal, (2) bisecting each other; (3) for the circle drawn around one of them will also pass through the limits of the other, (4) through <the property that> the parallelogram is right-angled].[29] (c) Let ΔΓ, ΔA be produced [to Z, H], (d) and let a small ruler be imagined, as ZBH, moved around some knob fixed at B, (d) and let

Catalogue: Plato
I avoid a full edition of this diagram. It is almost unique in the Archimedean corpus in offering a detailed three-dimensional perspective. Study of the nature of this three-dimensional representation will require attention to precise details of angles, which are very difficult to convey, and many lines can be named only by cumbersome expressions. To complicate further, scribes often had to erase and redraw parts of the diagram, making it much more complicated to ascribe anything to codex A. A facsimile of all figures, with discussion, is called for.   The diagram printed follows, for each line-segment drawn, the majority of codices, which is usually either the consensus of all codices, or the consensus of all codices but one. For the geometrical structure ABΓΔE: codex E has the line-segments in "correct" proportions (BΓ>BΔ>BE>BA) and, since codex E is on the whole the most conservative visually, it may perhaps be preferable.   Codex D has the geometrical

---

that this is a *geometrical* flight of fancy, momentarily more realistic with the reference to the axe-shaped grooves, but essentially a piece of geometry. This is a geometrical toy, and the language seems to suggest it is no more than a *hypothetical* geometrical toy: for indeed – for geometrical purposes – imagining the toy and producing it are equivalent.

[28] One version of this, that of the *Mechanical Introduction*, is preserved in Pappus' *Collection* (Hultsch [1886] I. 62–5, text and Latin translation). *The Construction of Missile-Throwing Machines* is an extant work (for text and translation, see Marsden [1971] 40–2). The following text agrees with both, though not in precise agreement; the differences are mainly minor, and the phenomenon is well known for ancient quotations in general. Hero was an Alexandrine, probably living not much before the year AD 100. Relatively many treatises ascribed to him are extant; some readers might feel too many. While a coherent individual seems to emerge from the writings (a competent but shallow popularizer of mathematics, usually interested in its more mechanical aspects), little is known about that individual, and perhaps no work may be ascribed to him with complete certainty.

[29] *Elements* III.22. Heiberg square-brackets Steps 1–4 here, as well as several other passages in this proof, because of their absence in the "original" of Hero. There are many possible scenarios (say, that we have here, in fact, the true original form of Hero, corrupted elsewhere; or that Hero had more than one version published . . . or that such questions miss the nature of ancient publication and quotation).

it be moved, until it cuts equal <lines drawn> from E, that is EH, HZ. (e) And let it <=the ruler> be imagined cutting <the lines> and having <its> position <as> ZBH, with the resulting EH, EZ being, as has been said, equal. [(f) So let a perpendicular EΘ be drawn from E on ΓΔ; (5) so it clearly bisects ΓΔ. (6) Now since ΓΔ is bisected at Θ, (7) and ΓZ is added, (8) the <rectangle contained> by ΔZΓ together with the <square> on ΓΘ is equal to the <square> on ΘZ.[30] (9) Let the <square> on EΘ be added in common; (10) therefore the <rectangle contained> by ΔZΓ together with the <squares> on ΓΘ, ΘE is equal to the <squares> on ZΘ, ΘE. (11) And the <squares> on ΓΘ, ΘE are equal to the <square> on ΓE,[31] (12) while the <squares> on ZΘ, ΘE are equal to the <square> on EZ];[32] (13) therefore the <rectangle contained> by ΔZΓ together with the <square> on ΓE is equal to the <square> on EZ. (14) So it shall be similarly proved that the <rectangle contained> by ΔHA, too, together with the <square> on AE, is equal to the <square> on EH. (15) And AE is equal to EΓ, (16) while HE <is equal> to EZ; (17) and therefore the <rectangle contained> by ΔZΓ is equal to the <rectangle contained> by ΔHA [(18) and if the <rectangle contained> by the extremes is equal to the <rectangle contained> by the means, the four lines are proportional];[33] (19) therefore it is: as ZΔ to ΔH, so AH to ΓZ. (20) But as ZΔ to ΔH, so ZΓ to ΓB (21) and BA to AH [(22) for ΓB has been drawn parallel to one <side> of the triangle ZΔH, namely to ΔH, (23) while AB <has been drawn> parallel to <another,> ΔZ];[34] (24) therefore as BA to AH, so AH to ΓZ and ΓZ to ΓB. (25) Therefore AH, ΓZ are two mean proportionals between AB, BΓ [which it was required to find].

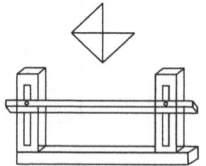

Plato (*cont.*) structure inside the mechanism, as in the thumbnail.

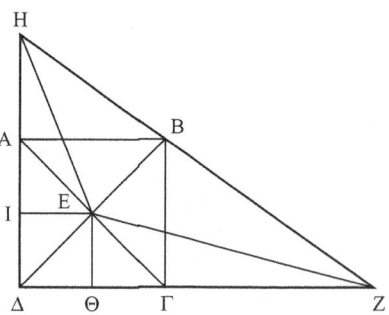

Catalogue: Hero Codices DE have AB greater than BΓ. Codex B omits I as well as the line IE.

---

[30] *Elements* II.6.

[31] *Elements* I.47. Original word order: "to the squares . . . is equal the square."

[32] *Elements* I.47.   [33] *Elements* VI.16.   [34] *Elements* VI.2.

As Philo the Byzantine[35]

Let the two given lines, whose two mean proportionals it is required to find, be AB, BΓ. (a) Let them be set out, so that they will contain a right angle, that at B, (b) and, having joined AΓ (c) let a semicircle be drawn around it, <namely> ABEΓ, (d) and let there be drawn: AΔ, in right <angles> to BA, (e) and ΓZ, <in right angles> to BΓ, (f) and let a moved ruler be set out as well, at the <point> B, cutting the <lines> AΔ, ΓZ (g) and let it be moved around B, until the <line> drawn from B to Δ is made equal to the <line> drawn from E to Z, (1) that is <equal> to the <line> between the circumference of the circle and ΓZ. (h) Now, let the ruler be imagined having a position as ΔBEZ has, (i) ΔB being equal, as has been said, to EZ. I say that AΔ, ΓZ are mean proportionals between AB, BΓ.

   (a) For let ΔA, ZΓ be imagined produced and meeting at Θ; (1) so it is obvious that (BA, ΘZ being parallel) (2) the angle at Θ is right, (b) and, the circle AEΓ being filled up, (3) it shall pass through Θ, as well.[36] (4) Now since ΔB is equal to EZ, therefore also the <rectangle contained> by EΔB is equal to the <rectangle contained> by BZE.[37] (5) But the <rectangle contained> by EΔB is equal to the <rectangle contained> by ΘΔA ((6) for each is equal to the <square> on the tangent <drawn> from Δ)[38] (7) while the <rectangle contained> by BZE is equal to the <rectangle contained> by ΘZΓ ((8) for each, similarly, is equal to the <square> on the tangent <drawn> from Z);[39] (9) so that, in turn, the <rectangle contained> by ΘΔA is equal to the <rectangle contained> by ΘZΓ, (10) and through this it is: as ΔΘ to ΘZ, so ΓZ to ΔA.[40] (11) But as ΘΔ to ΘZ, so both: BΓ to ΓZ, and ΔA to AB; (12) for BΓ has been drawn parallel to the <side> of the triangle ΔΘZ, <namely> ΔΘ (13) while BA <has been drawn> parallel to <its side> ΘZ;[41] (14) therefore it is: as BΓ to ΓZ, ΓZ to ΔA and ΔA to AB; which it was set forth to prove.

   And it should be noticed that this construction is nearly the same as that given by Hero; for the parallelogram BΘ is the same as that taken

---

   [35] Philo of Byzantium produced, in the fourth century BC, a collection of mechanical treatises, circulating in antiquity, but surviving now only in parts. Those parts reveal Philo as an original and brilliant author, probably one of the most important ancient mechanical authors. It appears that the solution quoted here was offered in a part of the work now lost. See Marsden (1971) 105–84.

   [36] *Elements* III.31.

   [37] *Elements* VI.1. That EΔ=BZ is a result of the construction ΔB=EZ (EB common).

   [38] *Elements* III.36.   [39] *Elements* III.36.   [40] *Elements* VI.16.

   [41] And then apply *Elements* VI.2 in addition to VI.16, to get Step 11.

in Hero's construction, as are the produced lines ΘA, ΘΓ and the ruler moved at B. They differ in this only: that there,[42] we moved the ruler around B, until the point was reached that the <lines drawn> from the bisection of AΓ, that is from K, on the <lines> ΘΔ, ΘZ, were cut off by it <=K> <as> equal, namely KΔ, KZ; while here, <we moved the ruler> until ΔB became equal to EZ. But in each construction the same follows. But the one mentioned here[43] is better adapted for practical use; for it is possible to observe the equality of ΔB, EZ by dividing the ruler ΔZ continuously into equal parts – and this much more easily than examining with the aid of a compass that the <lines drawn> from K to Δ, Z are equal.[44]

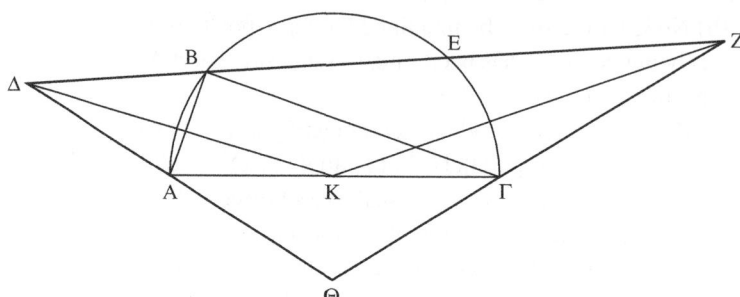

Catalogue: Philo Codex H has ΔZ parallel to AΓ. Codex D has Σ instead of E.

As Apollonius[45]

Let the two given lines, whose two mean proportionals it is required to find, be AB, AΓ (a) containing a right angle, that at A, (b) and with center B and radius AΓ let a circumference of a circle be drawn,

---

[42] I.e. Hero's solution.   [43] I.e. Philo's solution.

[44] The idea is this: we normally have an unmarked ruler, but we can mark it by continuous bisection, in principle a geometrically precise operation. The further we go down in the units by which we scale the ruler, the more precise the observation of equality. Since precise units are produced by continuous bisections from a given original length, there is a great advantage to having the two compared segments measured by units that both derive from the same original length. Hence the superiority of Philo's method, where the two segments lie on a single line, i.e. on a single ruler, or on a single scale of bisections. In other words, absolute units of length measurement were considered less precise than the relative units of measurement produced, geometrically, by continuous bisection.

[45] Apollonius is mainly known as the author of the *Conics* (originally an eight-book work, its first four books survive in Greek while its next three survive in Arabic, as do several other, relatively minor works.) The ancients thought, and the *Conics* confirm, that, as mathematician, he was second to Archimedes alone: not that you would guess it from the testimony included here.

<namely> KΘΛ, (c) and again with center Γ and radius AB let a circumference of a circle be drawn, <namely> MΘN, (d) and let it <=MΘN> cut KΘΛ at Θ, (e) and let ΘA, ΘB, ΘΓ be joined; (1) therefore BΓ is a parallelogram and ΘA is its diameter.[46] (f) Let ΘA be bisected at Ξ, (g) and with center Ξ let a circle be drawn cutting the <lines> AB, AΓ, after they are produced, (h) at Δ, E – (i) further, so that Δ, E will be along a line with Θ – (2) which will come to be if a small ruler is moved around Θ, cutting AΔ, AE and carried until <it reaches> such <a position> where the <lines drawn> from Ξ to Δ, E are made equal.

For, once this comes to be, there shall be the desideratum; for it is the same construction as that written by Hero and Philo, and it is clear that the same proof shall apply, as well.

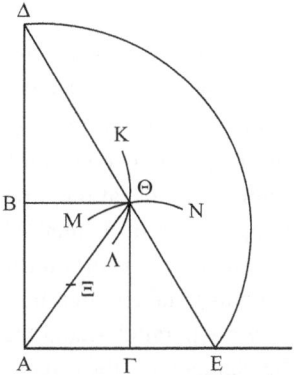

As Diocles in *On Burning Mirrors*[47]

Catalogue: Apollonius Codices BD have a quadrant for an arc. This is badly executed in codex D, where the arc falls short of E, falling instead on the line ΔE itself. Codex D has BΘ greater than ΓΘ. Codex G has ΔB equal to BA. Codex B has removed the continuation of line AE, and has added lines ΞΔ, ΞE.   Codex A had Ψ instead of E (corrected in codex B). Codex D omits Ξ.

In a circle, let two diameters be drawn at right <angles>, <namely> AB, ΓΔ, and let two equal circumferences be taken off on each <side> of B, <namely> EB, BZ, and through Z let ZH be drawn parallel to AB, and let ΔE be joined. I say that ZH, HΔ are two mean proportionals between ΓH, HΘ.

---

[46] By joining the lines ΘA, BΓ we can prove the congruity, first, of ΘBΓ, BΓA (*Elements* I.8), so the angle at Θ is right as well as that at A; and by another application of *Elements* I.8, we get the congruity of ΘAΓ, ΘAB, hence the angle at B = the angle at Γ, and ΘBAΓ must be a parallelogram.

[47] A work surviving in Arabic (published as Toomer [1976]) – Diocles' only work to survive. Probably active in the generation following Apollonius, Diocles belongs to a galaxy of brilliant mathematicians whose achievements are known to us only through a complex pattern of reflections.

(a) For let EK be drawn through E parallel to AB; (1) therefore EK is equal to ZH, (2) while KΓ <is equal> to HΔ. (b) For this will be clear once lines are joined from Λ to E, Z; (3) for the <angles contained> by ΓΛE, ZΛΔ will then be equal,[48] (4) and the <angles> at K, H are right; (5) therefore also all <=sides and angles> are equal to all,[49] (6) through ΛE's being equal to ΛZ;[50] (7) therefore the remaining ΓK, too, is equal to the <remaining> HΔ. (8) Now since it is: as ΔK to KE, ΔH to HΘ,[51] (9) but as ΔK to KE, EK to KΓ; (10) for EK is a mean proportional between ΔK, KΓ;[52] (11) therefore as ΔK to KE, and EK to KΓ, so ΔH to HΘ. (12) And ΔK is equal to ΓH, (13) while KE <is equal> to ZH, (14) and KΓ < is equal> to HΔ. (15) Therefore as ΓH to HZ, so ZH to HΔ and ΔH to HΘ. (c) So if equal circumferences – MB, BN – are taken on each side of B, (d) and, through N, NΞ is drawn parallel to AB, and ΔM is joined, (16) NΞ, ΞΔ, again, will be mean proportionals between ΓΞ, ΞO.

Now, producing in this way many continuous parallels between B, Δ; and, at the side of Γ, setting <circumferences> equal to those taken, by these <parallels>, from B; and joining lines from Δ to the resulting points (similarly to ΔE, ΔM) – then the parallels between B, Δ will be cut at certain points (in the diagram before us, at the <points> O, Θ), to which we join lines (by the application of a ruler <from one point to its neighbor>) – and then we shall have a certain line figured in the circle, on which: if a chance point is taken, and, through it, a parallel to AB is drawn, then: the drawn <parallel>, and the <line> taken by it <=the parallel> from the diameter (in the direction of Δ), will be mean proportionals between: the <line> taken by it <=the parallel> from the diameter (in the direction of the point Γ); and its <=the parallel's> part from the point in the line <=the line produced by the ruler> to the diameter ΓΔ.[53]

Having made these preliminary constructions, let the two given lines (whose mean proportionals it is required to find) be A, B, and let there be a circle, in which <let there be> two diameters in right <angles> to each other, ΓΔ, EZ, and let the line <produced> through the continuous points be drawn, as has been said, <namely> ΔΘZ, and

---

[48] *Elements* III.27.   [49] Referring to the triangles ΛKE, ΛHZ. *Elements* I.26.

[50] Two radii.   [51] *Elements* VI.2.   [52] *Elements* VI.8 Cor.

[53] This formulation is at least as opaque in the Greek as it is in my translation, and it is readable only by translating its terms to diagrammatic realities. This translation is effected in the ensuing proof. What must be understood at this stage is that we have repeated, virtually, the operation of the preceding argument a certain number of times (perhaps, infinitely many times), producing a line connecting many (or all) of the points of the type O, Θ.

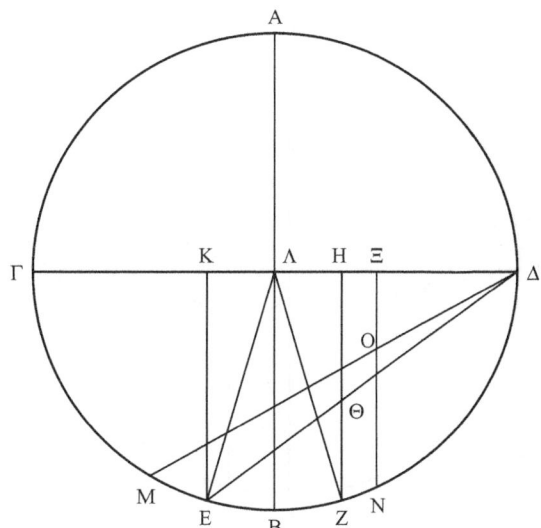

Catalogue: Diocles
Codex A had the letter
O on the intersection of
MΔ/HZ (corrected in
codices BDG).
Codex E has H (?)
instead of N.

let it come to be: as A to B, ΓH to HK,[54] and joining ΓK and producing it, let it cut the line at Θ, and let ΛM be drawn through Θ parallel to EZ; (1) therefore, through what has been proved above, MΛ, ΛΔ are mean proportionals between ΓΛ, ΛΘ. (2) And since it is: as ΓΛ to ΛΘ, so ΓH to HK,[55] (3) and as ΓH to HK, so A to B (a) if we insert means between A, B in the same ratio as ΓΛ, ΛM, ΛΔ, ΛΘ, e.g. N, Ξ,[56] (4) we shall have taken N, Ξ, mean proportionals between A, B, which it was required to find.

See diagram on
following page.

As Pappus in the *Introduction to Mechanics*[57]

Pappus put forth as his goal "to find a cube having a given ratio to a given cube," and while his arguments, too, proceeded towards such a goal, it is still clear that, finding this, the problem before us will be

---

[54] This defines the point K. *Elements* VI.12.

[55] *Elements* VI.2.

[56] This is not a *petitio principii*. Through *Elements* VI.12, it is possible to find the analogues of the series ΓΛ, ΛM, ΛΔ, ΛΘ, starting from the terms A, B – in fact, this is a mere change of scale.

[57] A reference to what we know as "Book 8" of Pappus' "Mathematical Collection," much of which is extant – his only surviving work in Greek (more may survive in Arabic, if we believe in the attribution of a commentary to Euclid's *Elements* Book X).

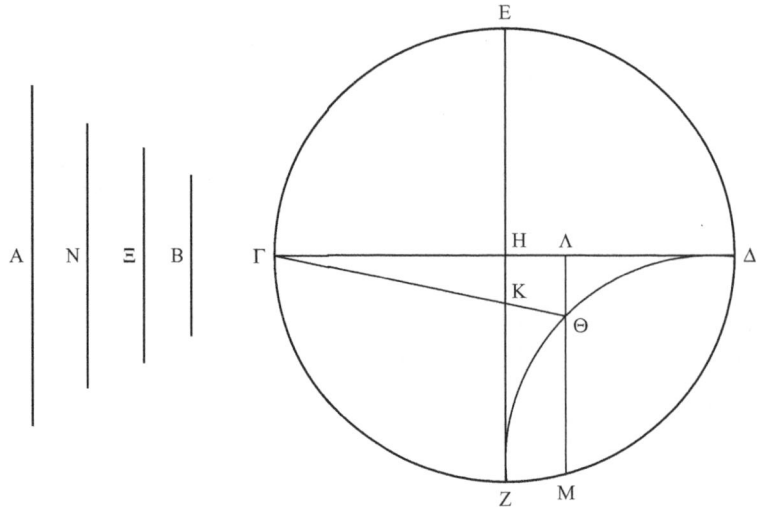

Catalogue: Diocles (second diagram) Codices GH had Ξ instead of Z (corrected later in Codex G).

found as well; for, given two lines, if the second of the two required means is found, the third will thereby be given as well.[58]

For let a semicircle ΑΒΓ be drawn (as he says himself, word by word),[59] and, from the center Δ, let ΔΒ be drawn at right <angles>,[60] and let a small ruler be moved around the point A, so that while one of its ends will be set to revolve around some small peg standing at the point A, the other end will be moved, between Β and Γ, as around the small peg.

Having constructed these, let it be demanded to find two cubes having to each other a demanded ratio.

Let the <ratio> of ΒΔ to ΔΕ be made the same as this <demanded> ratio, and joining ΓΕ let it be produced to Z. So let the ruler be moved along, between Β and Γ, until the <moment> when its part taken

---

[58] The idea is the following. To double a cube, it is necessary to find *one* of the middle terms in a four-terms geometrical progression. That is, if the side of the original cube is 1, and you want a cube with volume 2, the side of the new cube should be an A satisfying 1:A::A:X::X:2. So in a sense you do not need X, all you need is A – as far as doubling the cube is concerned. However, as Eutocius points out, A being given, X is already there: all we need to do is (for instance) to find the mean proportional of A and 2 (a simple Euclidean problem: *Elements* VI.13). Confusingly, Eutocius' ordinals here refer to the sequence of *four* terms in geometrical proportion. Hence the two mean proportionals are not "first and second" but "second and third."

[59] The meaning of this is that Eutocius has before him the original Pappic text (and not some second-hand report). That the quotation is not word-for-word is clear (and is to be expected given ancient practices), though the discrepancies are indeed minor.

[60] To the base of the semicircle, the line ΑΓ: the only straight line so far, hence a clear reference.

off between the lines ZE, EB becomes equal to the <part taken off> between the line BE and the circumference BKΓ (for this we will do easily by trial and error as we move the ruler). So let it have come to be and let it have a position <as> AK, so that HΘ, ΘK are equal. I say that the cube on BΔ has to the cube on ΔΘ the demanded ratio, that is the <ratio> of BΔ to ΔE.

(a) For let the circle be imagined completed, (b) and joining KΔ let it be produced to Λ, (c) and let ΛH be joined. (1) Therefore it <=ΛH> is parallel to BΔ (2) through KΘ's being equal to HΘ, (3) and KΔ's being equal to ΔΛ.[61] (d) So let both AΛ and ΛΓ be joined. (4) Now since the <angle contained> by AΛΓ is right ((5) for it is in a semicircle),[62] (6) and ΛM is a perpendicular, (7) therefore it is: as the <square> on ΛM to the <square> on MA, that is ΓM to MA,[63] (8) so the <square> on AM to the <square> on MH.[64] (9) Let the ratio of AM to MH be added as common; (10) therefore the ratio composed of both the <ratio> of ΓM to MA and the <ratio> of AM to MH, that is the ratio of ΓM to MH, (11) is the same as the <ratio> composed of both the ratio of the <square> on AM to the <square> on MH and the <ratio> of AM to MH. (12) But the ratio composed of both the <ratio> of the <square> on AM to the <square> on MH and the <ratio> of AM to MH is the same as the ratio which the cube on AM has to the <cube> on MH; (13) therefore the ratio of ΓM to MH, too, is the same as the ratio which the cube on AM has to the <cube> on MH. (14) But as ΓM to MH, so ΓΔ to ΔE,[65] (15) while as AM to MH, so AΔ to ΔΘ;[66] (16) therefore also: as BΔ to ΔE, that is the given ratio, (17) so the cube on BΔ to the cube on ΔΘ. (18) Therefore ΔΘ is the second of the two mean proportionals which it was required to find between BΔ, ΔE.

And if we make, as BΔ to ΔΘ, ΘΔ to some other <line>, the third will also be found.

And it must be realized that this sort of construction, too,[67] is the same as that discussed by Diocles, differing only in this: the other one <=Diocles'> draws a certain line through continuous points between the <points> A, B; on which <line> H was taken (ΓE being produced to cut the said line);[68] while here H is found by the

---

[61] Radii in a circle. 1 derives from 2 and 3 through *Elements* VI.2.

[62] *Elements* III.31.   [63] *Elements* VI.8 Cor.

[64] The angle at A is right, for the same reason that the angle at Λ is: both subtend a diameter (*Elements* III.31). Hence through *Elements* VI.8, 4, the claim follows.

[65] *Elements* VI.2.   [66] *Elements* VI.2.

[67] "Too:" i.e., the relation we see between Pappus and Diocles is the same as we saw above for the relation between Hero, Philo, and Apollonius.

[68] In an interesting move, Eutocius translates Diocles' argument to Pappus' diagram.

ruler AK's being moved around A. For we may learn as follows,
that H is the same, whether it is taken by the ruler (as in here) or
whether <it is taken> as Diocles said: (a) producing MH towards
N (b) let KN be joined. (1) Now since KΘ is equal to ΘH, (2) and
HN is parallel to ΘB, (3) KΞ is also equal to ΞN.[69] (3) And ΞB is
common and at right <angles>; (4) for KN is bisected, and at right
<angles>, by the <line drawn> through the center;[70] (5) therefore base
is equal to base, too,[71] (6) and through this the circumference KB <is
equal> to the <circumference> BN, too.[72] Therefore H is on Diocles'
line.

And the proof, too, is the same. For Diocles has said that (1) it is:
as ΓM to MN, so MN to MA and AM to MH.[73] (2) And NM is equal
to MΛ; (3) for the diameter cuts it at right <angles>;[74] (4) therefore
it is: as ΓM to MΛ, so ΛM to MA and AM to MH. (5) Therefore ΛM,
MA are mean proportionals between ΓM, MH. (6) But as ΓM to MH,
ΓΔ to ΔE,[75] (7) while as ΓM to MΛ, AM to MH[76] (8) that is ΓΔ to
ΔΘ;[77] therefore ΔΘ is also the second of the means between ΓΔ, ΔE,
that which Pappus found as well.

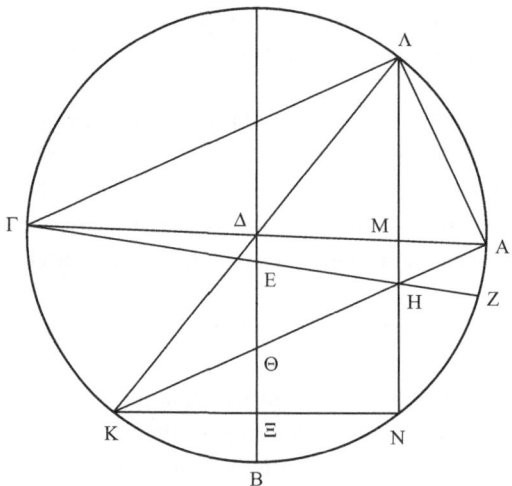

Catalogue: Pappus
Codex D has omitted Z.

---

[69] *Elements* VI.2.   [70] A close formulaic reference to *Elements* III.3.

[71] *Elements* I.4.   [72] *Elements* III.28.

[73] A translation of Diocles' first proof, Step 11, into Pappus' diagram.

[74] *Elements* III.3.   [75] Pappus' Step 14.

[76] Two successive applications of *Elements* III.31, VI.8 Cor., parallel in this respect
to Steps 7, 8 in Pappus' proof.

[77] A variation on Step 15 of Pappus' proof.

As Sporus[78]

Let the two given unequal lines be AB, BΓ; so it is required to find two mean proportionals in a continuous proportion[79] between AB, BΓ.

(a) Let ΔBE be drawn from B at right angles to AB, (b) and with a center B, and a radius BA, let a semicircle be drawn, <namely> ΔAE, (c) and let a line joined from E to Γ be drawn through to Z, (d) and let a certain line be drawn through from Δ in such a way, that HΘ will be equal to ΘK; (1) for this is possible;[80] (e) and let perpendiculars be drawn from H, K on ΔE, <namely> HΛ, KNM. (2) Now since it is: as KΘ to ΘH, MB to BΛ,[81] (3) and KΘ is equal to ΘH, (4) therefore MB, too, is equal to BΛ; (5) so that the remainder, too, ME,[82] <is equal> to ΛΔ. (6) Therefore the whole ΔM, too, is equal to ΛE, (7) and through this it is: as MΔ to ΔΛ, ΛE to EM.[83] (8) But as MΔ to ΔΛ, KM to HΛ,[84] (9) while as ΛE to EM, HΛ to NM.[85] (10) Again, since it is: as ΔM to MK, KM to ME,[86] (11) therefore as ΔM to ME, so the <square> on ΔM to the <square> on MK,[87] (12) that is the <square> on ΔB to the <square> on BΘ,[88] (13) that is the <square> on AB to the <square> on BΘ;[89] (14) for ΔB is equal to BA.[89] (15) Again, since it is: as MΔ to ΔB, ΛE to EB,[90] (16) but as MΔ to ΔB, KM to ΘB,[91] (17) while as ΛE to EB, HΛ to ΓB,[92] (18) therefore also: as KM to ΘB, HΛ to ΓB; (19) and alternately, as KM to HΛ, ΘB to ΓB.[93] (20)

---

[78] Apparently Sporus wrote a book called *Honeycombs* (or *Aristotelean honeycombs*?), probably in late antiquity (third century AD?). Our knowledge is a surmise based on indirect evidence from Pappus, in other words our knowledge is minimal. Perhaps he is to be envisaged as a collector of remains from ancient times, mathematical, philosophical and others? In this case, the lack of originality in his solution should not come as a surprise. As usual, consult Knorr (1989) 87–93.

[79] "Mean proportionals in a continuous proportion" is an expanded way of saying "mean proportionals."

[80] Cf. Eutocius' comment on Philo's solution. Perhaps this sentence, too, is a Eutocian comment, a brief intrusion into the Sporian text.

[81] AB is perpendicular on ΔE, just as KM and HΛ are. Hence through *Elements* I.28, VI.2 the set of proportions ΔK:ΔΘ:ΔH::ΔM:ΔB:ΔΛ can be derived, from which, through *Elements* V.17, it is possible to derive, *inter alia*, KΘ:ΘH::MB:BΛ.

[82] "Remainder" after MB is taken away from the radius EB.

[83] The reasoning is similar to *Elements* V.7.   [84] *Elements* VI.2, 4.

[85] *Elements* VI.2, 4. Steps 7–9 seem to lead to the conclusion KM:HΛ::HΛ:MN. This conclusion however is not asserted, and is not required in the proof. Steps 7–9 are thus a false start. Is this text an uncorrected draft? Mistaken? Corrupt?

[86] *Elements* III.31, VI.8.   [87] *Elements* VI. 20 Cor. 2.

[88] *Elements* VI.2, 4.   [89] Radii in circle.

[90] The same kind of reasoning as in Step 7.   [91] *Elements* VI.2, 4.

[92] *Elements* VI.2, 4.   [93] *Elements* V.16.

But as KM to HΛ, MΔ to ΔΛ,[94] (21) that is ΔM to ME,[95] (22) that is the <square> on AB to the <square> on ΘB;[96] (23) therefore also: as the <square> on AB to the <square> on ΘB, BΘ to BΓ. (f) Let a mean proportional be taken between ΘB, BΓ, <namely> Ξ. (24) Now since, as the <square> on AB to the <square> on BΘ, ΘB to BΓ, (25) but the <square> on AB has to the <square> on BΘ a ratio duplicate of AB to BΘ, (26) while ΘB has to BΓ a ratio duplicate of ΘB to Ξ,[97] (27) therefore also: as AB to BΘ, BΘ to Ξ. (28) But as ΘB to Ξ, Ξ to BΓ; (29) therefore also: as AB to BΘ, ΘB to Ξ and Ξ to BΓ.

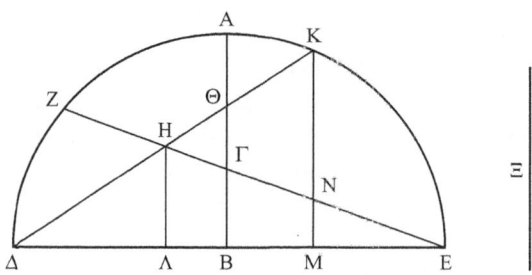

Catalogue: Sporus
Codex D has omitted
Θ. Codex E has H
instead of N.

And it is obvious that this, too, is the same as that proved by both Pappus and Diocles.

As Menaechmus[98]

Let the two given lines be A, E; so it is required to find two mean proportionals between A, E.

(a) Let it come to be,[99] and let <the mean proportionals> be B, Γ, (b) and let a line be set out, <given> in position, <namely> ΔH

---

[94] *Elements* VI.2, 4.   [95] Step 5.

[96] Step 13.   [97] *Elements* V. Def. 10.

[98] A fourth-century BC mathematician, apparently involved, *inter alia*, with the origins of Conics. Perhaps a student of Eudoxus and an acquaintance of Plato, he seems to have had wide, philosophical interests. Whether he has produced both this solution and its alternative is, however, another question: Toomer has argued in (1976) 169–70 that the alternative solution was, in fact, by Diocles, as a closely related proof is preserved in the Arabic translation of Diocles' *On Burning Mirrors*.

[99] That is, assume the problem solved. The following passage is in an analysis/synthesis structure.

(limited at Δ),[100] (c) and, at Δ, let ΔZ be set equal to Γ, (d) and let
ZΘ be drawn at right <angles =to ΔH>, (e) and let ZΘ be set equal
to B.[101] (1) Now since three lines <are> proportional, A, B, Γ, (2)
the <rectangle contained> by A, Γ is equal to the <square> on B;[102]
(3) therefore the <rectangle contained> by a given <line> A, and by Γ,
that is ΔZ, (4) is equal to the <square> on B, (5) that is to the <square>
on ZΘ. (6) Therefore Θ is on a parabola drawn through Δ.[103] (f) Let
parallels ΘK, ΔK be drawn as parallels.[104] (7) And since the <rectangle
contained> by B, Γ is given; (8) for it is equal to the <rectangle
contained> by A, E;[105] (9) therefore the <rectangle contained> by
KΘZ is given as well. (10) Therefore Θ is on a hyperbola in KΔ, ΔZ as
asymptotes.[106] (11) Therefore Θ is given;[107] (12) so that Z <is given>,
too.

So it will be constructed like this. Let the two given lines be A, E, and
<let> ΔH <be given> in position, limited at Δ, (a) and, through Δ, let
a parabola be drawn, whose axis is ΔH, while the *latus rectum*[108] of the
figure is A, and let the lines drawn down <from the parabola> in a right
angle on ΔH, be in square the rectangular areas applied along A,[109]
having as breadths the <lines> taken by them <from the line ΔH>

---

[100] The line ΔH is not so much a magnitude, as a position: it is the line on which Z
is situated, ΔK is erected, etc. Hence the strange description, "Given in position, limited
at Δ."

[101] ZΘ begins as a position at (d) and becomes a magnitude at (e).

[102] *Elements* VI.17. From this point onwards, A and Δ are consistently inverted in
the manuscript's text. It would seem that in Eutocius' original the two given lines were
Δ, E and the vertex of the parabola was A. Eutocius inverted A and Δ, in his diagram
and at the beginning of his text, but here he forgot about this and just went on copying
from his original: let him who has never switched labels in his diagrams cast the first
stone. I follow Heiberg's homogenization, keeping Eutocius' inversion (Torelli, following
Moerbeke, chose the other way around).

[103] *Conics* I.11. To paraphrase algebraically, Menaechmus notes that there is a constant A satisfying A*ΔZ=ZΘ², so that Θ is on a parabola whose vertex is Δ, its *latus
rectum* (see below) A.

[104] ΘK, ΔK are parallel not to each other, but to the already drawn lines, ZΔ, ZΘ.

[105] A, E are the original, *given* lines, so the rectangle they contain is given as well.

[106] *Conics* II.12. Menaechmus notes that the point Θ determines a constant rectangle
intercepted by the two lines KΔ, ΔH, which is a property of the hyperbola.

[107] Θ is now given as the intersection of a given parabola and a given hyperbola.

[108] *Latus rectum*: a technical term. In every parabola, there is a line X such that, for
every point on the axis of the parabola (e.g. Z in our case) the rectangle contained by the
line from the point to the vertex (e.g. ΔZ in our case) and by the line X, is always equal
to the square on what is known as the "ordinate" on that point (e.g. ZΘ in our case). This
line X is known as the *Latus rectum*.

[109] That is, rectangles whose one side is A . . .

towards the point Δ.[110] (b) Let it be drawn and let it be the <parabola> ΔΘ, (c) and <let> ΔK <be> right,[111] (d) and let a hyperbola be drawn in KΔ, ΔZ <as> asymptotes, (1) on which <hyperbola>, the <lines> drawn parallel to KΔ, ΔZ make the rectangular area <contained by them> equal to the <rectangle contained> by A, E;[112] (2) so it <=the hyperbola> will cut the parabola. (e) Let it cut <it> at Θ, (f) and let KΘ, ΘZ be drawn as perpendiculars. (3) Now since the <square> on ZΘ is equal to the <rectangle contained> by A, ΔZ,[113] (4) it is: as A to ZΘ, ΘZ to ZΔ.[114] (5) Again, since the <rectangle contained> by A, E is equal to the <rectangle contained> by ΘZΔ, (6) it is: as A to ZΘ, ZΔ to E.[115] (7) But as A to ZΘ, ZΘ to ZΔ; (8) and therefore as A to ZΘ, ZΘ to ZΔ and ZΔ to E. (g) Let B be set equal to ΘZ, (h) while Γ <be> set equal to ΔZ; (9) therefore it is: as A to B, B to Γ and Γ to E. (10) Therefore A, B, Γ, E are continuously proportional; which it was required to find.

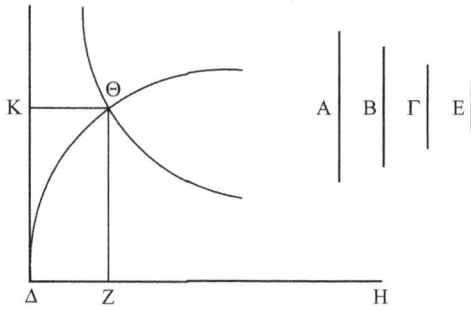

Catalogue:
Menaechmus
Codex H has omitted
H, has Λ for A.

In another way

Let the two given lines be (at right <angles> to each other) AB, BΓ, (a) and let their means come to be <as> ΔB, BE, so that it is: as ΓB to BΔ, so BΔ to BE and BE to BA, (b) and let ΔZ, EZ be drawn at right <angles>.[116] (1) Now since it is: as ΓB to BΔ, ΔB to BE, (2) therefore the <rectangle contained> by ΓBE, that is the <rectangle contained>

---

[110] . . . and whose other side are lines such as ΔZ. This lengthy description unpacks the property of the *latus rectum* – the property of the parabola. The construction of a parabola, given its *latus rectum*, is provided at *Conics* I.52.

[111] That is, ΔK is at right angles to the axis of the parabola.

[112] *Conics* II.12.   [113] *Conics* I.11.

[114] *Elements* VI.17.   [115] *Elements* VI.16.

[116] "At right angles:" Both to each other and to the original lines EB, BΔ.

by a given <line> and by BE (3) is equal to the <square> on BΔ,[117] (4) that is <to the square on> EZ.[118] (5) Now since the <rectangle contained> by a given <line> and by BE is equal to the <square> on EZ, (6) therefore Z touches[119] a parabola, <namely that> around the axis BE.[120] (7) Again, since it is: as AB to BE, BE to BΔ, (8) therefore the <rectangle contained> by ABΔ, that is the <rectangle contained> by a given <line> and by BΔ, (9) is equal to the <square> on EB,[121] (10) that is <to the square on> ΔZ;[122] (11) therefore Z touches a parabola, <namely that> around the axis BΔ; (12) but it has touched another given <parabola, namely that> around BE; (13) therefore Z is given. (14) And ZΔ, ZE are perpendiculars; (15) therefore Δ, E are given.

And it will be constructed like this. Let the two given lines be (at right <angles> to each other) AB, BΓ, (a) and let them be produced, from B, without limit, (b) and let a parabola be drawn around the axis BE, so that the lines drawn down <from the parabola> on BE are in square the <rectangles applied> along BΓ.[123] (c) Again let a parabola be drawn around the axis ΔB, so that the lines drawn down <from the parabola on the axis> are in square the <rectangles applied> along AB; (1) so the parabolas will cut each other. (d) Let them cut <each other> at Z, (e) and let ZΔ, ZE be drawn from Z as perpendiculars. (2) Now since ZE, that is ΔB[124] (3) has been drawn down in a parabola, (4) therefore the <rectangle contained> by ΓBE is equal to the <square> on BΔ;[125] (5) therefore it is: as ΓB to BΔ, ΔB to BE.[126] (6) Again, since ZΔ, that is EB[127] (7) has been drawn in a parabola, (8) therefore the <rectangle contained> by ΔBA is equal to the <square> on EB;[128] (9) therefore it is: as ΔB to BE, BE to BA.[129] (10) But as ΔB to BE, so ΓB to BΔ; (11) and therefore as ΓB to BΔ, BΔ to BE and EB to BA; which it was required to find.

---

[117] *Elements* VI.17.   [118] *Elements* I.28, 33.

[119] "Touches:" a somewhat strange verb to use for a *point*. The claim is that Z is on the parabola.

[120] *Conics* I.11. Around a given axis there can be an infinite number of parabolas. Our parabola, however, is uniquely given, since Step 5 effectively defines its *latus rectum*. The same situation is found in Step 11 below.

[121] *Elements* VI.17.   [122] *Elements* I.28, 33.

[123] To spell this out: for whatever point Z, EZ^2 shall be equal to $EB*B\Gamma$. The text lapses here into the peculiar dense formulae of the theory of conic sections. The construction itself is provided at *Conics* I.52.

[124] *Elements* I.28, 33.   [125] *Conics* I.11.

[126] *Elements* VI.17.   [127] *Elements* I.28, 33.

[128] *Conics* I.11.   [129] *Elements* VI.17.

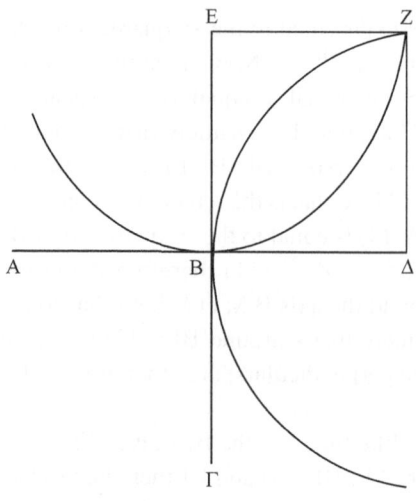

Catalogue:
Menaechmus, second
diagram
Codex G has BΔ equal
to BE, while D has BΔ
greater than BE.
Related to this, codex
D has the right-hand
curved line composed
of two arcs, not one, as
in the thumbnail.

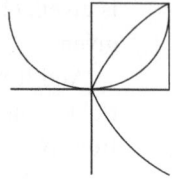

[And the parabola is drawn by the compass invented by our teacher, Isidore the Milesian mechanician, this being proved by him in the commentary which he produced to Hero's *On Vaulting*.][130]

Archytas' solution, according to Eudemus' *History*[131]

Let the two given lines be AΔ, Γ; so it is required to find two mean proportionals between AΔ, Γ.

(a) Let a circle, <namely> ABΔZ, be drawn around the greater <line>, AΔ, (b) and let AB, equal to Γ, be fitted inside, (c) and, produced, let it meet, at Π, the tangent to the circle <drawn> from Δ, (d) and let BEZ be drawn parallel to ΠΔO, (e) and let a right semicylinder be imagined on the semicircle ABΔ, (f) and <let> a right semicircle <be imagined> on the <line> AΔ, positioned in the parallelogram of the semicylinder;[132] (1) so, when this semicircle is rotated, as from Δ to

---

[130] Probably an intrusion by the same person as the scholiast writing at the end of the commentaries to Book I and II (see notes there) about whose identity little is known; certainly he belongs to the same general period as that of Eutocius himself. Neither *On Vaulting* nor its commentary are extant.

[131] Archytas was a central figure in early fourth-century BC intellectual life, clearly, among other things, a mathematician. Eudemus, Aristotle's pupil, wrote, near the end of the same century, a history of mathematics. Eutocius writes about a thousand years later, and it is an open question: what did he have as direct evidence for the works of Archytas, or of Eudemus?

[132] "Right" semicircle – i.e., in right angles to the plane of the original circle (the plane of the page, as it were). "In the parallelogram of the semicylinder" – a semicylinder consists of a half of a cylinder, together with a parallelogram where the original cylinder

B (the limit A of the diameter remaining fixed), it will cut the cylindrical surface in its rotation, and will draw in it a certain line.[133] (2) And again, if the triangle AΠΔ is moved in a circular motion (AΔ remaining fixed), opposite <in direction> to that of the semicircle, it will produce, by the line AΠ, a conical surface; which <line>, rotated, will meet the cylindrical line at a certain point;[134] (3) and at the same time B, too,

---

was cut. If a cylinder is a circle, extended into space, then a semicylinder is a semicircle – bound by a semi-circumference and a diameter – extended into space, and "the parallelogram of the semicylinder" is the diameter, extended into space. Effectively, what we are asked to do is to take the semicircle ABΔ, and lift it up in space (keeping the line AΔ in place), until it has rotated 90 degrees. More of such spatial thinking is to come.

[133] Here verbal and two-dimensional representations almost break down. I will make an effort: keep in mind the semicircle we have just lifted at right angles into space – the semicircle on top of the diameter AΔ. We now detach it from the diameter, keeping however the point A fixed. We rotate it, gliding along the surface of the diagram, keeping its upright position. Learn to do this; glide it in your imagination; imagine it skating along the ice-rink of the original circle (the ice-rink of the diagram, of the page), always keeping one foot firmly on the point A. I shall soon return to this choreography. Now, when your mind is used to this operation, evoke another imaginary object, the semicylinder on top of the original semicircle ABΔ. So we have two objects: the rotating semicircle, and the semicylinder. As the semicircle glides along, it is possible to identify the point where it cuts the semicylinder. For instance, at its Start position, it cuts the semicylinder at Δ. And at the other position depicted in our diagram, it cuts the semicylinder at K. Notice that K is higher than Δ. At its Start position, the semicircle fits snugly inside the semicylinder. As it glides further, parts of it begin to project out of the semicylinder – the second foot is no longer inside the semicylinder – indeed the semicircle will completely emerge out of the semicylinder after a quarter rotation. So look at it, at that other position, AKΔ (another Δ now; this point is allowed – almost uniquely in Greek mathematics – to keep its name while in movement): now the semicircle projects out of the semicylinder, and the point where it cuts the semicylinder is not right at the bottom of the semicylinder (as with the original position of Δ), but a bit higher, K. Move the semicircle further, and the intersection is again a bit higher. So we can imagine the line composed of such points of intersection – and this is finally the line which this Step 1 calls into existence.

[134] Here the three-dimensional construction is slightly redundant. We do not require the *cone* as such, but we merely use it as scaffolding for the line AΠ. This line is to be rotated around the diameter AΔ, keeping its head at A and keeping its distance from the diameter AΔ constant. Think of it now as three-dimensional, gravity-free ballet. We have two dancers, a ballerina and a male dancer. The ballerina is the curved line, the arc of the moving semicircle ("Tatiana"). I have discussed Tatiana in the preceding note. We now also have a male dancer, the straight line AΠ ("Eugene"). Both glide effortlessly in space: Tatiana with her two feet on the ground, one foot firmly kept at A, the other rotating; Eugene, even more acrobatically, holds Tatiana at her firm foot A and rotates in space, going round and round, always keeping the same distance from the line AΔ (which happens to be the base position of Tatiana's movement). Even more fantastically, our dancers can intersect with each other and with the stage-props, and go on dancing. Now Eugene, in his movement, keeps intersecting with the (static) semicylinder. At first, he intersects with the semicylinder at the point B. As he moves higher into space,

will, rotating, draw a semicircle in the surface of the cone.[135] (g) So let
the moved semicircle[136] have its position at the place of the meeting of
the lines,[137] as the <position> of △KA, (h) and <let> the contrariwise
rotated triangle[138] <have> the <position> of △ΛA, (i) and let the said
point of intersection be K, (j) and, also, let the semicircle drawn by B
be BMZ, (k) and let the common section of it <=semicircle BMZ>
and of the circle BΔZA be the line BZ, (l) and let a perpendicular be
drawn from K on the plane of the semicircle BΔA;[139] (4) so it will
fall on the circumference of the circle,[140] (5) through the cylinder's
being set up right.[141] (m) Let it <=the perpendicular> fall and let it
be KI, (n) and let the <line> joined from I to A meet BZ at Θ, (o) and
let AΛ meet the semicircle BMZ at M,[142] (p) and let KΔ, MI, MΘ be
joined. (6) Now since each of the semicircles △KA, BMZ is right to the

---

his intersections with the semicylinder move higher, too, as they also move away from
the point A. Thus Eugene, too, draws a line of intersections – "Eugene's line," the
line drawn by the intersection of the rotating line and the original semicylinder (just
as Tatiana had produced her own, "Tatiana's Line," made of her intersections with the
cylindrical surface, in the preceding step). We shall soon look, finally, at intersection of
those lines of intersections – a second-order intersection – between Tatiana's and Eugene's
lines.

[135] If we look at a point along the line AΠ, and plot its circular movement as the line
AΠ keeps rotating, we will see a circle (or a semicircle if we concentrate, as Archytas
does, on the part "above" the original circle, above the plane of the page). Archytas
concentrated on the point B, and on the circle BMZ it traces in its movement. This will
become important later on in the proof.

Objects, moved, leave a trace, a virtual object. This is the heart of this solution.

[136] That is "Tatiana."

[137] That is the meeting-point of "Tatiana's and Eugene's lines" – the lines drawn on
the semicylinder. This is the second-order intersection between lines of intersections,
mentioned in n. 134. That the point of intersection exists, and that it is unique, can be
shown by the following topological intuitive argument: Eugene's line starts at B and
moves continuously upwards as it moves towards Δ, reaching finally a point above Δ.
Tatiana's line starts at Δ and moves continuously upwards as it moves towards B, reaching
finally a point above B. This chiastic movement must have a point of intersection.

[138] Instead of the line AΠ rotated, we now imagine the entire triangle AΠΔ rotated,
its side AΔ remaining fixed as the cone is drawn. Its motion is "contrariwise" to Tatiana's.

[139] That is, on the original plane of the page.

[140] That is the original circle ABΔZ.

[141] The point K is on the surface of the (right) semicylinder, projecting upwards from
the original semicircle ABΔ, and therefore the perpendicular drawn directly downwards
from K is simply a line on the semicylinder, and must fall on the circumference of ABΔ.

[142] △ΛA is the position of the rotating triangle △ΠA when it reaches the point of
intersection K. We take two snapshots of the line AΠ: once, resting on the plane of the
page, when it is ABΠ; again, when it is stretched in mid-air, passing through the point
K. Now Π has become Λ, B has become M, while A remained fixed. The import of Step
o is the identification of M as the mapping of B into the line AΛ.

underlying plane,[143] (7) therefore the common section <=of the two semicircles>, too, MΘ, is at right <angles> to the plane of the circle <=ABΔZ>;[144] (8) so that MΘ is right to BZ, as well.[145] (9) Therefore the <rectangle contained> by BΘZ, that is the <rectangle contained> by AΘI,[146] (10) is equal to the <square> on MΘ;[147] (11) therefore the triangle AMI is similar to each of the <triangles> MIΘ, MAΘ, and the <angle contained> by IMA is right;[148] (12) and the <angle contained> by ΔKA is right, too;[149] (13) therefore KΔ, MI are parallel,[150] (14) and it will be proportional: as ΔA to AK, that is KA to AI,[151] (15) so IA to AM, (16) through the similarity of the triangles.[152] (17) Therefore four <lines>, ΔA, AK, AI, AM are continuously proportional, (18) and AM is equal to Γ, (19) since <it is> also <equal> to AB; (20) therefore two mean proportionals have been found, between the given <lines> AΔ, Γ <namely> AK, AI.

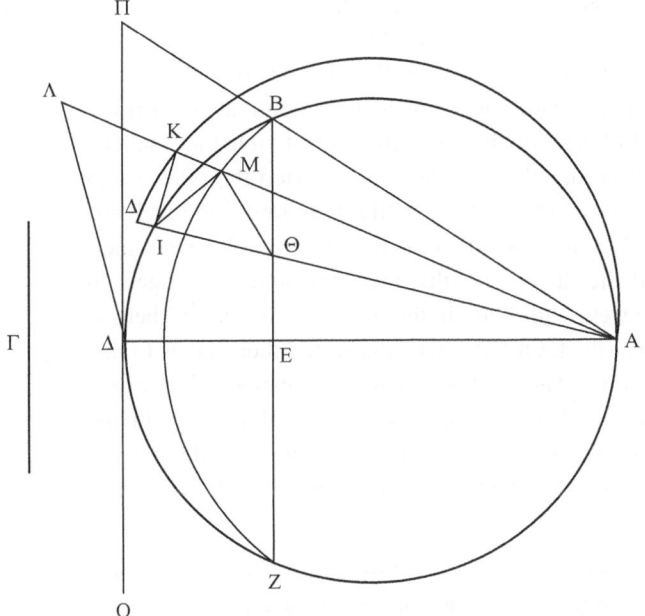

Catalogue: Archytas Codex E extends the line ΠO in both directions, beyond the points Π, O. Something went wrong in codex A with the letter Γ. Codices DE have Ϲ (!) instead of Γ, while Codex B may perhaps have omitted it altogether to begin with (mystified by a very unfamiliar Greek character?). Most likely, this was a very badly executed Γ. Codex H positions K on the intersection of ΠO/ΛA.

---

[143] "Underlying plane:" the plane of the original circle; the plane of the diagram/page.

[144] *Elements* XI.19.   [145] *Elements* XI. Def. 3.   [146] *Elements* III.35.

[147] *Elements* III.31, VI.8. It should be seen that BZ is the diameter of the circle traced by the point B in its rotating movement.

[148] *Elements* VI.8.

[149] *Elements* III.31. Remember, once again: ΔKA is a semicircle, and ΔA is a diameter.

[150] *Elements* I.28.   [151] *Elements* VI.8, 4.   [152] *Elements* VI.8, 4.

As Eratosthenes[153]

Eratosthenes to king Ptolemy, greetings.

They say that one of the old tragic authors introduced Minos, building a tomb to Glaucos, and, hearing that it is to be a hundred cubits long in each direction, saying:

> *You have mentioned a small precinct of the tomb royal;*
> *Let it be double, and, not losing its beauty,*
> *Quickly double each side of the tomb.*

He seems, however, to have been mistaken; for, the sides doubled, the plane becomes four times, while the solid becomes eight times. And this was investigated by the geometers, too: in which way one could double the given solid, the solid keeping the same shape; and they called this problem "duplication of a cube:" for, assuming a cube, they investigated how to double it. And, after they were all puzzled by this for a long time, Hippocrates of Chios was the first to realize that, if it is found how to take two mean proportionals, in continuous proportion, between two straight lines (of whom the greater is double the smaller), then the cube shall be doubled, so that he converted the puzzle into another, no smaller puzzle.[154] After a while, they say, some Delians, undertaking to fulfil an oracle demanding that they double one of their altars, encountered the same difficulty, and they sent messengers to the geometers who were with Plato in the Academy, asking of them to find that which was asked. Of those who dedicated themselves to this diligently, and investigated how to take two mean proportionals between two given lines, it is said that Archytas of Tarentum solved this with the aid of semicylinders, while Eudoxus did so with the so-called curved lines;[155] as it happens, all of them wrote demonstratively, and it was

---

[153] A third-century BC polymath, the librarian in the Alexandria library; we shall get to see him again in Volume 3 of this translation, as the addressee to one of Archimedes' works, the *Method*. The genuineness of the following letter has been doubted by Wilamowitz (1894). I myself follow Knorr (1989) in thinking this is by Eratosthenes. The treatise is dedicated to Ptolemy III Euergetes (reign 246–221 BC).

[154] Notice that converting X into something, not smaller than X, is the theme of the problem of duplication itself. Eratosthenes' text is shot through with this kind of intelligent play.

[155] This "it is said" is lovely. The line of myth starts with Minos and tragedy, is stressed by the repeated vague allusions ("one of the tragic authors . . ."), then is reinforced through the Delian oracle; so that now even the fully historical, relatively recent Archytas and Eudoxus may acquire the same literary–mythical aura ("it is said;" and the vague, deliberately tantalizing descriptions: "semicylinders . . . so-called [!] curved lines." Clearly, as the rest of the letter shows, Eratosthenes knew the constructions in full mathematical detail). Eratosthenes writes of mathematics, within *literary* Greek culture.

impossible practically to do this[156] by hand (except Menaechmus, by the *shortness*[157] – and this with difficulty). But we have conceived of a certain easy mechanical way of taking proportionals through which, given two lines, means – not only two, but as many as one may set forth – shall be found. This thing found, we may, generally: reduce a given solid (contained by parallelograms) into a cube, or transform one solid into another, both making it[158] similar[159] and, while enlarging it, maintaining the similitude, and this with both altars and temples;[160] and we can also reduce into a cube, both liquid and dry measures (I mean, e.g., a *metertes*[161] or a *medimnos*[162]), and we can then measure how much the vessels of these liquid or dry materials hold, using the side of the cube.[163] And the conception will be useful also for those who wish to enlarge catapults and stone-throwing machines; for it is required to augment all – the thicknesses and the magnitudes and the apertures and the *choinikids*[164] and the inserted strings – if the throwing-power is to be proportionally augmented, and this can not be done without finding the means. I have written to you the proof and the construction of the said machine.[165]

For let there be given two unequal lines, <namely> AE, ΔΘ, between which it is required to find two mean proportionals in continuous proportion, (a) and, on a certain line, <namely> EΘ, let AE be set at right <angles>, (b) and let three parallelograms, <namely> AZ, ZI, IΘ, be constructed on EΘ, (c) and, in them, let diagonals be drawn: AZ, ΛH, IΘ; (1) so they themselves will be parallel.[166] (d) So, the middle parallelogram (ZI) remaining in place, let AZ be pushed above the middle <parallelogram>, <and let> IΘ <be pushed> beneath it, as in the second figure, until A, B, Γ, Δ come to be on a <single> line,[167] (e) and let a line be drawn through the points A, B, Γ, Δ,

---

[156] I.e. duplicating the cube.

[157] Another tantalizing, vague description. I therefore keep the manuscripts' reading against Heiberg's (possible) emendation, "except, to some small extent, Menaechmus."

[158] I.e. the created solid.   [159] I.e. to the original solid.

[160] So we round the theme of the Minos/Delos myths.

[161] A liquid measure.   [162] A dry measure.

[163] In other words, the two mean proportions will allow us, given a vessel containing X measures, to construct a vessel containing Y measures.

[164] Boxes containing the elastic strings of the throwing machines.

[165] Note the abruptness of the down-to-business move, here between the rhetorical introduction and the mathematical proof. We suddenly see the style of Greek mathematics vividly set against another Greek discourse.

[166] This is true only if EZ=ZH=HΘ (and then *Elements* I.28, I.4), an assumption which is nowhere stated. An oversight by Eratosthenes? A textual corruption?

[167] This is extremely confusing, especially since Eratosthenes (who assumes the reader is acquainted with a model of the machine) did not bother to explain to us that the configuration is, in a way, three dimensional. We must imagine the three parallelograms

(f) and let it meet the <line> EΘ, produced, at K; (2) so it will be: as AK to KB, EK to KZ (in the parallels AE, ZB),[168] (3) and ZK to KH (in the parallels AZ, BH).[169] (4) Therefore as AK to KB, EK to KZ and KZ to KH. (5) Again, since it is: as BK to KΓ, ZK to KH (in the parallels BZ, ΓH),[170] (6) and HK to KΘ (in the parallels BH, ΓΘ), (7) therefore as BK to KΓ, ZK to KH and HK to KΘ. (8) But as ZK to KH, EK to KZ; (9) therefore also: as EK to KZ, ZK to KH and HK to KΘ. (10) But as EK to KZ, AE to BZ, (11) and as ZK to KH, BZ to ΓH, (12) and as HK to KΘ, ΓH to ΔΘ;[171] (13) therefore also: as AE to BZ, BZ to ΓH and ΓH to ΔΘ. (14) Therefore two means have been found between AE, ΔΘ, <namely> both BZ and ΓH.[172]

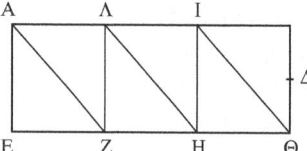 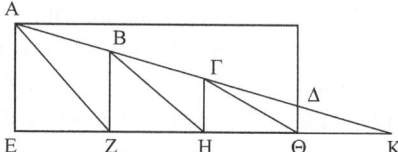

So these are proved for geometrical surfaces. But so as we may also take the two means by a machine, a box is fixed (made of wood, or ivory, or bronze), holding three equal tablets, as thin as possible. Of these, the middle is fitted in its place, while the other two are moveable along grooves (the sizes and the proportions may be as anyone wishes them; for the arguments of the proof will yield the conclusion in the same way). And, for taking the lines in the most precise way, it must be done with great art, so that when the tablets are simultaneously

Catalogue:
Eratosthenes
Codex D has EZ greater than ZH, both smaller than HΘ, in the left-hand rectangle; a neat trisection in the right-hand rectangle. Codex E has EZ smaller than ZH, both smaller than HΘ, in the left-hand rectangle, a neat trisection in the right-hand triangle.

AZ, ΛH, IΘ as three sliding doors, each on a different groove (they are on the wall of the tatami room), AZ nearest to us, IΘ farthest from us, ΛH midway between the two. We now slide these sliding doors: AZ to the right ("covering" part of the door ΛH), IΘ to the left (partly covered, then, by the door ΛH). We look throughout at the following two points: the point where the diagonal IΘ (painted on the door IΘ) meets IH (the edge of the door ΛH); and the point where the diagonal ΛH meets ΛZ, (the edge of the door AZ). At first these are the points I, Λ, respectively. As we slide the doors, slowly, through some trial and error, we shall reach a point where the two points (now christened Γ, B) both lie on the line AΔ (itself constantly changing as we slide the doors!). Here we stop. Notice this one crucial point: by sliding the doors to the left or to the right, the painted diagonals remain parallel to each other, as do the edges of the doors. Essentially, before us is a parallelism-preserving machine.

[168] *Elements* VI.2, 4.   [169] *Elements* VI. 2, 4.

[170] *Elements* VI.2, 4.   [171] Steps 10–12: *Elements* VI.2, 4.

[172] To signal the end of the strictly mathematical discourse, Eratosthenes now redundantly speaks of "both" BZ and ΓH – a redundancy that throws us back into the rhetorical world of the introduction of the letter.

moved they all remain parallel[173] and firm[174] and touching each other throughout.[175]

In the dedication, the machine is made of bronze, and is fitted with lead below the crown of that pillar, and the proof below it (phrased more succinctly), and the figure, and with it the epigram.[176] So let these be written below as well, for you, so that you have, also, just as in the dedication. (Of the two figures, the second is inscribed in the pillar.)

Given two lines, to find two mean proportionals in continuous proportion. Let AE, $\Delta\Theta$ be given.[177] (a) So I move the tables in the machine together, until the points A, B, Γ, Δ come to be on a <single> line. ((b) So let it be imagined, as in the second figure.) (1) Therefore it is: as AK to KB, EK to KZ (in the parallels AE, BZ),[178] (2) and ZK to KH (in the <parallels> AZ, BH);[179] (3) therefore as EK to KZ, KZ to KH. (4) But as they themselves are to each other, so are both: AE to BZ and BZ to ΓH.[180] (5) And we shall prove in the same way that, also, as ZB to ΓH, ΓH to $\Delta\Theta$; (6) therefore AE, BZ, ΓH, $\Delta\Theta$ are proportional. Therefore two means have been found between the two given <lines>.

And if the given <lines> will not be equal to AE, $\Delta\Theta$, then, after we make AE, $\Delta\Theta$ proportional to them, we shall take the means between them <=AE, $\Delta\Theta$>, and return to those <given lines>,[181] and we shall have the task done. And if it is demanded to find several means: we shall insert tablets in the machine, <so that their total is> always more by one than <the number of> the means to be taken; and the proof is the same.

---

[173] That is, no tilting to the left or to the right.

[174] That is, no tilting backwards or forwards.

[175] That is, as one tablet slides along another, the two remain constantly in close touch, the three separate planes simulating a single plane to the greatest possible degree.

[176] The sense is clear enough: the machine (almost two dimensional) is fixed, very much like a plaque, right below the crown of a certain pillar, and then, going downwards, are, inscribed: a brief proof, a diagram and an epigram. The sentence is difficult, because of the way in which it assumes three hitherto unmentioned objects as part of the universe of discourse: the dedication (which dedication?), the pillar (which pillar?), and the epigram (which epigram?). In short, the author assumes we have seen the pillar.

[177] This is supposed to be a report of an inscription, though obviously not an exact one (consider e.g. Step b, referring to the right-hand, preceding diagram, clearly meaningless in the dedication). Of course my lettering and numbering ought to be ignored when the original inscription is envisaged. Anyway, one wonders how tall the pillar was, and how large the letters were (could this be why Ptolemy asked for an explanatory letter?).

[178] *Elements* VI.2, 4.   [179] *Elements* VI.2, 4.   [180] *Elements* VI.2, 4.

[181] Suppose the greater given line is twice AE; we take half the smaller given line as our $\Delta\Theta$, and then we double the obtained means.

If you plan, of a small cube, its double to fashion,
Or – dear friend – any solid to change to another
In nature: it's yours. You can measure, as well:
Be it byre, or corn-pit, or the space of a deep,
Hollow well.[182] As they run to converge, in between
The two rulers – seize the means by their boundary-ends.[183]
Do not seek the impractical works of Archytas'
Cylinders; nor the three conic-cutting Menaechmics;
And not even that shape which is curved in the lines
That Divine Eudoxus[184] constructed.
By these tablets, indeed, you may easily fashion –
With a small base to start with – even thousands of means.
O Ptolemy, happy! Father, as youthful as son:
You have given him all that is dear to the muses
And to kings.[185] In the future – O Zeus! – may *you* give him,
From your hand, this, as well: a sceptre.[186]
May it all come to pass. And may him, who looks, say:
"Eratosthenes, of Cyrene, set up this dedication."

As Nicomedes[187] in *On Conchoid Lines*

And Nicomedes, too, writes (in the book on conchoids which is written by him[188]) of the construction of a machine accomplishing the
same service. From the book the man seems to have prided himself
immensely, while making great fun of the solutions of Eratosthenes, as

---

[182] A georgic touch, this mention of rural measures. Eratosthenes does much more
than put geometry in metre; he brings it inside poetic genres.

[183] The rural imagery – now transformed into the shadow of a hunting scene.

[184] Can I have a quadrisyllabic here, please (E-ou–dok-sus, stress on the 'dok' sound)?
More important: the poem modulates into the invocation-of-myth theme, and so past
mathematicians shade (as they did in the prose letter) into mythical heroes.

[185] Eratosthenes is possibly writing now in the capacity of a tutor to the prince.

[186] Ptolemy gave his son a good education; Zeus would give him the rule over Egypt
(note, incidentally, that Eratosthenes is quite proper. There is nothing regicidal in wishing
a king to be outlived by his son. But the ground is a bit shaky, hence the "as youthful as
son" above).

[187] Otherwise virtually unknown: he may have lived, during the third/second century
BC (as the mathematical interests and polemics suggest), in Asia Minor (as the name
suggests). Eutocius' text is closely related (especially towards its end) to Pappus, Book
IV 26–8 (pp. 242–50).

[188] The "writes"/"written by him" dissonance is in the original.

impractical[189] and at the same time devoid of geometrical skill. So, for the sake of not missing any of those who troubled over this problem, and for comparison with Eratosthenes, we add him too to what we have written so far.[190] He writes, in essence, the following:

(a) One must imagine: two rulers conjoined at right <angles> to each other, in such a way that a single surface keeps hold of them, as are AB, ΓΔ, (b) and, in the <ruler> AB, an axe-shaped groove, inside which a *chelonion*[191] can run freely, (c) and, in the <ruler> ΓΔ, a small cylinder – at the part next to Δ and the middle line dividing the width of the ruler – which is fitted to the ruler and protrudes slightly from the higher surface of the same ruler,[192] (d) and another ruler, as EZ, (e) which has <the following> (<beginning> at some small distance from <its> limit at Z): a cut, as HΘ, which may be mounted on the small cylinder at Δ,[193] (f) and a rounded hole next to E,[194] which will be inserted in a certain axle attached to the freely running *chelonarion*[195] in the axe-shaped groove which is in the ruler AB.[196] (1) So, the ruler EZ fitted (first in the cut HΘ, over the small cylinder next to Δ, and second in the hole E, over the axle attached to the *chelonarion*), if one, taking hold of the K end of the ruler, moves it in the direction of A, then

---

[189] "Impractical:" in the original, "amechanical," with a nice pun (the solution is bad as a piece of theoretical mechanics, while being impractical).

[190] The author of this passage (probably Eutocius, though possibly an earlier compiler) does not approve of polemics in mathematics, in an interesting example of the change of intellectual mores from ancient to late ancient times. As Eutocius implies, this is the concluding solution in this catalogue. The overall structure is clear: Eratosthenes referred to many of the preceding solutions, and so he had to be penultimate; Nicomedes, who referred to Eratosthenes, had to be last. The unintended result is that polemics mark the end of this catalogue: the actors leave the stage bathetically, in a loud, vulgar quarrel.

[191] "Chelonion:" a very polysemic noun. The basic meaning is "tortoise-shell" but the Greeks, apparently, saw tortoise-shells everywhere, in parts of the body and in various artificial objects. "Knob" is probably the best stab at what is meant here.

[192] In other words: imagine a small cylinder – another knob – this time not freely running (as the knob in the ruler AB), but fixed at Δ.

[193] We take the ruler, and cut an internal rectangle away from it, producing a rectangular hole; so that we may now fix this cut ruler, loosely, with its hole upon the cylinder at Δ.

[194] Now remove the ruler, and cut it again, this time just with a rounded hole, a hole to be fixed firmly on the knob at E.

[195] A variation on the word *chelonion*, clearly meaning the very same object as in Step b.

[196] So now all the items of the construction come together. The characters are: three rulers AB, ΓΔ, EZ; an axe-shaped groove in the ruler AB (upon which, a *chelonion*; upon which, again, an axle); a cylinder on the ruler ΓΔ; and the third ruler, EZ, with two holes, one mounted on the axle on the ruler AB, the other mounted on the cylinder on the ruler ΓΔ. None of this can be grasped without the diagram.

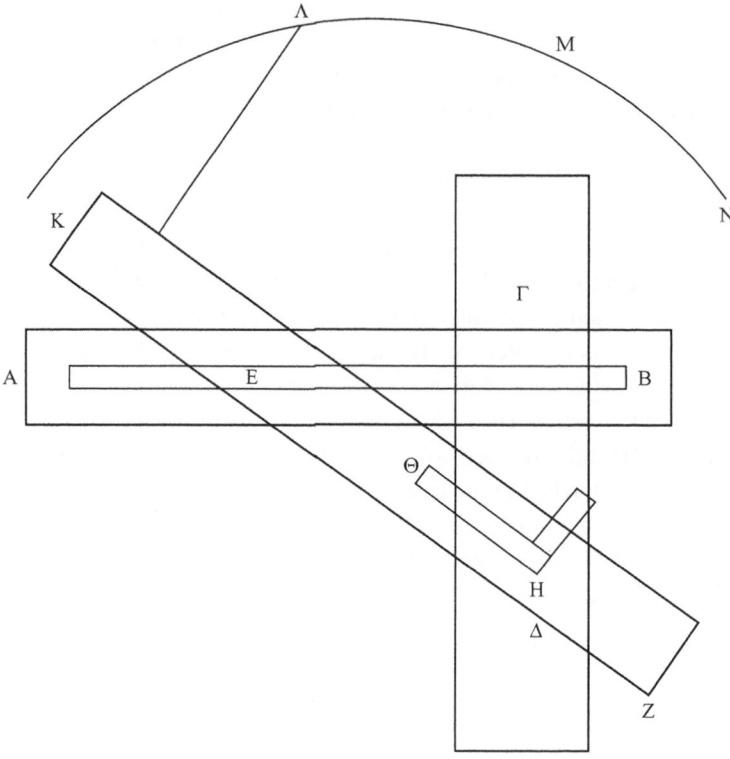

Catalogue: Nicomedes Codices GH4 have the upper angle between the rectangle KZ and the line to Λ acute. Codex 4 does not close the small line near Z of the rectangle KZ. Codex A had M instead of H.   Codex D has omitted N, and has Λ instead of A, Θ instead of B, and A instead of Δ.   Codices EH have omitted Λ.

in that of B, the point E[197] will always be carried on the ruler AB, (2) while the cut HΘ will always be moved on the small cylinder next to Δ (the middle line of the ruler EZ imagined to pass, in its movement, through the axis of the cylinder at Δ[198]), (3) the projection of the ruler, EK,[199] remaining the same.[200] (4) So if we conceive of some writing-tool at K, fixed on the base,[201] a certain line will be drawn, such as ΛMN, which Nicomedes calls "First Conchoid Line," and <he calls>

---

[197] We have moved from imperative to indicative, from construction to argument; to an extent, we have moved from mechanics to geometry, hence the sudden "point."

[198] EZ is fitted in such a way, that its middle is exactly on the center of the cylinder at Δ. Also, it is so firmly attached so as not to sway as it moves, always keeping its middle exactly on the center of Δ. That geometrical precision is never perfectly instantiated is acknowledged by the verb "imagined."

[199] That is, a projection beyond the ruler AB.

[200] As the ruler runs along the groove, its rigidity is unaffected and its internal distances are kept, including that between the points E and K.

[201] "The base of the ruler:" The writing-tool is imagined to move on, say, a piece of papyrus, which is *beneath* the machine.

the magnitude of the ruler EK "Radius[202] of the Line," and <he calls> Δ "Pole."[203]

So he proves for this line: that it has the property that it draws nearer and nearer to the ruler AB; and that if any straight line is drawn between the line and the ruler AB, it will always cut the line. And the first of the properties is best seen on another diagram. Imagining a ruler, AB, and a pole Γ, and a radius ΔE, and a conchoid line ZEH, let the two <lines> ΓΘ, ΓZ be drawn forward from Γ, (1) the resulting <lines> KΘ, ΛZ being, obviously, equal. I say that the perpendicular ZM is smaller than the perpendicular ΘN.[204]

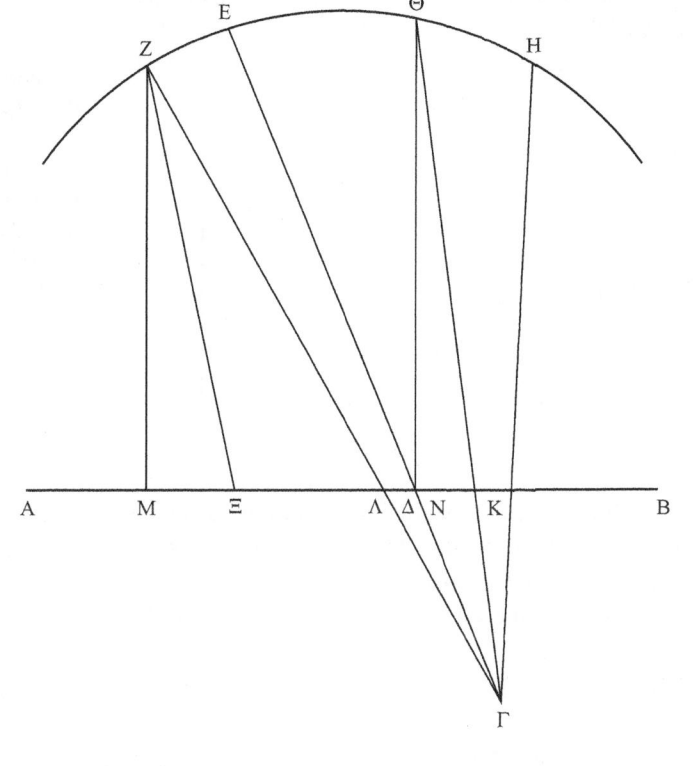

Catalogue: Nicomedes, second diagram Codex B draws ΘN so that N does not coincide with Δ, but falls on the line AB between the points K, Δ.   Codex G has the line ΓH not perpendicular to the line AB, but tilted to the left, the rest following with an even stronger leftward tilt. Codex E had an obvious mistake, corrected perhaps immediately by the same scribe, with the distribution of the letters: at first, it had A instead of M, M instead of Ξ, Ξ instead of Λ. Curiously, codex 4 has a somewhat similar mistake, corrected apparently by the same hand: A instead of M, B instead of Ξ. Codex H has Z instead of Ξ, A instead of Δ. Heiberg has strangely omitted B.

---

[202] *Diastema*, literally, "interval:" one of the expressions used by the Greeks for our "radius."

[203] Abstracting away the machinery, the conchoid is the locus of points K where a given length EK is on a line passing through a given point Δ, and E is on a given line AB. (There is a further condition, that K is on the other side of AB than Δ.) With the limiting case of Δ being on the line AB itself, the conchoid becomes a circle, accounting for Nicomedes' metaphor of EK as "radius."

[204] It appears that, in the figure, the points Δ, N are taken to coincide (so as to save space and avoid clutter), though – so as to keep the case general – they preserve their separate names and identities.

(2) For, the angle <contained> by MΛΓ being greater than the <angle contained> by NKΓ,[205] (3) the remainder, (being short of the two right <angles>), the <angle contained> by MΛZ, is smaller than the remainder <angle contained> by NKΘ,[206] (4) and through this, the <angles> at M, N being right, (5) the <angle> at Z, too, shall be greater than the <angle> at Θ.[207] (a) And if we construct the <angle contained> by MZΞ equal to the <angle> at Θ, (6) KΘ, that is ΛZ, (7) shall have to ΘN the same ratio, which ΞZ <has> to ZM;[208] (8) so that ZΛ shall have to ΘN a smaller ratio than <it has> to ZM, (9) and through this ΘN is greater than ZM.[209]

And the second <property> was that the line drawn between AB and the <conchoid> line cuts the <conchoid> line; and this is made understood as follows:[210]

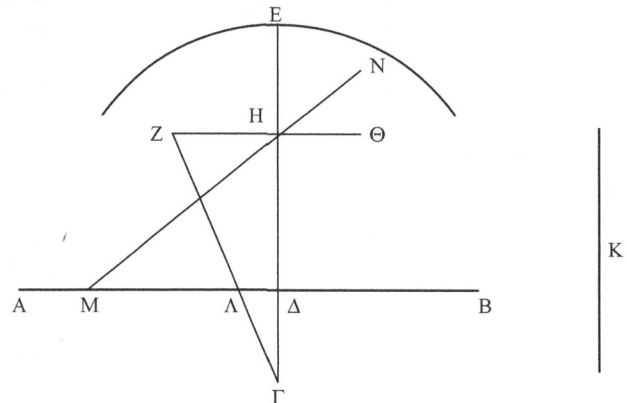

(1) For the drawn <line> is either parallel to AB or not. (a) First let it be parallel, as ZHΘ, (b) and let it come to be: as ΔH to HΓ, so ΔE to some other <line>, K, (c) and when a circumference <of a circle> is drawn, with Γ <as> center and K as radius, (d) let it cut ZH at Z,[211]

---

[205] *Elements* I.16.

[206] *Elements* I.13: the sum of two angles on a single straight line is two right angles (what we call 180 degrees). Hence MΛZ is the "the remainder, (being short of the two right <angles>)," as the angle MΛΓ is taken away from the sum MΛΓ+MΛZ.

[207] *Elements* I.32.

[208] *Elements* III.32, VI.4. Further, Ξ is between M, Λ.

[209] *Elements* V.10.

[210] It is now understood that Γ is still the pole, AB still in the same role it had in all previous constructions (the line beyond which the radius projects), and the conchoid passes through the point E. Nicomedes, or Eutocius, or some intermediate source, omit all this; the last omission, in particular, makes the argument very confusing.

[211] That we may assume that the line and the circle cut each other, can be shown, following Heiberg, from K's being greater than ΓH (from ΔH:HΓ::ΔE:K, get ΔH:ΔE::HΓ:K, and ΔH<ΔE, hence HΓ<K).

(e) and let ΓZ be joined; (2) therefore it is: as ΔH to HΓ, so ΛZ to ZΓ.[212] (3) But as ΔH to HΓ, so was ΔE to K, (4) that is to ΓZ; (5) therefore ΔE is equal to ΛZ;[213] (6) which is impossible; (7) for Z must be on the line.[214]

(e) But then, let the <drawn> line not be parallel, and let it be as MHN,[215] (f) and let ZH be drawn through H parallel to AB. (8) Therefore ZH shall meet the line; (9) so that, much more, MN <shall meet the line as well>.[216]

These being the concomitant results following through the machine, its usefulness for the <problem> put forth is shown as follows:

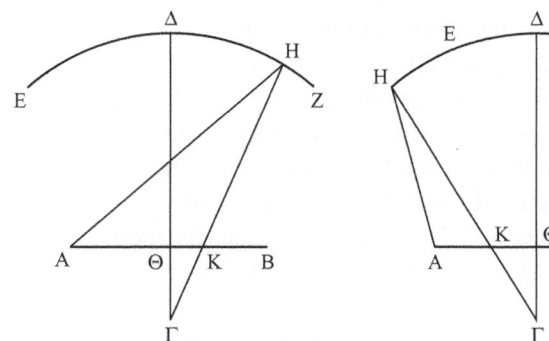

Catalogue: Nicomedes, fourth diagram
Codex B adds the letter Z on the right end of the right arc.
Codex E continues the right arc ΔH beyond H.

---

[212] *Elements* VI.2.

[213] And so Z is a point on the locus of the conchoid line; for it is the end of a straight line whose start is Γ, and whose length beyond AB, namely the length ZΛ, is equal to ΔE, which is the radius of the conchoid (in the simplest point of the conchoid – the point directly above Γ – the part projecting beyond AB is ΔE).

[214] Heiberg dislikes the proof, because the words "assume that the parallel line does not cut the conchoid" were not explicitly said, and because Step 7 draws out not a premise of the argument (which is refuted by reductio), but its conclusion (which itself provides the reductio). The text is indeed jarring, but not an impossibility.

[215] In another simplification of the diagram, similar to the congruence of Δ/N in the previous diagram, here the point M is made to settle on the line AB.

[216] The proof for the parallel case is obviously symmetrical, so that the parallel line must cut the conchoid at two points, in either direction of the line ΓE. So we have a parallel, ZHΘ, cutting the line in two points. We now tilt it with its fixed point at H, one part of it going upwards (away from AB), the other downwards (towards AB). The part going downwards will indeed, at some stage, miss the conchoid. But, for the part going upwards, the "much more" clause is entirely appropriate – it will meet the conchoid even before the parallel ZΘ does! Always within Greek standards of proof, where topology is mostly intuitive.

Again, given an angle, A,[217] and an external point, Γ,[218] to draw ΓH
and to make KH[219] equal to a given <line>.[220]

(a) Let a perpendicular, <namely> ΓΘ, be drawn from the point Γ
on AB, (b) and let it be produced, and let ΔΘ be equal to the given
<line>, (c) and with the pole Γ, and the given radius ΔΘ, and the ruler
AB, let a first conchoid line be drawn, EΔZ. (1) Therefore AH cuts it
((2) through what was proved). (d) Let it cut it at H, (e) and let ΓH be
joined; (3) therefore KH is equal to the given <line>.

These things proved, let two lines be given at right <angles> to each
other, ΓΛ, ΛA, between whom it is required to find continuous two
mean proportionals (a) and let the parallelogram ABΓΛ be completed,
(b) and let each of AB, BΓ be bisected by the points Δ, E, (c) and,
having joined ΔΛ, let it be produced and let it meet ΓB, produced <as
well>, at H, (d) and <let> EZ <be drawn> at right <angles> to BΓ,
(e) and let ΓZ be produced, being equal to AΔ,[221] (f) and let ZH be
joined, (g) and <let> ΓΘ <be drawn> parallel to it <ZH>, (1) and,
there being an angle, (the <one contained> by KΓΘ), (h) let ZΘK
be drawn through, from Z, (a given <point>), making ΘK equal to
AΔ or to ΓZ; (2) for it was proved through the conchoid that this is
possible;[222] (i) and, having joined KΛ, let it be produced and let it meet
AB, produced <as well>, at M. I say, that it is: as ΓΛ to KΓ, KΓ to
MA and MA to AΛ.

(1) Since BΓ has been bisected by E, (2) and KΓ is added to it, (3)
therefore the <rectangle contained> by BKΓ with the <square> on
ΓE is equal to the <square> on EK.[223] (4) Let the <square> on EZ
be added <as> common; (5) therefore the <rectangle contained> by
BKΓ with the <squares> on ΓEZ,[224] that is <with> the <square> on
ΓZ[225] (6) is equal to the <squares> on KEZ, (7) that is the <square>
on KZ.[226] (8) And since, as MA to AB, MΛ to ΛK,[227] (9) but as MΛ to
ΛK, so BΓ to ΓK,[228] (10) therefore also: as MA to AB, so BΓ to ΓK.

---

[217] "An angle, A:" by this Nicomedes refers to the angle BAH.

[218] Γ is *external* in the sense that it does not fall in the section of the plane intercepted
by the angle BAH. It is external to this angle.

[219] KH is implicitly defined roughly as follows: "the section of the line drawn from
Γ, which is intercepted by the angle BAH."

[220] Confusingly, the text does not move on directly to the problem of finding two
mean proportionals, but adds a final lemma, this time a problem solved with the
conchoid.

[221] Step e is where the point Z is completely determined (Step d merely set the line
on which it is located).

[222] In the preceding lemma.   [223] *Elements* II.6.

[224] "The <squares> on ΓEZ:" A way of saying "the squares on ΓE, EZ."

[225] *Elements* I.47.   [226] *Elements* I.47.

[227] *Elements* VI.2.   [228] *Elements* VI.2.

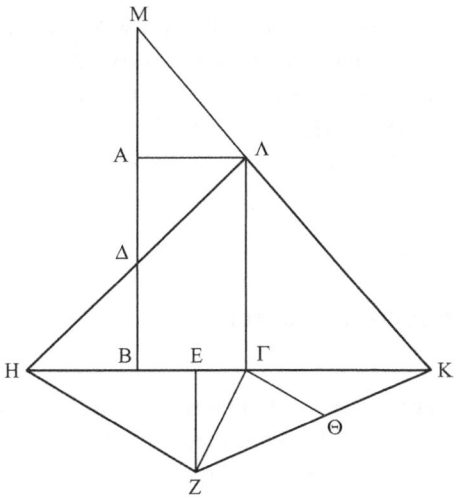

Catalogue: Nicomedes, fifth diagram
Codex D has △B greater than △A; codex G has △A greater than △B.

(11) And AΔ is half AB, (12) while ΓH is twice BΓ ((13) since ΓΛ, too, <is twice> ΔB[229]); (14) therefore it shall also be: as MA to AΔ, so HΓ to KΓ.[230] (15) But as HΓ to ΓK, so ZΘ to ΘK ((16) through the parallels HZ, ΓΘ[231]); (17) therefore compoundly, too: as MΔ to ΔA, ZK to KΘ.[232] (18) But AΔ, in turn, is assumed equal to ΘK, (19) since AΔ is equal to ΓZ, as well;[233] (20) therefore MΔ, as well, is equal to ZK; (21) therefore the <square> on MΔ, too, is equal to the <square> on ZK. (22) And the <rectangle contained> by BMA with the <square> on ΔA is equal to the <square> on MΔ,[234] (23) while the <rectangle contained> by BKΓ, with the <square> on ΓZ, was proved equal to the <square> on ZK, (24) of which, the <square>

---

[229] Step 12 derives from Step 13 through *Elements* VI.2.

[230] The move from Steps 10–12 to Step 14 is interesting. The intuition is that, halving the consequent of a ratio, and doubling the antecedent of a ratio, are equivalent operations. If you take 17:12, you may "double" it in two different ways: either by doubling the antecedent (so you get 34:12), or by halving the consequent (so you get 17:6). Ratios yield a binary structure, a Noah's ark of paired actions and counter-actions.

[231] And then apply *Elements* VI.2. From Step 14 (MA:AΔ::HΓ:KΓ) and 15 (HΓ:ΓK::ZΘ:ΘK) one expects: Step 16* (MA:AΔ::ZΘ:ΘK). This is not asserted, but is understood as the starting-point for Step 17.

[232] *Elements* V.18.

[233] That the text derives Step 18 – whose claim was explicitly stated in Step h – is very jarring. (Most probably, the final clause in Step h, "or to ΓZ," is a later gloss added on the basis of Step e, anticipating Step 18.)

[234] *Elements* II.6. The original reads "to the square . . . is equal the rectangle with the square," so the topic of the Greek sentence is the square on MΔ, connecting all Steps 20–2.

ΑΔ[235] is equal to the <square> on ΓΖ;[236] (25) for ΑΔ was assumed equal to ΓΖ; (26) therefore the <rectangle contained> by BMA, too, is equal to the <rectangle contained> by BKΓ. (27) Therefore as MB to BK, KΓ to AM.[237] (28) But as BM to BK, ΓΛ to ΓK;[238] (29) therefore also: as ΛΓ to ΓK, ΓK to AM. (30) And it is also: as ΛΓ to ΓK, MA to ΑΛ;[239] (31) therefore also: as ΛΓ to ΓK, ΓK to AM and AM to ΑΛ.

To the second theorem

Arch. 192   "And compoundly, as ΔΘ to ΘΓ, ΓΑ to ΑΕ, that is the <square> on ΓΒ to the <square> on ΒΕ." For, as on the same diagram in the said:[240] (1) since, in a right angled triangle – ΓΒΑ – a perpendicular has been drawn from the right <angle> on the base, (2) the triangles next to the perpendicular are similar both to the whole and to each other,[241] (3) and through this it is: as ΓΑ to ΑΒ, ΒΑ to ΑΕ (4) and ΓΒ to ΒΕ; (5) so that also: as the <square> on ΓΑ to the <square> on ΑΒ, so the <square> on ΓΒ to the <square> on ΒΕ. (6) But as the <square> on ΓΑ to the <square> on ΑΒ, so ΓΑ to ΑΕ;[242] (7) for as the first to the third, so the <square> on the first to the <square> on the second.[243] (8) Therefore as ΓΑ to ΑΕ, so the <square> on ΓΒ to the <square> on ΒΕ.

Arch. 193   Through the same it is proved, that it is: as ΓΑ to ΓΕ, so the <square> on ΑΒ to the <square> on ΒΕ.[244] (1) For through the similarity of the triangles, (2) it is, again: as ΑΓ to ΓΒ, so ΒΓ to ΓΕ,[245] (3) that is, as the <square> on ΑΓ to the <square> on ΓΒ, so ΑΓ to ΓΕ;[246] (4) while as the <square> on ΑΓ to the <square> on ΓΒ, so

---

[235] The preposition "on" is omitted in the manuscripts. Probably it should be reinstated (as in Heiberg, following Moerbeke), but I keep the manuscripts' reading, to point at the possibility that someone along the chain of transmission thought of this square in an abstract, very modern way.

[236] Starting from the implicit result of Steps 21–3: ((rect. BMA)+(sq. ΑΔ))=((rect. BKΓ)+(sq. ΓΖ)) we further notice in Step 24 that (sq. ΑΔ)=(sq. ΓΖ), whence Step 26 is obvious.

[237] *Elements* VI.16.   [238] *Elements* VI.2.   [239] *Elements* VI.2, applied twice.

[240] Proposition 2 has two diagrams so one has to distinguish between the two. Eutocius explains he refers to the diagram in the text from which the preceding quotation is taken.

[241] *Elements* VI.8.

[242] Can be deduced from *Elements* VI.8 Cor., 20 Cor. 2.

[243] The ordinals are terms in a continuous proportion first:second::second:third. The reference is to *Elements* VI.20 Cor. 2.

[244] *SC* II.2, Step 31.   [245] *Elements* VI.8 Cor.   [246] *Elements* VI. 20 Cor. 2.

the <square> on AB to the square on BE;[247] (5) therefore also: as AΓ
to ΓE, the <square> on AB to the <square> on BE.

Later on, following this, as he works on proving that the cone BKZ
is equal to the segment of the sphere BAZ (after he has set out a cone
N, having a base equal to the surface of the segment BAZ, and a height

Arch. 193   equal to the radius of the sphere), he says that "the cone N is equal
to the solid sector ZABΘ, as has been proved in the first book." But
it must be understood that in the first book he did not demonstrate
that the sector of *this* kind is equal to a cone thus taken, but rather
the <sector> contained by the surface of the cone and by a spherical
surface smaller than a hemisphere – which is also what he seemed, in
the definitions, to call, in the strict sense, "a solid sector." For he has
said: "When a cone cuts a sphere, having the vertex at the center of the
sphere, I call the figure contained by the surface of the cone and by the
<surface> inside the cone a 'solid sector' ".[248] But the figure set forth
now is contained by a conical surface having the vertex at the center
of the sphere, and by a spherical surface – but not by the one taken
inside the cone. And it will be proved in the following way, through the
<results> proved in the first book, that this kind of figure, too, comes
to be equal to: the cone having a base equal to the spherical surface
containing the segment, while <its> height is equal to the radius of the
sphere.

(a) Let a sphere be imagined separately,[249] (b) and let it be cut
by some plane not <passing> through the center, <namely> by the
circle around the diameter BΔ, (c) and <let> A <be> the center of the
sphere, (d) and let a cone be imagined, <namely> that having <as>
base the circle around the diameter BΔ, and, <as> vertex, the point
A, (e) and let a cone, E, be set out, (f) and let its base be equal to
the surface of the sphere, while its height is the radius of the sphere;
(1) therefore the cone E is equal to the sphere; (2) for it is four times
the cone having <as> base the great circle, and the same height;[250]
(3) of which same <cone> the sphere was also proved <to be> four
times.[251] (g) And also let two other cones be set out, Z, H, (h) of whom,
let Z have a base equal to the surface at the segment BΓΔ, and, <as>
height, the radius of the sphere, (h) and let H have a base equal to the
surface at the segment BΘΔ, and the same height; (4) therefore the
cone Z is equal to the sector, whose vertex is A, and <its> spherical

---

[247] *Elements* VI. 8. (the claim is proved directly from the similarity of the triangles).

[248] Amazingly, this quotation of *SC* I. Def. 5 – otherwise only moderately different
from the text of the definition as we have it – is in a (lightly touched) Doric dialect.

[249] That is, apart from the original diagram of *SC* II.2.

[250] *SC* I.33, *Elements* XII.11.   [251] *SC* I.34.

surface is the <surface> at BΓΔ.[252] (5) Now since the base of the cone
E is equal to the bases of the cones Z, H, (6) and they are under the
same height, (7) therefore the cone E, that is the sphere (8) is equal to
the cones Z, H.[253] (9) But the <cone> Z was proved equal to the solid
sector at BΓΔ, having A <as> vertex; (10) therefore the remaining
cone H is equal to the remaining segment,[254] having, <as> base, the
surface at the segment BΘΔ, and the radius <as> height.

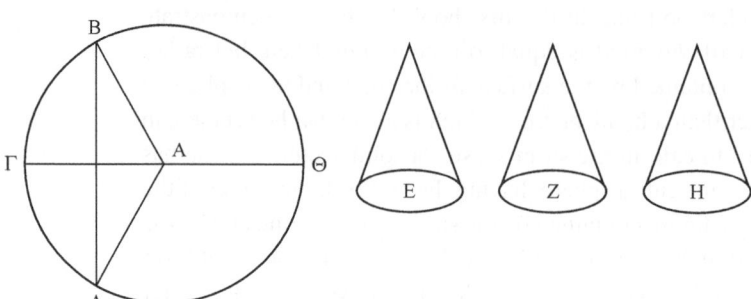

In II.2
Codex G has the
implied diameters of E,
Z, H at the same height
as ΓΘ; H has them
even higher.

Arch. 193   Later again he says: "therefore the cone N, that is the sector BΘZA,
is equal to the figure BΘZK." (1) For since the conclusion was reached
for the cone N that it is equal to a cone, whose base is the circle around
the diameter BZ, while <its> height is ΘK, (2) but the cone, whose
base is the same, while <its> height is EK, is equal to the said cone[255]
and to the <cone> having the same base, and EΘ <as> a height;
(3) for they are to each other as the heights;[256] (4) taking away, <as>
common, the cone having the same base, and EΘ <as> a height, (5) the
remaining figure BΘZK is equal to the cone having the circle around
the diameter BZ <as> base, and ΘK <as> height, (6) that is to the
cone N, (7) that is to the sector BAΘZ.

       After attaching, at the end of theorem, the corollary from the con-
clusions, he then derives the last part of the theorem (that is that the
segment of the sphere ABZ is equal to the cone BKZ), by another proof,
Arch. 195 and, as he sets out to do so, he says: "therefore as KΘ to ΘΓ, ΘΔ to
ΔΓ, and the whole KΔ is to ΔΘ, as ΔΘ to ΔΓ." (1) For since it is:
as KΘ to ΘΓ, ΘΔ to ΔΓ, (2) alternately, too: as KΘ to ΘΔ, ΘΓ to
ΓΔ,[257] (3) compoundly, too: as KΘ to ΔΘ, ΘΔ to ΔΓ,[258] (4) "that is
KΘ to ΘA";[259] (5) for it was: as KΘ to ΘΓ, ΘΔ to ΔΓ, (6) and ΘΓ
is equal to ΘA.

---

[252] *SC* I.44.   [253] *Elements* XII.11.

[254] "Segment:" Eutocius is still unhappy about calling this a "sector."

[255] "The said cone:" not cone N itself, but the cone to which N is equal.

[256] *Elements* XII.11.   [257] *Elements* V.16.   [258] *Elements* V.18.

[259] Step 4 is the direct continuation of the lemma in Archimedes' text. The Eutocian
text here is part commentary, part an expanded re-enactment of the original Archimedean
formulation.

Arch. 195   And a bit later: "therefore as KΘ to ΔΘ, so AE to EΓ; therefore also: as the <square> on KΔ to the <rectangle contained> by KΘΔ, so the <square> on AΓ to the <rectangle contained> by AEΓ." For let the <lines> KΔ, AΓ be imagined set separately, and let it be: as KΘ to ΘΔ, so AE to EΓ. I say that it is also: as the <square> on KΔ to the <rectangle contained> by KΘΔ, so the <square> on AΓ to the <rectangle contained> by AEΓ.

(1) For since it is: as KΘ to ΘΔ, so AE to EΓ, (2) It is also, compoundly: as KΔ to ΔΘ, so AΓ to ΓE;[260] (3) so that also: as the <square> on KΔ to the <square> on ΔΘ, so the <square> on AΓ to the <square> on EΓ. (4) Again, since it is: as KΘ to ΘΔ, so AE to EΓ, (5) but as KΘ to ΘΔ, so the <rectangle contained> by KΘΔ to the <square> on ΘΔ, (6) ΘΔ taken <as> a common height,[261] (7) and as AE to EΓ, so the <rectangle contained> by AEΓ to the <square> on EΓ, (8) EΓ taken, again, <as> a common height,[262] (9) therefore also: as the <rectangle contained> by KΘΔ to the <square> on ΘΔ, so the <rectangle contained> by AEΓ to the <square> on EΓ. (10) And it was proved: as the <square> on ΘΔ to the <square> on ΔK, so the <square> on EΓ to the <square> on ΓA; (11) Therefore also, through the equality: as the <rectangle contained> by KΘΔ to the <square> on KΔ, so the <rectangle contained> by AEΓ to the <square> on AΓ.[263] (12) inversely, also;[264] which it was required to prove.

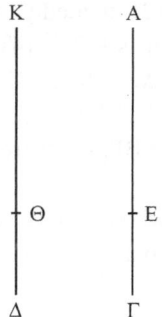

K   A

     Θ   E

Δ   Γ

In II.2 Second diagram Codex D has KΔ appreciably greater than AΓ; so also codices EG, but by a tiny amount. Codex G also has ΔΘ slightly greater than ΓE.

To 3

Arch. 199   "And as the said circles to each other, the <square> on AΔ to the <square> on ΔB, that is AΓ to ΓB." For as in the same diagram as

---

[260] *Elements* V.18.   [261] *Elements* VI.1.
[262] *Elements* VI.1   [263] *Elements* V.22.
[264] *Elements* V.7 Cor., yielding the implicit conclusion (sq. KΔ):(rect. KΘΔ):: (sq. AΓ):(rect.AEΓ).

the said (1) since a perpendicular (<namely>, ΔΓ) has been drawn in a right-angled triangle, AΔB; (2) and <it has been drawn> from the right <angle>, (3) it is a mean proportional between the segments of the base,[265] (4) and the triangles next to the perpendicular are similar to the whole <triangle> and to each other;[266] (5) so that it is: as BΓ to ΔΓ, BΔ to ΔA;[267] (6) therefore also the <squares> on them; (7) but as the <square> on BΓ to the <square> on ΓΔ, so the first BΓ to the third ΓΑ;[268] (8) therefore also: as BΓ to ΓΑ, the <square> on BΔ to the <square> on ΔA.[269]

Arch. 199   "And a given ratio, of ΑΓ to ΓB; so that the point Γ is given." (1) For since the sphere is assumed given, (2) therefore its diameter, AB, is given as well. (3) And the ratio of ΑΓ to ΓB is given; (4) and if a given magnitude is divided into a given ratio, each of the segments is given;[270] (5) so that ΑΓ is given. (6) and A is given; (7) for it is on the common section of lines given in position;[271] (8) therefore Γ is given as well.[272]

To 4

Arch. 202   "And through the same <arguments> as before, through the construction, as ΛΔ to ΚΔ, KB to BP and ΔX to XB." For in the <proposition> preceding this one, it was thus concluded: (1) Since it is, as ΚΔ, ΔX taken together to ΔX, so PX to XB,[273] (2) dividedly: as ΚΔ to ΔX, PB to BX;[274] (3) alternately: as ΚΔ, that is KB[275] (4) to BP, ΔX to XB.[276] (5) Again, since it is: as ΛX to XΔ, so KB, BX taken together to XB,[277] (6) dividedly and alternately: as ΛΔ to ΔK, ΔX to XB.[278] (7) And it was also: as ΔX to XB, KB to BP; (8) therefore as ΛΔ to ΔK, ΔX to XB and KB to BP.

Arch. 202   "And therefore the whole PΛ to the whole KΛ is as KΛ to ΛΔ." For as one to one, so all the antecedents to all the consequents.[279]

---

[265] *Elements* VI.8 Cor.

[266] *Elements* VI.8.   [267] *Elements* VI.4.

[268] The reference of "first" and "third" is to three terms in a continuous proportion, first:second::second:third. That the lines in question form such a continuous proportion was asserted at Step 3 above. *Elements* VI.20 Cor. 2.

[269] The manuscripts have "as BΓ to ΓΑ, the <line> BΔ to ΔA." (No indication of noun for ΔA.) This is most likely to be mere textual corruption, but a more interesting possibility is that a more abstract representation of the square – directly through the diagrammatic symbols, without the word "square" – is being approached.

[270] *Data* 7.   [271] *Data* 25.   [272] *Data* 27.

[273] Archimedes' construction.   [274] *Elements* V.17.   [275] Both radii.

[276] *Elements* V.16.   [277] Archimedes' construction.   [278] *Elements* V.16, 17.

[279] *Elements* V.12. In modern terms: given $(a_1 : b_1 :: a_2 : b_2 :: \ldots :: a_n : b_n)$ it can be concluded, for any k between 1 and n: $(a_k : b_k :: (a_1 + a_2 + \cdots + a_n) : (b_1 + b_2 + \cdots + b_n))$.

Arch. 203   "Therefore as PΛ to ΛΔ, the <square> on KΛ to the <square> on ΛΔ." (1) For since it is: as PΛ to ΛK, KΛ to ΛΔ, (2) therefore also: as the first to the second, so the <square> on the first to the <square> on the second;[280] (3) therefore it is: as PΛ to ΛΔ, so the <square> on PΛ to the <square> on ΛK. (4) But as the <square> on PΛ to the <square> on ΛK, so the <square> on ΛK to the <square> on ΛΔ; (5) for they are proportional;[281] (6) therefore as PΛ to ΛΔ, so the <square> on ΛK to the <square> on ΛΔ.

Arch. 203   "Let BZ be set equal to KB; (for it is clear that it will fall beyond P)." (1) For since it is: as XΔ to XB, so KB to BP[282] (2) and ΔX is greater than XB,[283] (3) therefore KB, as well, is greater than BP. (4) Therefore Z falls beyond P.

Arch. 203   "And since <the> ratio of ΔΛ to ΛX is given, as well as the ratio of PΛ to ΛX, therefore <the> ratio of PΛ to ΛΔ, too, is given." (1) For since it is: as KBX taken together to BX, that is ZX to XB,[284] (2) so ΛX to XΔ,[285] (3) Convertedly: as XZ to ZB, so XΛ to ΛΔ,[286] (4) also inversely: as BZ to ZX, ΛΔ to ΛX.[287] (5) And the ratio of BZ to ZX is given, (6) since ZB is equal to the radius of the given sphere, (7) while BX is given (8) as its limits B, X are given by hypothesis,[288] (9) the sphere being cut by the plane ΑΓ and by the <line> ΔB being at

---

[280] *Elements* VI.20 Cor. 2.

[281] Here is one of those moments when I get seriously excited and people watch on bemused as I cry aloud and foam, but please, please pay attention! Eutocius speaks about "they," meaning the "they" which are independently known to be proportional, i.e. the line segments PΛ, ΛK, ΛΔ. This use of "they" shows that these line segments are understood to be the logical subjects of this Step 4 itself. I.e., Step 4 is understood to be not on squares, but on segments of lines. This can be immediately grasped by our own expression of the type $a^2:b^2::c^2:d^2$, where clearly the expression is felt to be about (a, b, c, d), rather than about (a^2, b^2, c^2, d^2): the "^{2}" symbol is merely something we do with the main protagonists (just as we manipulate them with the ":" symbol, yet no one would think the expression is about ":"). This is at the heart of what constitutes the symbolic nature of our "^{2}," which does not yield an object so much as transforms another, separately present object. The opposite is usually the case with the fully geometrical Greek "<square> on," which is not a symbolic manipulation, but is a real geometrical expression, yielding an object completely distinct from the line segment from which we started: a square, distinct from a line. Thus the enormous significance of the expression before us: Greek "<square> on" acts and feels like a purely symbolic, modern "^{2}." This is typical of the commentary, second-order position of Eutocius: extensions of symbolism towards the fully second-order symbolism of algebra are often suggested, though never fully followed. More of this to come below.

[282] See Eutocius' first comment above.

[283] An assumption made explicit in Archimedes' synthesis, and at this stage – the analysis – based on the diagram alone.

[284] "KBX taken together" is equal to ZX, since by Archimedes' construction ZB=KB.

[285] Archimedes' construction.   [286] *Elements* V.19 Cor.

[287] *Elements* V.7 Cor.   [288] *Data* 26.

right <angles> to the <line> ΑΓ,[289] (10) and through this the whole XZ, too, <is given>, (11) as well as the <ratio> of XZ to ZB;[290] (12) so that the ratio of XΛ to ΛΔ is given as well. (13) Again, since the ratio of the segments is given, (14) the ratio of the cone ΛΑΓ to the cone ΑΡΓ is given, as well.[291] (15) So that the <ratio> of ΛX to XP <is given>, too; (16) for they are to each other as the heights;[292] (17) therefore <the> ratio of the whole PΛ to ΛX is given.[293] (18) Now since the ratio of each of PΛ, ΛΔ to ΛX is given, (19) therefore the ratio of PΛ to ΛΔ is given as well. (20) For the <magnitudes> which have a given ratio to the same, have also a given ratio to each other.[294]

Arch. 203   "Now since the ratio of PΛ to ΛX is combined of both: the <ratio> which PΛ has to ΛΔ, and <of> ΛΔ to ΛX." It is obvious that, once ΛΔ is taken as a mean, the synthesis of ratios is taken (as this is taken in the *Elements*, too[295]). Since, however, the discussion of the subject has been somewhat confused, and not such as to make the concept satisfactory, (as can be found reading Pappus[296] and Theon[297] and Arcadius[298] who, in many treatises, present the operation not by arguments, but by examples), there will be no incongruity if we linger briefly on this subject so as to present the operation more clearly.[299]

So: I say that if some middle term is taken between two numbers (or magnitudes), the ratio of the initially taken numbers[300] is composed of

---

[289] *Data* 25.   [290] *Data* 1.   [291] *SC* II.2.   [292] *Elements* XI.14.

[293] *Data* 22.   [294] *Data* 8.   [295] Reference to *Elements* VI.23.

[296] The reference must be to a work, or works, no longer extant.

[297] Theon of Alexandria, late fourth century AD, Hypatia's father, known especially through his (extant) commentary on the Almagest. The reference here is to this commentary, pp. 61 ff. Basil. and possibly, to other, lost, works (Eutocius' plural "many treatises" is very emphatic).

[298] Known only through this reference. One wonders if the sequence Pappus – Theon is not meant to be chronological, in which case Arcadius is probably a very late author, not much earlier than Eutocius himself – which could help to explain how Eutocius knows him but we don't. See Knorr (1989) 166, however, for a suggestion linking Arcadius with a known (and unattributed) *Introduction to the Almagest*, containing a passage on the composition of ratios.

[299] What Eutocius says is that as far as the mathematical consensus is concerned, Archimedes' argument is clear and even obvious. However, since the mathematical consensus itself seems to be at fault here, a commentary is required. First we had a spirit of philological enterprise, in the catalogue of two mean proportionals, and now a mathematical independence. Eutocius has grown considerably since the commentary to the first book. The composition of ratios is indeed a sore point in Greek mathematics: let's see how much sense he will make out of it (Eutocius himself clearly was happy with his own discussion, and he has recycled it in his later commentary to Apollonius' *Conics*, II. pp. 218 ff.).

[300] So the immediately preceding "or magnitudes" is an afterthought.

the ratio, which the first has to the mean and of the <ratio> which the mean has to the third.

So first it ought to be recalled how a ratio is said to be composed of ratios. For as in the *Elements*: "when the quantities of the ratios, multiplied, produce a certain <quantity>,"[301] where "quantity" clearly stands for "the number" whose cognate is the given ratio[302] (as say several authors as well as Nicomachus[303] in the first book of *On Music* and Heronas[304] in the commentary to the *Arithmetical Introduction*[305]), which is the same as saying: "the number which, multiplied on[306] the consequent term of the ratio, produces the antecedent as well." And the quantity would be taken in a more legitimate way in the case of multiples,[307] while in the case of superparticulars, superpartients,[308] it is no longer possible for the quantity to be taken with the unit remaining undivided;[309] so that in these cases the unit must be divided – which, even if this does not belong to what is proper in arithmetic, yet it does belong to what is proper in calculation. And the unit is divided by the part or by the parts by which the ratio is called,[310] so that (to say this in a clearer way), the quantity of the half-as-large-again is, added to the unit, half the unit; and <the quantity of the> four-thirds is, added to the unit, one third the unit, so that, as has been said above as well, the quantity of the ratio, multiplied on the consequent term, produces the

---

[301] *Elements* V. Def. 5, bracketed in Heiberg's edition of the *Elements* but apparently in Eutocius' own text.

[302] The idea is that a typical ratio is, for instance, the multiplicative "twice," whose cognate is the cardinal "two." So the term for ratio "twice" is the cognate of the term for number "two."

[303] A first–second centuries AD Pythagorean philosopher-mathematician. His *On Music* does not survive.

[304] Known only from this reference.

[305] An extant treatise written by Nicomachus.

[306] Standard English usage has X multiplied by Y, not on Y. I prefer to stick to the literal translation of the Greek particle ἐπί since, a few pages below, in a geometrical context, Archimedes and Eutocius are about to employ the same expression so as to suggest a similar calculation, applied to geometry.

[307] I.e. integer multiplicatives such as "twice," "three times," etc.

[308] Kinds of what we call non-integer, positive rational numbers. Here they are mentioned as kinds of ratio, not as kinds of numbers. See Eutocius' comment to *SC* I.2 for the terms.

[309] The quantity of non-integer ratios is not a cardinal. This seemingly trivial point must be stressed by Eutocius, since, in trying to make sense of "ratios as quantities" he starts from the relationship between ratios and their cognate number, which exists only in the case of integers.

[310] For instance: 2 is a sixth of 12, therefore it is a "part" of 12 (namely, a sixth part). 9 to 12 cannot be expressed by such a single term. It is three quarters of 12, three parts. Therefore it is "parts" of 12 (*Elements* VII. Deff. 3–4).

antecedent. For the quantity of nine to six, being the unit and the half, multiplied on 6, produces 9, and it is possible to observe the same in the other cases as well.

Having clarified these first, let us return to the enunciated proposition. For let the two given numbers[311] be A, B, and let a certain mean be taken between them, Γ. So it is to be proved that the ratio of A to B is combined of the <ratio> which A has to Γ, and Γ to B.

(a) For let the quantity of the ratio A, Γ[312] be taken, <namely> Δ, (b) and <let the quantity> of the <ratio> Γ, B <be taken, namely> E; (1) therefore Γ, multiplying Δ, produces A, (2) while B, multiplying E, <produces> Γ. (c) So let Δ, multiplying E, produce Z. I say that Z is a quantity of the ratio of A to B, that is, that Z, multiplying B, produces A. (d) For let B, multiplying Z, produce H. (3) Now since B, multiplying Z, has produced H, (4) and, multiplying E, <it has produced> Γ, (5) it is therefore: as Z to E, H to Γ.[313] (6) Again, since Δ, multiplying E, has produced Z (7) while, multiplying Γ, it has produced A, (8) it is therefore: as E to Γ, Z to A. (9) Alternately: as E to Z, Γ to A,[314] (10) inversely also: as Z to E, so A to Γ.[315] (11) But as Z to E, H was proved <to be> to Γ; (12) therefore also: as H to Γ, A to Γ; (13) therefore A is equal to H.[316] (14) But B, multiplying Z, has produced H; (15) therefore B, multiplying Z, produces A as well; (16) therefore Z is a quantity of the ratio of A to B.[317] (17) And Z is: Δ, multiplied on E, (18) that is: the quantity of the ratio A, Γ, <multiplied> on the quantity of the ratio Γ, B; (19) therefore the ratio of A to B is composed of both: the <ratio>, which A has to Γ, and Γ to B; which it was required to prove.[318]

---

[311] "Numbers:" any pretence at generality is by now dropped. In what follows, Eutocius consistently uses the masculine article, referring to "number." My translation "A," abbreviated to avoid excessive tedium, thus stands for the original "the <number> A."

[312] "The ratio A, Γ:" a revolutionary expression.

[313] This can be related to *Elements* VII.17 – which is apparently what Eutocius conceives of as the basis for his own argument. Eutocius seems to discuss the subject matter neither of *Elements* V (geometrical magnitudes), nor that of *Elements* VII (integers), but the subject matter of what he has called "calculation," which we call positive rational numbers. Since, however, his tool box is so heavily based on Euclid, he probably finds it natural to deal with (what we call) positive rational numbers as if they were integers.

[314] *Elements* V.16 (magnitudes) or VII.13 (numbers)?

[315] *Elements* V.7 Cor. (No separate *Elements* proof for numbers).

[316] *Elements* V.9.

[317] Eutocius does not stop here – the interim definition of goal – but goes on to obtain the original goal, referring explicitly to the composition of ratios.

[318] The proof is valid for the rational numbers Eutocius has in mind. But its limited generalizability is fundamental. While it is easy to provide a clear sense of the-ratio-of-two-rational-numbers as a rational number, there is no such simple way of defining, say, the-ratio-of-two-lines as, say, a line, let alone as any numerical magnitude of the

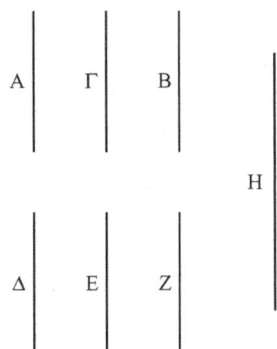

In II.4
Codex D has A greater
than Γ, in turn greater
than B; and Δ greater
than E, in turn greater
than Z. Codex E has A
greater than B, in turn
equal to Γ; and Δ
greater than E, in turn
equal to Z. Codex G
has the three lines Δ,
E, Z (equal to each
other) greater than the
three lines A, B, Γ
(equal to each other).

And so that the thing said shall be clarified with an example, too,[319] let some mean number, <namely> 4, be inserted between 12 and 2. I say that the ratio of 12 to 2, that is the six-times, is composed of the thrice (12 to 4) and of the twice (4 to 2).

For if we multiply the quantities of the ratios on each other, that is 3 on 2, 6 results, being the quantity of the ratio of 12 to 6, and it is six-times, which is also what was put forth to prove.

And even if the inserted mean happens not to be (first) smaller than the greater and (second) greater than the smaller, but is instead the opposite of that,[320] or it is greater than both, or smaller than both; even so the composition mentioned above follows. Let some mean be inserted between 9 and 6, greater than both, <namely> 12. I say that from both: the converse-of-a-third-as-large-again ratio (9 to 12) and from the twice (12 to 6), the half-as-large-again is composed (9 to 6).

(1) For the quantity of the ratio 9 to 12 is three fourths, (2) that is half and a fourth, (3) and the quantity of 12 to 6 is 2. (4) Now if we multiply 2 on half and fourth, a one unit and a half results, (5) which is a quantity of the half-as-large-again ratio, (6) which 9 has to 6, as well. And similarly, if 4, as well, is inserted <as> a mean between 9 and 6: from the 9 to 4 (twice-cum-converse-of-four-times) and <from> the 4 to 6 (converse-of-half-as-large-again), the half-as-large-again is composed. For again, when we multiply the quantity of the

---

kinds the Greeks knew. The ratio of lines is just that, a ratio. What we need however for Eutocius' purposes (who after all deals here with the geometrical magnitudes of Archimedes' treatise) is to see any ratio whatsoever as some sort of single object, not just as a relation between two objects. In other words, we need modern mathematics which, for better or worse, is not what Eutocius is offering us.

[319] Unlike his predecessors, Eutocius explicates the concept with a proof, rather than an example. Having done so, he now goes on to add the examples.

[320] That is, smaller than the smaller and greater than the greater. This, as Heiberg notes in his textual apparatus, is not possible. Heiberg suggests Eutocius may have nodded off here, but I would be even happier to believe this is some sort of an attempt at humor.

twice-cum-converse-of-four-times, <namely> 2¼,[321] on the quantity of the converse-of-half-as-large-again, that is the two thirds, we shall get the one <and> a half, the quantity of the half-as-large-again ratio, as has been said. And the same argument will apply similarly in all other cases.

From the things said it is also clear that if not one mean term is inserted between two given numbers or magnitudes, but many, the ratio of the extremes is composed of all the ratios which the terms have, arranged in sequence, starting from the first and terminating in the last one in the order of the <terms> standing in ratios.[322]

For, there being two terms, A, B, let more than one <terms, namely> Γ, Δ, be inserted. I say that the ratio of A to B is composed of the <ratio> which A has to Γ, and Γ to Δ, and Δ to B.

(1) For since the <ratio> of A to B is composed of the <ratio>, which A has to Δ, and Δ to B, (2) as has been said above, (3) and the ratio of A to Δ is composed of the <ratio> which A has to Γ, and Γ to Δ, (4) therefore the ratio of A to B is combined of the <ratio> which A has to Γ, and Γ to Δ, and Δ to B. And similarly it will be proved in the remaining <cases? means?>.

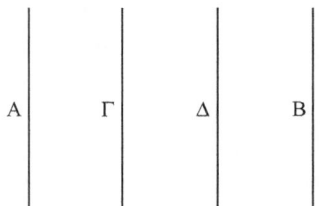

A   Γ   Δ   B

In II.4 Second diagram

Arch. 203

Further in the text he says:[323] "but as PΛ to ΛΔ, <so> the <square> on ΔB was proved to be to the <square> on ΔX." (1) For since it has been proved: as PΛ to ΛΔ, the <square> on ΛK to the <square> on ΛΔ,[324] (2) and as the <square> on KΛ to the <square> on ΛΔ, so the <square> on BΔ to the <square> on ΔX ((3) for it was proved: as KΛ to ΛΔ, BΔ to ΔX, through the "compoundly"[325]); (4) therefore as PΛ to ΛΔ, the <square> on BΔ to the <square> on ΔX.

---

[321] Greek strictly speaking does not have the symbol "¼." Instead, the symbol "4'" is used to mean "fourth part."

[322] Eutocius is at pains to clarify that the order need not be that of quantity (from great to small) but can be any order whatsoever.

[323] The word "text" is a free translation of what means roughly "what is said." Eutocius' point is that the lemma immediately follows the preceding lemma: the commentary follows one Archimedean sentence, and then the next.

[324] Step 15 in Archimedes' analysis.

[325] Step 17 of Archimedes' analysis. This time "compoundly" refers not to the composition-of-ratios operation, but to the "compoundly" proportion argument, *Elements* V.18.

Arch. 203   "And let it be made, as PΛ to ΛX, BZ to ZΘ." Wherever the
point Θ be positioned, as far as the logical consequence of the proof is
concerned, no obstacle to the argument may arise. But it shall be clear
that it always falls (just as it is positioned in the diagram) between B,
P, as follows: (1) for since it is, as ΛK to ΔK, that is to KB,[326] (2) so
KP to PB,[327] (3) and therefore also as one to one, so all to all:[328] (4) as
ΛP to PK, KP to PB. (5) But ΛP has to PX a greater ratio than ΛP to
PK;[329] (6) therefore ΛP has to PX a greater ratio than KP to BP, too,
(7) that is ZB to BP. (8) Conversely, PΛ has to ΛX a smaller ratio than
BZ to ZP.[330] (9) Therefore if we make: as PΛ to ΛX, so BZ to some
other <line>, it shall be to a <line> greater than ZP.

And it is at once apparent that ZΘ is greater than ΘB.[331] (1) For
since it has been proved: as ΛΔ to ΔK, ΔX to XB (2) and KB to BP,[332]
(3) and ΔX is greater than XB, (4) therefore ΛΔ, too, is greater than
ΔK, (5) and KB <is greater> than BP; (6) so that ΛΔ <is greater>
than BP, as well.[333] (7) Therefore the whole ΛX is greater than XP, as
well; (8) so that ΘZ <is greater> than ΘB, as well.

Arch. 204   "Remaining, therefore, it is: as the <square> on BΔ, that is a given,
to the <square> on ΔX, so ZX to ZΘ." (1) For since the <ratio>
composed of the <square> on BΔ to the <square> on ΔX and of
BZ to ZX was proved to be the same as the ratio of BZ to ΘZ,[334]
(2) and the same ratio (of BZ to ZΘ) is the same also as the <ratio>
composed of the <ratio> of BZ to ZX and of XZ to ZΘ, (3) therefore,
also, the ratio composed of the <ratio> of the <square> on BΔ to the
<square> on ΔX and of the <ratio> of BZ to ZX is the same as the
<ratio> composed of the <ratio> of BZ to ZX and of the <ratio>
of XZ to ZΘ. (4) Now if we take away the <ratio> common to both
ratios, <namely> the <ratio> of BZ to XZ, the remaining ratio of the
<square> on BΔ to the <square> on ΔX is the same as the <ratio>
of XZ to ZΘ.

Arch. 204   And "So it is required to cut a given line, ΔZ, at the <point> X,
and to produce: as XZ to a given <line>" (that is ZΘ) "so the given
<square>" (that is the <square> on BΔ) "to the <square> on ΔX.
And this, said in this way – without qualification – is soluble only given
certain conditions, but with the added qualification of the specific characteristics of the problem at hand" (that is, both that ΔB is twice BZ

---

[326] Both radii.

[327] Step 10 of Arcihmedes' analysis, plus an implicit use of *Elements* V.18.

[328] *Elements* V.12.   [329] *Elements* V.8.

[330] See Eutocius' commentary to *SC* I.2.

[331] So we get an even firmer grasp of where the point Θ is.

[332] Steps 10–11 in Archimedes' analysis.

[333] Remember ΔK=KB (radii).   [334] Step 34 in Archimedes' analysis.

and that ZΘ is greater than BZ – as is seen in the analysis) "it is always soluble; and the problem will be as follows: given two lines ΔB, BZ (and ΔB being twice BZ), and a point on BZ, <namely> Θ; to cut ΔB at X, and to produce: as the <square> on ΔB to the <square> on ΔX, XZ to ZΘ; and these <problems> will be, each, both analyzed and constructed at the end." While he promised to prove the aforementioned claim at the end, it is impossible to find the promised thing in any of the manuscripts. Which is why, as we found, Dionysodorus, too, failing to get to the same proofs – being unable to lay hands on the lost lemma – went on another route to the entire problem, which we shall write down in the following. And Diocles too, in the book he composed *On Burning Mirrors*, also in the belief that Archimedes promised, but had not delivered the promise, attempted to fill the gap himself; and we shall write down the attempt in the following. (Indeed, this again has nothing resembling the lost argument, but, similarly to Dionysodorus, he constructs the problem through a different proof.)

But – in a certain old book (for we did not cease from the search for many books), we have read theorems written very unclearly (because of the errors), and in many ways mistaken about the diagrams. But they had to do with the subject matter we were looking for, and they preserved in part the Doric language Archimedes liked using, written with the ancient names of things: the parabola called "section of a right-angled cone," the hyperbola "section of an obtuse-angled cone." From which things we began to suspect, whether these may not in fact be the things promised to be written at the end. So we read more carefully the content itself (since we have found – as had been said – that it has been an uneasy piece of writing, because of the great number of mistakes), taking apart the ideas one by one.[335] We write it down, as far as possible, word-for-word (but in a language that is more widely used, and clearer).

The first theorem is proved for the general case, so that his claim, concerning the limits on the solution, becomes clearer. Then it is also applied to the results of the analysis in the problem.[336]

---

[335] Note how narrative form is kept throughout, beginning from the romantic quest for books, following the commentator in his study – the moment of sudden conversion, and then the long work of taking the treatise apart.

[336] The textual and mathematical commentary on the following passage – on Archimedes' problem in the lost appendix and on its later solutions – could not be contained within the boundaries of this volume. I publish them separately in Netz (forthcoming b). Within this volume, I limit myself to immediate comments on the details of the text.

Given a line, AB, and another, AΓ, and an area, Δ: let it first be put forth:[337] to take a point on AB, such as E, so that it is: as AE to AΓ, so the area Δ to the <square> on EB.

(a) Let it come to be, (b) and let AΓ be set at right <angles> to AB, (c) and, having joined ΓE, (d) let it be drawn through to Z, (e) and let ΓH be drawn through Γ, parallel to AB, (f) and let ZBH be drawn through B, parallel to AΓ, meeting each of the <lines> ΓE, ΓH, (g) and let the parallelogram HΘ be filled in, (h) and let KEΛ be drawn through E parallel to either ΓΘ or HZ, (i) and let the <rectangle contained> by ΓHM be equal to the <area> Δ.

(1) Now since it is: as EA to AΓ, so the <area> Δ to the <square> on EB,[338] (2) but as EA to AΓ, so ΓH to HZ,[339] (3) and as ΓH to HZ, so the <square> on ΓH to the <rectangle contained> by ΓHZ,[340] (4) therefore as the <square> on ΓH to the <rectangle contained> by ΓHZ, so the <area> Δ to the <square> on EB, (5) that is to the <square> on KZ;[341] (6) alternately also: as the <square> on ΓH to the <area> Δ, that is to the <rectangle contained> by ΓHM, (7) so the <rectangle contained> by ΓHZ to the <square> on ZK.[342] (8) But as the <square> on ΓH to the <rectangle contained> by ΓHM, so ΓH to HM;[343] (9) therefore also: as ΓH to HM, so the <rectangle contained> by ΓHZ to the <square> on ZK. (10) But as ΓH to HM, so (HZ taken as a common height) the <rectangle contained> by ΓHZ to the <rectangle contained> by MHZ;[344] (11) therefore as the <rectangle contained> by ΓHZ to the <rectangle contained> by MHZ, so the <rectangle contained> by ΓHZ to the <square> on ZK; (12) therefore the <rectangle contained> by MHZ is equal to the <square> on ZK.[345] (13) Therefore if a parabola is drawn through H around the axis ZH, so that the lines drawn down <to the axis> are in square the <rectangle applied> along HM,[346] it shall pass through K,[347] (14) and it <=the parabola> shall be given in position, (15) through HM's being given in magnitude (16) as it contains, together with the given HΓ, the given <area> Δ;[348] (17) therefore K touches a parabola given in position.

---

[337] I.e. "let the geometrical task be."   [338] The assumption of the analysis.

[339] *Elements* VI.2, 4, and I.34.   [340] *Elements* VI.1.   [341] *Elements* I.34.

[342] *Elements* V.16.   [343] *Elements* VI.1.

[344] *Elements* VI.1.   [345] *Elements* V.7.

[346] For any point Z on the axis, the square on the line drawn from the parabola to the point Z, i.e. the square on KZ, is equal to the rectangle contained by ZH (i.e. the line to the vertex of the parabola) and by the constant line HM (the *latus rectum*) – i.e. to the rectangle ZHM.

[347] Converse of *Conics* I.11.

[348] For Step 15 to derive from Step 16, *Data* 57 is required. Step 14 derives from Step 15 in the sense that there is a unique parabola given an axis, a vertex and a *latus*

(j) Now, let it <=the parabola> be drawn, as has been said, and let it be as HK. (18) Again, since the area ΘΛ is equal to the <area> ΓB,[349] (19) that is the <rectangle contained> by ΘKΛ is equal to the <rectangle contained> by ABH, (20) if a hyperbola is drawn through B, around the asymptotes ΘΓ, ΓH, it shall pass through K, (21) through the converse of the 8th theorem of the second book of the *Conic Elements* of Apollonius,[350] (22) and it shall be given in position (23) through <the fact> that each of ΘΓ, ΓH is <given in position>, as well, (24) further yet – <through the fact> that B is given in position, too. (k) Let it be drawn, as has been said, and let it be as KB; (25) therefore K touches a hyperbola given in position; (26) and it also touched a parabola given in position. (27) Therefore K is given. (28) And KE is a perpendicular drawn from it to a <line> given in position, <namely> to AB; (29) therefore E is given.[351] (30) Now since it is: as EA to the given <line> AΓ, so the given <area> Δ to the <square> on EB: (31) two solids, whose bases are the <square> on EB and the <area> Δ, and whose heights are EA, AΓ, have the bases reciprocal to the heights; (32) so the solids are equal;[352] (33) therefore <the solid produced by> the <square> on EB, on EA <as the solid's height> is equal to <the solid produced by> the given <area> Δ on the given <line> ΓA <as the solid's height>.[353] (34) But <the solid produced by> the <square> on BE on EA <as the solid's height> is the greatest of all similarly taken <solids> on BA, when BE is twice EA, as shall be proved;[354] (35) therefore <the solid produced by> the given <area> on the given <line as the solid's height> must be not greater than <the solid produced by> the <square> on BE on EA <as the solid's height>.[355]

---

rectum, once again through an obvious converse of *Conics* I.11 (any other conic section must yield two unequal lines drawn to the axis, both producing an equal rectangle when applied to the same *latus rectum*).

[349] *Elements* I.43.

[350] What we have as *Conics* II.12. Notice that this type of reference is most probably due to Eutocius.

[351] *Data* 30.   [352] *Elements* XI.34.

[353] The expression "plane on line" has here a geometrical significance, yet it can be also interpreted as the multiplicative "on" used in the examples of calculation earlier, where we had "number on number." For this ambiguity of meaning, see Netz (forthcoming b).

[354] For the modern reader: the maximum of $x^2(a - x)$ for $a>x>0$ is at $x = 2/3a$. Archimedes indeed proves this below, obviously, as we shall see, following a geometrical route.

[355] That is, assuming BE is twice EA. The idea is the following. You take the original line BA, divide it at the point where BE is twice EA, derive the solid BE^2*EA, and now you've got a maximum for the solid $Δ*ΓA$. Since both Δ and ΓA are independently given, they could theoretically be given in such a way that $BE^2*EA<Δ*ΓA$. This is the limit on the conditions of solubility.

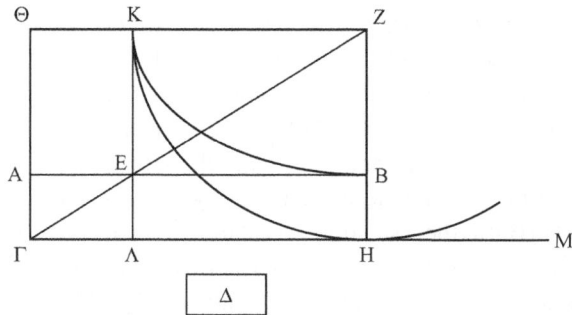

And it will be constructed like this: let the given line be AB, and some other given <line> AΓ, and the given area Δ, and let it be required to cut AB, so that it is: as one segment to the given AB, so the given <area> Δ to the <square> on the remaining segment.

(a) Let AE be taken, a third part of AB; (1) therefore the <area> Δ, on the <line> AΓ is either greater than the <square> on BE, on EA, or equal, or smaller.

(2) Now then, if it is greater, the problem may not be constructed, as has been proved in the analysis; (3) and if it is equal, the point E produces the problem. (4) For, the solids being equal, (5) the bases are reciprocal to the heights,[356] (6) and it is: as the <line> EA to the <line> AΓ, so the <area> Δ to the <square> on BE.

(7) And if the <area> Δ, on AΓ, is smaller than the <square> on BE, on EA, it shall be constructed like this:

(a) Let AΓ be set at right <angles> to AB, (b) and let ΓZ be drawn through Γ parallel to AB, (c) and let BZ be drawn through B parallel to the <line> AΓ, (d) and let it meet ΓE (<itself> being produced) at H, (e) and let the parallelogram ZΘ be filled in, (f) and let KEΛ be drawn through E parallel to ZH. (8) Now, since the <area> Δ, on AΓ, is smaller than the <square> on BE, on EA, (9) it is: as EA to AΓ, so the <area> Δ to some <area> smaller than the <square> on BE,[357] (10) that is, <smaller> than the <square> on HK.[358] (g) So let it be: as EA to AΓ, so the <area> Δ to the <square> on HM, (h) and let the <rectangle contained> by ΓZN be equal to the <area> Δ.[359] (11) Now since it is: as EA to AΓ, so the <area> Δ, that is the <rectangle contained> by ΓZN (12) to the <square> on HM,

In II.4 Third diagram Codices DH have the rectangle Δ to the right of the main figure, as in the thumbnail. In these two codices, it is also a near square.
Codex D has a redundant line parallel to ΘΓ, KΛ, between them; codex 4 had a redundant line AZ, erased (perhaps by the same scribe, immediately correcting a trivial error).
Codex E has the line KΛ slightly slanted, so that K is to the right of Λ.

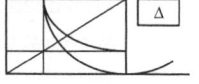

---

[356] *Elements* XI.34.

[357] The closest foundation in Euclid is *Elements* VI.16, proving that if $a*b = c*d$, then $a{:}d{:}{:}c{:}b$ (for a, b, c and d being lines).

[358] Steps b, e, f, *Elements* I.34.

[359] Steps g and h define the points M, N respectively, by defining areas that depend upon those points.

(13) but as EA to AΓ, so ΓZ to ZH,[360] (14) and as ΓZ to ZH, so
the <square> on ΓZ to the <rectangle contained> by ΓZH,[361] (15)
therefore also: as the <square> on ΓZ to the <rectangle contained>
by ΓZH, so the <rectangle contained> by ΓZN to the <square> on
HM; (16) alternately also: as the <square> on ΓZ to the <rectangle
contained> by ΓZN, so the <rectangle contained> by ΓZH to the
<square> on HM.[362] (17) But as the <square> on ΓZ to the <rectangle
contained> by ΓZN, ΓZ to ZN,[363] (18) and as ΓZ to ZN, (taking
ZH as a common height) so is the <rectangle contained> by ΓZH
to the <rectangle contained> by NZH;[364] (19) therefore also: as the
<rectangle contained> by ΓZH to the <rectangle contained> by NZH,
so the <rectangle contained> by ΓZH to the <square> on HM; (20)
therefore the <square> on HM is equal to the <rectangle contained>
by HZN.[365] (21) Therefore if we draw, through Z, a parabola around
the axis ZH, so that the lines drawn down <to the axis> are, in square,
the <rectangle applied> along ZN – it shall pass through M.[366] (i) Let
it be drawn, and let it be as the <parabola> MΞZ. (22) And since the
<area> ΘΛ is equal to the <area> AZ,[367] (23) that is the <rectangle
contained> by ΘKΛ to the <rectangle contained> by ABZ,[368] (24) if
we draw, through B, a hyperbola around the asymptotes ΘΓ, ΓZ, it shall
pass through K[369] (through the converse of the 8th theorem of <the
second book of> Apollonius' *Conic Elements*). (j) Let it be drawn,
and let it be as the <hyperbola> BK, cutting the parabola at Ξ, (k)
and let a perpendicular be drawn from Ξ on AB, <namely> ΞOΠ, (l)
and let the <line> PΞΣ be drawn through Ξ parallel to AB. (25) Now,
since BΞK is a hyperbola (26) and ΘΓ, ΓZ are asymptotes,[370] (27) and
PΞΠ[371] are drawn parallel to ABZ, (28) the <rectangle contained> by
PΞΠ is equal to the <rectangle contained> by ABZ;[372] (29) so that
the <area> PO, too, <is equal> to the <area> OZ. (30) Therefore if
a line is joined from Γ to Σ, it shall pass through O.[373] (m) Let it pass,
and let it be as ΓOΣ. (31) Now, since it is: as OA to AΓ, so OB to
BΣ,[374] (32) that is ΓZ to ZΣ,[375] (33) and as ΓZ to ZΣ (taking ZN as a

---

[360] Steps b, e, f, *Elements* I.29, 32, VI.4.   [361] *Elements* VI.1.

[362] *Elements* V.16.   [363] *Elements* VI.1.   [364] *Elements* VI.1.

[365] *Elements* V.7.   [366] The converse of *Conics* I.11.

[367] Based on *Elements* I.43.

[368] As a result of Step a (the angle at A right), all the parallelograms are in fact
rectangles.

[369] Converse of what we call *Conics* II.12.   [370] Steps 25–6: based on Step j.

[371] An interesting way of saying "the <lines> PΞ, ΞΠ."   [372] *Conics* II.12.

[373] Step 30 is better put as: "The diagonal of the parallelogram PΣZΓ passes through
O," which can then be proved as a converse of *Elements* I.43.

[374] *Elements* I.29, 32, VI.4.   [375] *Elements* VI.2.

common height) the <rectangle contained> by ΓZN to the <rectangle contained> by ΣZN,[376] (34) therefore as OA to AΓ, too, so the <rectangle contained> by ΓZN to the <rectangle contained> by ΣZN. (35) And the <rectangle contained> by ΓZN is equal to the area Δ,[377] (36) while the <rectangle contained> by ΣZN is equal to the <square> on ΣΞ, (37) that is to the <square> on BO,[378] (38) through the parabola.[379] (39) Therefore as OA to AΓ, so the area Δ to the <square> on BO. (40) Therefore the point O has been taken, producing the problem.

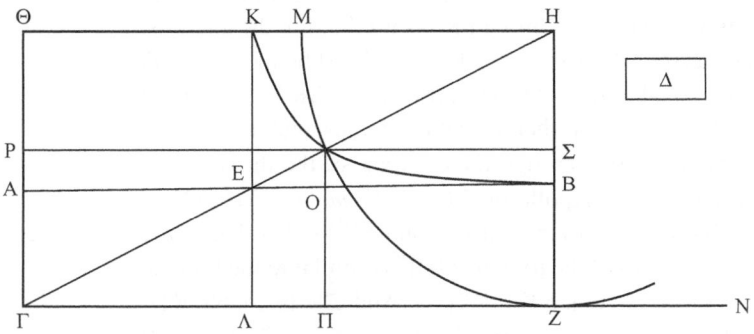

And it will be proved like this that, BE being twice EA, the <square> on BE, on EA, is <the> greatest of all <magnitudes> similarly taken on BA.[380]

For let there be, as in the analysis, again: (a) a given line, at right <angles> to AB, <namely> AΓ, (b) and, having joined ΓE, (c) let it be produced and let it meet at Z the <line> drawn through B parallel to AΓ, (d) and, through the <points> Γ, Z, let ΘZ, ΓH be drawn parallel to AB, (e) and let ΓA be produced to Θ, (f) and, parallel to it, let KEΛ be drawn through E, (g) and let it come to be: as EA to AΓ, so the <rectangle contained> by ΓHM to the <square> on EB; (1) therefore the <square> on BE, on EA, is equal to the <rectangle contained> by ΓHM, on AΓ, (2) through the <fact> that the bases of the two solids are reciprocal to the heights.[381] Now I say that the <rectangle

In II.4 Fourth diagram Codices BDG have the line AB parallel to ΓZ. Codex D has Δ as nearly a square. It also fails to have the points H, Σ, B, Z aligned on a single line, and does not draw a line HZ. Codex E has the lines ΛK, ΠΞ divergent so that K is to the left of Λ, Ξ is to the right of Π. Codex G has a straight line instead of the arc segment MΞ; H has an arc segment instead of the straight line ΞO. The letter N is omitted in codex A. Heiberg restores it on the line HZ, between the points Σ, B (he also removes the line segment that continues from ΓZ beyond Z).

---

[376] *Elements* VI.1.

[377] Step h. The original Greek is literally: "To the <rectangle contained> by ΓZN is equal the area Δ" (with the same syntactic structure, inverted by my translation, in the next step).

[378] Steps a, e, k, l, *Elements* I.34.

[379] A reference to *Conics* I.11 – the "symptom" of the parabola.

[380] Here we reach the proof for the limits of solubility, promised at the end of the analysis.

[381] *Elements* XI.34.

contained> by ΓHM, on AΓ, is <the> greatest of all <magnitudes> similarly taken on BA.[382]

(h) For let a parabola be drawn through H, around the axis ZH, so that the <lines> drawn down <to the axis> are in square the <rectangle applied> along HM;[383] (3) so it will pass through K, as has been proved in the analysis, (4) and, produced, it will meet ΘΓ (5) since it is parallel to the diameter of the section, ((6) through the twenty-seventh theorem of the first book of Apollonius' *Conic Elements*[384]). (i) Let it <=the parabola> be produced and let it meet <the line ΓΘ produced> at N, (j) and let a hyperbola be drawn through B, around the asymptotes NΓH; (7) therefore it will pass through K, as was said in the analysis. (k) So let it pass, as the <hyperbola> BK, (l) and, ZH being produced, (m) let HΞ be set equal to it <=to ZH>, (n) and let ΞK be joined, (o) and let it be produced to O; (8) therefore it is obvious, that it <=ΞO> will touch the parabola, (9) through the converse of the thirty-fourth theorem of the first book of Apollonius' *Conic Elements*.[385] (10) Now since BE is double EA ((11) for so it is assumed[386]) (12) that is ZK <is twice> KΘ,[387] (13) and the triangle OΘK is similar to the triangle ΞZK,[388] (14) ΞK, too, is twice KO.[389] (15) And ΞK is double KΠ, as well, (16) through the <facts> that ΞZ, too, is double KH,[390] (17) and that ΠH is parallel to KZ;[391] (18) therefore OK is equal to KΠ. (19) Therefore OKΠ, being in contact with the hyperbola, and lying between the asymptotes, is bisected <at the point of contact with the hyperbola>; (20) therefore it touches the hyperbola[392] (21) through the converse of the third theorem of the second book of Apollonius' *Conic Elements*. (22) And it touched the parabola, too, at the same <point> K. (23) Therefore the parabola touches the hyperbola at K.[393] (p) So let the

---

[382] The point E is taken implicitly to satisfy the relation mentioned in the introduction to the proof: EB is equal to twice EA.

[383] For every point taken on the parabola (say, in this diagram, K): (sq.(KZ) = rect.(ZH, HM)). (The point Z is obtained by KZ being, in this case, at right angles to the axis of the parabola and, in general, by its being parallel to the tangent of the parabola at the vertex of the diameter considered for the property.)

[384] What we call *Conics* I.26.   [385] What we know as *Conics* I.33.

[386] This is the implicit assumption of the entire discussion.

[387] Step d, *Elements* I.30, 34.   [388] Step c, *Elements* I.29, 32.

[389] *Elements* VI.4.   [390] Step m.

[391] Step d, *Elements* I.30. Finally, Step 15 derives from 16, 17 through *Elements* VI.2.

[392] In the sense of "being a tangent."

[393] As far as the extant corpus goes, this is a completely intuitive statement. Not only in the sense that we do not get a proof of the implicit assumption ("if two conic sections have the same tangent at a point, they touch at that point"), but also in a much more fundamental way, namely, we never have the concept of two conic sections being tangents even *defined*.

hyperbola, produced, as towards P, be imagined as well,[394] (q) and let a chance point be taken on AB, <namely> Σ, (r) and let TΣY be drawn through Σ parallel to KΛ, (s) and let it meet the hyperbola at T, (t) and let ΦTX be drawn through T parallel to ΓH. (24) Now since (through the hyperbola and the asymptotes)[395] (25) the <area> ΦY is equal to the <area> ΓB; (26) taking the <area> ΓΣ away <as> common, (27) the <area> ΦΣ is then equal to the <area> ΣH, (28) and through this, the line joined from Γ to X will pass through Σ.[396] (u) Let it pass, and let it be as ΓΣX. (29) And since the <square> on ΨX is equal to the <rectangle contained> by XHM[397] (30) through the parabola,[398] (31) the <square> on TX is smaller than the <rectangle contained> by XHM.[399] (v) So let the <rectangle contained> by XHΩ come to be equal to the <square> on TX.[400] (32) Now since it is: as ΣA to AΓ, so ΓH to HX,[401] (33) but as ΓH to HX (taking HΩ as a common height), so the <rectangle contained> by ΓHΩ to the <rectangle contained> by XHΩ,[402] (34) and <the rectangle contained by ΓHΩ> to the <square> on XT (which is equal to it <=to the rectangle contained by XHΩ>)[403] (35) that is to the <square> on BΣ,[404] (36) therefore the <square> on BΣ, on ΣA, is equal to the <rectangle contained> by ΓHΩ, on ΓA.[405] (37) But the <rectangle contained> by ΓHΩ, on ΓA, is smaller

---

[394] In Step k it has been drawn only as far as K.

[395] Refers to *Conics* II.12, already invoked in setting-up the hyperbola. For the theorem to apply in the way required here, it is important that the asymptotes are at right angles to each other (as indeed provided by the setting-out of the theorem).

[396] Converse of *Elements* I.43.

[397] The point Ψ is the intersection of the parabola with the line ΦX. Since this line had not yet come to existence when the parabola was drawn, this point could not be made explicit then, and it is left implicit now, to be understood on the basis of the diagram – this, the most complex of diagrams!

[398] *Conics* I.11.

[399] Archimedes effectively assumes that, inside the "box" KZHΛ, the hyperbola is always "inside" the parabola. This is nowhere proved by Apollonius. Greeks could prove this, e.g. on the basis of *Conics* IV.26.

[400] This step does not construct a rectangle (this remains a completely virtual object). Rather, it determines the point Ω.

[401] Steps c, d, *Elements* I.29, 30, 32, VI.4.   [402] *Elements* VI.1.

[403] Step v.   [404] Steps r, t, *Elements* I.34.

[405] *Elements* XI.34. The structure of Steps 32–6 being somewhat involved, I summarize their mathematical gist: (32) ΣA:AΓ::ΓH:HX, but (33) ΓH:HX::rect.(ΓHΩ):rect.(XHΩ), (34) rect.(XHΩ) = sq.(XT) hence (from 33–4) the result (not stated separately): (34′) ΓH:HX::rect.(ΓHΩ):sq.(XT); then (35) sq.(XT)=sq.(BΣ) hence the result (not stated separately): (35′) ΓH:HX::rect.(ΓHΩ):sq.(BΣ) and, with 32 back in the argument, the result (not stated separately): (35″) ΣA:AΓ::rect.(ΓHΩ):sq.(BΣ) whence finally: (36) sq.(BΣ) on ΣA = rect.(ΓHΩ) on ΓA.

than the <rectangle contained> by ΓHM, on ΓA;[406] (38) therefore the <square> on BΣ, on ΣA, is smaller than the <square> on BE, on EA.

(39) So it will be proved similarly also in all the points taken between the <points> E, B.

But then let a point be taken between the <points> E,A, <namely> ϛ. I say that like this, too, the <square> on BE, on EA, is greater than the <square> on Bϛ, on ϛA.[407]

(w) For, the same being constructed, (x) let ϙϛP be drawn through ϛ parallel to KΛ, (y) and let it meet the hyperbola at P; (40) for it meets it, (41) through its being parallel to the asymptote;[408] (z) and, having drawn A'PB' through P, parallel to AB, let it meet HZ (being produced), at B'. (42) And since, again, through the hyperbola, (43) the <area> Γ'ϙ is equal to <the area> AH,[409] (44) the line joined from Γ to B' will pass through ϛ.[410] (a') Let it pass, and let it be ΓϛB'. (45) And since, again, through the parabola, (46) the <square> on A'B' is equal to the <rectangle contained> by B'HM.[411] (47) Therefore the <square> on PB' is smaller than the <rectangle contained> by B'HM.[412] (b') Let the <square> on PB' come to be equal to the <rectangle contained> by B'HΩ.[413] (48) Now since it is: as ϛA to AΓ, so ΓH to HB',[414] (49) but as ΓH to HB' (taking HΩ as a common height), so the <rectangle contained> by ΓHΩ to the <rectangle contained> by B'HΩ,[415] (50) that is to the <square> on PB', (51) that is to the <square> on Bϛ,[416] (52) therefore the <square> on Bϛ, on ϛA, is equal to the <rectangle contained> by ΓHΩ, on ΓA.[417] (53) But the <rectangle contained> by ΓHM is greater than the <rectangle contained> by ΓHΩ;[418] (54) therefore the <square> on BE on EA is greater than the <square> on Bϛ, on ϛA, as well.

(55) So it shall be proved similarly in all the points taken between the <points> E, A, as well. (56) And it was also proved for all the <points>

---

[406] Step v, *Elements* XI.32.

[407] In Netz (1999) I suggest that this part of the argument may be due to Eutocius, rather than Archimedes.

[408] *Conics* II.13.   [409] *Conics* II.12.

[410] Converse to *Elements* I.43.   [411] *Conics* I.11.

[412] Steps 40–7 retrace the ground covered earlier at 24–31. Step 47 is unargued, like its counterpart 31.

[413] This is a very strange moment: an already determined point (Ω, determined at Step v above) is now being re-determined.

[414] *Elements* I.29, 32, VI.4.   [415] *Elements* VI.1.

[416] Steps w, x, z, *Elements* I.30, 34.   [417] *Elements* XI.34.

[418] Step b', *Elements* VI.1. The implicit result of: (52) sq.(Bϛ) on ϛA = rect.(ΓHΩ) on ΓA and (53) rect.(ΓHM) > rect.(ΓHΩ) is (53') sq.(Bϛ) on ϛA < rect.(ΓHM) on ΓA. This implicit Step 53' (together with Step 1!) is the basis of the next, final step.

between the <points> E, B; (57) therefore, of all the <magnitudes> taken similarly on AB, the greatest is the <square> on BE on EA, when BE is twice EA.

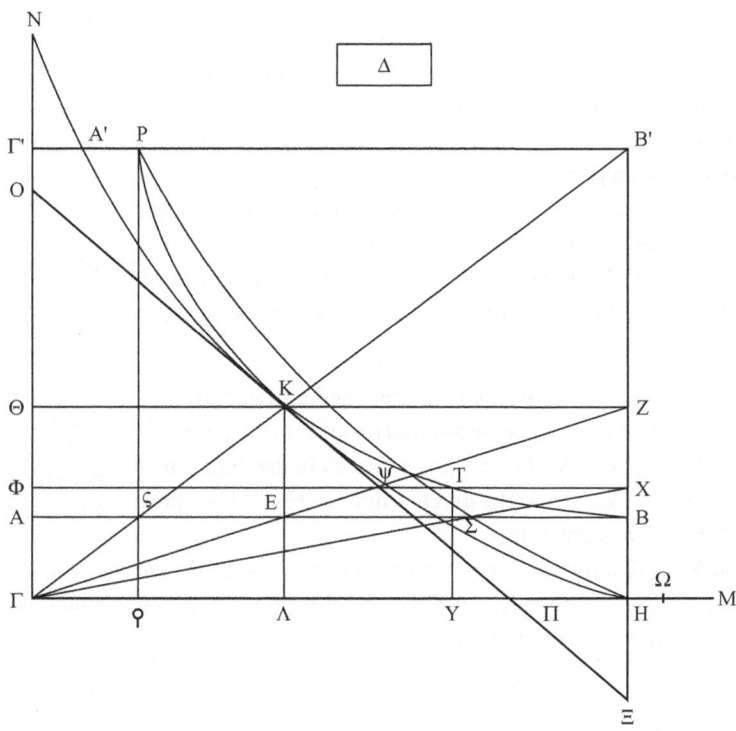

In II.4 Fifth diagram Codex A had the two curves – parabola and hyperbola – meet at a point higher, and to the left of P, as in the thumbnail. This is preserved in all copies. Codices E4 have the lower segment of the hyperbola (below K) drawn twice to the point B, once passing correctly at the point T in its correct position (the intersection TY/XΦ), once passing above it. Codex H has a similar arrangement, only it has the higher branch of the hyperbola terminate not at the point B, but above it, between B and X; codex D has only the higher branch, drawn to the point B. In all these codices DEH4, the letter T is consequently repositioned to be at the intersection of the higher branch and the line TY. It appears likely that codex A had the arrangement of E4. In all copies, the parabola passes rather near the point Σ (and not, as in the printed diagram, somewhat below the point). Perhaps it is intended to pass through that point. In codices DG, the rectangle Δ has the sides somewhat longer

Now one must understand also the consequences of the diagram above.[419] For since it has been proved that the <square> on BΣ, on ΣA, and the <square> on Bς, on ςA, are smaller than the <square> on BE, on EA: <therefore> it is possible to produce the task assigned by the original problem, by cutting the <line> AB at two points (when the given area on the given <line> is smaller than the <square> on BE on EA).

(a) And this comes to be, if we imagine a parabola drawn around the diameter XH, so that the lines drawn down <to the diameter> are in square the <rectangle applied> along HΩ; (1) for such a parabola certainly passes through the <point> T.[420] (2) And since it <=the parabola> must meet ΓN ((3) being parallel to the diameter[421]), (4) it is clear that it cuts the hyperbola at some point above K, ((b) as, here,[422] at P), (5) and <it is clear that> a perpendicular drawn from P on AB

---

[419] Here we almost certainly have Eutocius, rather than Archimedes speaking.
[420] From Step v of the proof, and then the converse to *Conics* I.11.
[421] Step 2 derives from Step 3, through *Conics* I.26.   [422] I.e. "in this diagram."

((c) as, here, Pς), cuts AB at ς, so that the point ς produces the task assigned by the problem, (6) and so that the <square> on BΣ on ΣA is then equal to the <square> on Bς on ςA (7) as is self-evident from the preceding proofs.

So that – it being possible to take two points on BA, producing the required task – one may take whichever one wishes, either the <point> between the <points> E, B, or the <point> between the <points> E, A. For if <one takes> the <point> between the <points> E, B, then, as has been said, one draws a parabola through the points H, T, which cuts the hyperbola at two points. <Of these two points,> the <point> closer to H, that is to the axis of the parabola, will procure[423] the <point> between the <points> E, B (as here T has procured Σ), while the point more distant <from the diameter will procure> the <point> between the <points> E, A (as here P procures ς).

Now, <taken> generally, the problem is analyzed and constructed like this. But, in order that it may also be applied to Archimedes' text,[424] let the diameter of the sphere ΔB be imagined (just as in the diagram of the text), and the radius BZ,[425] and the given <line> ZΘ.[426] Therefore he says the problem comes down to:

"To cut ΔZ at X, so that it is: as XZ to the given <line> so the given <square> to the <square> on ΔX. This, said in this way – without qualification – is soluble only given certain conditions."

In II.4 (*cont.*) than the base. Codex G has the rectangle Δ to the left of the main figure.   Many lines do not appear parallel. Horizontals: codex D has AB, ΦX climb to the right, ΘZ fall a little to the right; codex E has ΦX fall a little to the right; codex H has ΦX climb a little to the right; codex 4 has AB fall a little to the right. Verticals: codex E has P rather to the left of $\subset$, K slightly to the left of Λ.   Codex A had Φ instead of Ψ (so in all copies). Codex E omits Λ and Γ', Codex G omits P. I think I might see a η where Heiberg (whom I follow) prints a Γ'. The character is so rare either way that no real decision is possible.

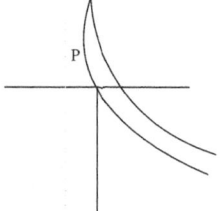

---

[423] The verb *heurisko*, better known to mathematicians for its first perfect singular used by an animate subject (*hêurêka*, translated "I have found," "I've got it"), commonly used in the infinitive with an animate logical subject understood (in the definition of goal inside problems: "*dei heurein* . . ." translated "it is required to find . . ." i.e. by the mathematician). Here, a third-person present/future with an *inanimate* subject, the translation must be different, and mine is only one of many possible guesses.

[424] I.e., the text of *SC* II.4 (all this, after all, is a commentary to that proposition!). Archimedes has promised (*SC* II.4, passage following Step 40) to analyze and construct "both problems," meaning 1. The general problem, given any two lines and an area, 2. The problem required in this proposition (the given lines and area are limited within certain parameters). The lost text found by Eutocius contained only the first, general problem. Perhaps we have lost the particular case. (It is clear that Eutocius' source was not another text of the *SC* – he would have told us that – but rather, some compilation of mathematical results. In such a context, the particular problem would have been of no interest.) Perhaps Archimedes never did give a particular solution; perhaps he meant it to be implicit in the general solution. It is so, in a sense, and Eutocius' business here is to make this implicit particular solution explicit.

[425] Another case where identity and equality are not distinguished. Eutocius' intention is not that BZ is the radius, but that it is equal to the radius.

[426] Note that Eutocius' new diagram does not come directly from the original diagram of *SC* II.4. Eutocius produced a mirror-inversion of the original diagram, putting the greater segment to the right. This is done in order to make this new diagram fit the diagrams for the solution of the problem.

For if the given <area>, on the given <line> turns out to be greater than the <square> on ΔB, on BZ, the problem would be impossible, as has been proved.[427] And if it is equal, the point B would produce the task assigned by the problem, and in this way, too, the solution would have no relevance to what Archimedes originally put forward; for the sphere would not be cut according to the given ratio.[428] Therefore, said in this way, without qualification, it was only soluble given certain added conditions. "But with the added qualification of the specific characteristics of the problem at hand" (that is, both that ΔB is twice BZ and that BZ is greater than ZΘ), "it is always soluble." For the given <square> on ΔB, on ZΘ, is smaller than the <square> on ΔB, on BZ (through BZ's being greater than ZΘ), and, when this is the case, we have shown that the problem is possible, and how it then unfolds.

It should also be noticed that Archimedes' words fit with our analysis. For previously (following his analysis) he stated, in general terms, that which the problem came down to, saying: "it is required to cut a given <line>, ΔZ, at the <point> X, and to produce: as XZ to a given <line>, so the given <square> to the <square> on ΔX." Then he says that, in general, said in this way, this is soluble only given certain conditions, but with the addition of specific characteristics of the problem that he has obtained (that ΔB is twice BZ, and that BZ is greater than ZΘ), it is always soluble. And so he takes this problem in particular, and says this: "And the problem will be as follows: given two lines ΔB, BZ (and ΔB being twice BZ), and <given> a point on BZ, <namely> Θ; to cut ΔB at X . . ." – and no longer saying, as previously, that it is required to cut ΔZ, but <to cut> ΔB, instead – because he knew (as we ourselves have proved above) that there are two points which, taken on ΔZ, produce the task assigned by the problem, one between the <points> Δ, B, and another between the <points> B, Z. Of these, the <point> between the <points> Δ, B would be of use for what Archimedes put forward originally.[429]

---

[427] Since BZ is equal to the radius, and BΔ is the diameter, obviously BΔ is twice BZ. Hence the point B is the maximum for the solid, as shown in the lemma to the analysis. If the solid required by the terms of the problem is greater than this maximum, it simply cannot be constructed.

[428] What we are looking for is a point at which to cut the sphere, so that its two segments then have a given ratio. The point B, on the surface of the sphere, can be said to produce no cutting into two segments at all. (Or, if it is said to cut the sphere, the two "segments" – one a sphere, one a point – do not have a ratio.)

[429] A phrasing reminiscent of the point made above, why B would not do as a solution (it does not produce a cut in the sphere). The same consideration applies here: we require only that solution which picks a point inside the sphere.

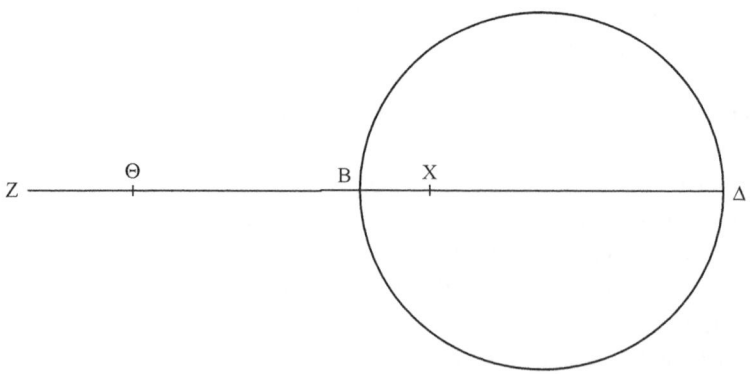

In II.4 Sixth diagram

So we have copied this down, in conformity with Archimedes' words, as clearly as possible.[430]

Dionysodorus, too, as has been said above, did not get to read what Archimedes promised to have written at the end, and was incapable of discovering again, as it were, the unpublished proofs. Taking another route to the whole problem,[431] the method of solution he uses in his treatise is not without grace. We therefore thought it incumbent upon us to add him to the above, correcting the text as best we could. For with him, too, in all the manuscripts we had come across, as a result of men's massive carelessness, much of the proofs was difficult to understand, with the sheer number of mistakes.

As Dionysodorus

To cut the given sphere by a plane, so that its segments will have to each other the given ratio.

Let there be the given sphere, whose diameter is AB, and <let> the given ratio be that which ΓΔ has to ΔE. So it is required to cut the sphere by a plane, right to AB, so that the segment whose vertex is A has to the segment whose vertex is B the ratio which ΓΔ has to ΔE.

(a) Let BA be produced to Z, (b) and let AZ be set <as> half of AB, (c) and let ZA have to AH <that> ratio which ΓE has to EΔ, (d) and let AH be at right <angles> to AB, (e) and let AΘ

---

[430] At face value, this seems to suggest that so far we had only Archimedes' words. But of course this is not the meaning. For the sake of the transition, from Archimedes to Dionysodorus, Eutocius lumps together all the preceding text as "Archimedes." It is always salutary to realize how careless are ancient commentators in signposting their text and dividing lemmas from commentary.

[431] Meaning now the main problem of *SC* II.4.

be taken <as> a mean proportional between ZA, AH; (1) therefore AΘ is greater than AH.[432] (f) And let a parabola be drawn
through the <point> Z around the axis ZB, so that the <lines> drawn
down <on the axis> are in square <the rectangles applied> along
AH;[433] (2) therefore it shall pass through Θ, (3) since the <rectangle
contained> by ZAH is equal to the <square> on AΘ.[434] (g) So let
it be drawn, and let it be as the <line> ZΘK, (h) and let BK be
drawn down through B, parallel to AΘ, (i) and let it cut the parabola
at K, (j) and let a hyperbola be drawn through H, around ZBK <as>
asymptotes; (4) so it cuts the parabola between the <points> Θ, K.[435]
(k) Let it cut <the parabola> at Λ, (l) and let ΛM be drawn <as> a
perpendicular from Λ on AB, (m) and let HN, ΛΞ be drawn through
H, Λ parallel to AB. (5) Now since HΛ is a hyperbola, (6) and ABK
are asymptotes, (7) and MΛΞ are parallel to AHN, (8) the <rectangle
contained> by AHN is equal to the <rectangle contained> by MΛΞ,
(9) through the 8th theorem of the second book of Apollonius' *Conic
Elements*.[436] (10) But HN is equal to AB,[437] (11) while ΛΞ <is equal>
to MB; (12) therefore the <rectangle contained> by ΛMB is equal to
the <rectangle contained> by HAB, (13) and through the <fact> that
the <rectangle contained> by the extremes is equal to the <rectangle
contained> by the means, (14) the four lines are proportional;[438] (15)
therefore it is: as ΛM to HA, so AB to BM; (16) therefore also: as the
<square> on ΛM to the <square> on AH, so the <square> on AB
to the <square> on BM. (17) And since (through the parabola), the

---

[432] AΘ is greater than AH, because it is the mean proportional in the series ZA –
AΘ – AH (Step e). ZA is greater than AH, because ZA, AH have the same ratio as ΓE,
EΔ (Step c), and ΓE is greater than EΔ – which, finally, we know from the diagram.

[433] Notice that the *latus rectum* is here not at the vertex of the parabola.

[434] Step e, *Elements* VI.17. Step 2 derives from Step 3 on the basis of the converse
of *Conics* I.11.

[435] The key insight of Archimedes' solution was that the parabola contained the
hyperbola in the relevant "box." The key insight of Dionysodorus' solution is that the
hyperbola cuts the parabola at the relevant "box." Both insights are stated without proof,
typical for such topological insights in Greek mathematics. Dionysodorus' understanding
of the situation may have been like this. Concentrate on the wing of the hyperbola to
the right of AΘ. It must get closer and closer to the line BK, without ever touching
that line (BK is an asymptote to the hyperbola: the relevant proposition is *Conics* II.14).
So the hyperbola cannot pass wholly below or above the point K; at some point, well
before reaching the line BK, it must pass higher than the point K. Since at the stretch
ΘK, the parabola's highest point is K (this can be shown directly from the construction
of the parabola, *Conics* I.11), what we have shown is that the hyperbola, starting below
the parabola (H below Θ), will become higher than the parabola, well before either
reaches the line BK. Thus they must cut each other.

[436] What we call *Conics* II.12.   [437] *Elements* I.34.   [438] *Elements* VI.16.

<square> on ΛM is equal to the <rectangle contained> by ZM, AH,[439] (18) therefore it is: as ZM to MΛ, so MΛ to AH;[440] (19) therefore also: as the first to the third, so the <square> on the first to the <square> on the second and the <square> on the second to the <square> on the third;[441] (20) therefore as ZM to AH, so the <square> on ΛM to the <square> on HA. (21) But as the <square> on ΛM to the <square> on AH, so the <square> on AB was proved <to be> to the <square> on BM; (22) therefore also: as the <square> on AB to the <square> on BM, so ZM to AH. (23) But as the <square> on AB to the <square> on BM, so the circle whose radius is equal to AB to the circle whose radius is equal to BM;[442] (24) therefore also: as the circle whose radius is equal to AB to the circle whose radius is equal to BM, so ZM to AH; (25) therefore the cone having the circle whose radius is equal to AB <as> base, and AH <as> height, is equal to the cone having the circle whose radius is equal to BM <as> base, and ZM <as> height;[443] (26) for such cones, whose bases are in reciprocal proportion to the heights, are equal.[444] (27) But the cone having the circle whose radius is equal to AB <as> base, and ZA <as> height, is to the cone having the same base, but <having> AH <as> height, as ZA to AH,[445] (28) that is ΓE to EΔ ((29) for, being on the same base, they are to each other as the heights[446]); (30) therefore the cone, too, having the circle whose radius is equal to AB <as> base, and ZA <as> height, is to the cone having the circle whose radius is equal to BM <as> base, and ZM <as> height, as ΓE to EΔ. (31) But the cone having the circle whose radius is equal to AB <as> base, and ZA <as> height, is equal to the sphere,[447] (32) while the cone having the circle whose radius is equal to BM <as>

---

[439] *Conics* I.11.   [440] *Elements* VI.17.   [441] *Elements* VI.20 Cor. 2.

[442] *Elements* XII.2. Those are curious circles. We are not quite given them, since we do not know their exact radii. (We know what their radii are *equal to*, but this is not yet knowing what they *are*.) On the other hand, these are fully fledged individuals: they are "the" circles of their kind, not just "a" circle whose radius is equal to a given line. Over and above the semi-reality of the diagram, we are asked to invent another toy reality, where certain unnamed circles subsist. More of this to follow.

[443] The toy circles, introduced in Step 23, spring out of their boxes, now as cones. These cones have a particularly funny spatial location: their heights are not merely equal to certain lines, but *are* in fact those certain lines. Hence they are half in the toy universe of the circles, half in the more tangible universe of the diagram. Or more precisely: the sense of location has been eroded, and we are faced with purely hypothetical geometrical objects.

[444] *Elements* XII.15.   [445] *Elements* XII.14.

[446] This belated explicit reference to *Elements* XII.14 is meant to support Step 27, not Step 28. It is probably Eutocius' contribution and, if so, so are probably the other references to the *Elements* and the *Conics*.

[447] *SC* I.34.

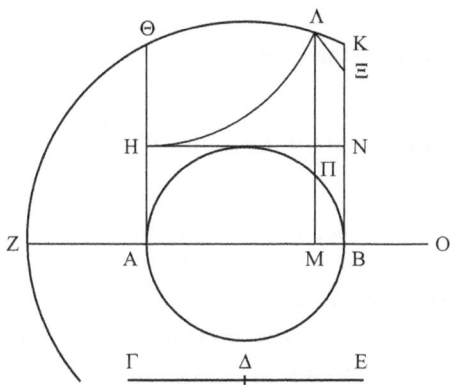

In II.4 Seventh diagram Codex E has the line ΛΞ drawn to the point K, so that the points Ξ, K coalesce; meanwhile, it repositions the letter Ξ so that it appears to belong to the point Λ (!), right beneath it. The letter Ξ is similarly, at least ambiguously positioned in codices GH. All codices omit the letter N.

base, and ZM <as> height, is equal to the segment of the sphere, whose vertex is B, and <whose> height is BM, (33) as shall be proved further on; (34) therefore the sphere, too, has to the said segment the ratio which ΓE has to EΔ; (35) dividedly, also: the segment, whose vertex is A, and <whose> height is AM, has to the segment, whose vertex is B, and <whose> height is BM, this ratio, which ΓΔ has to ΔE.[448] (36) Therefore the plane produced through ΛM, right to AB, cuts the sphere according to the given ratio; which it was required to do.

And it shall be proved like this, that the cone having the circle whose radius is equal to BM <as> base, and ZM <as> height, is equal to the segment of the sphere whose vertex is B, and whose height is BM:

(n) For let it come to be: as ZM to MA, so OM to MB;[449] (37) therefore the cone having the same base as the segment, and OM <as> height, is equal to the segment.[450] (38) And since it is: as ZM to MA, so OM to MB, (39) alternately also: as ZM to MO, so AM to MB,[451] (40) but as AM to MB, so the <square> on ΠM to the <square> on MB,[452] (41) and as the <square> on ΠM to the <square> on MB, so the circle whose radius is equal to ΠM, to the circle whose radius is equal to MB,[453] (42) therefore as the circle whose radius is equal to ΠM, to the circle whose radius is equal to MB, so MZ to MO. (43) Therefore the cone having the circle whose radius is equal to MB <as> base, and ZM <as> height, is equal to the cone having the circle whose radius is equal to ΠM <as> base, and MO <as> height; (44) for their bases are in reciprocal proportion to the heights;[454] (45) so

---

[448] *Elements* V.17.   [449] This is the definition of the point O.

[450] *SC* II.2.   [451] *Elements* V.16.

[452] The point Π is defined by the diagram alone. It is the intersection of the line ΛM with the circle. The claim is based on *Elements* III.31, VI.8 Cor., VI.20 Cor. 2.

[453] *Elements* XII.2.   [454] *Elements* XII.15.

that it <=the cone having the circle whose radius is equal to MB as
base, and ZM as height> is equal to the segment, too.

As Diocles in *On Burning Mirrors*[455]

And Diocles, too, gives a proof, following this introduction:

Archimedes proved in *On Sphere and Cylinder* that every segment
of a sphere is equal to a cone having the same base as the segment, and,
<as> height, a line having a certain ratio to the perpendicular <drawn>
from the vertex of the segment on the base: <namely, the ratio> that:
the radius of the sphere, and the perpendicular of the alternate segment,
taken together, have to the perpendicular of the alternate segment.[456]
For instance, if there is a sphere ABΓ, and if it is cut by a certain plane,
<namely> the circle around the diameter ΓΔ,[457] and if (AB being
diameter, and E center) we make: as EA, ZA taken together to ZA, so
HZ to ZB, and yet again, as EB, BZ taken together to ZB, so ΘZ to
ZA, it is proven: that the segment of the sphere ΓBΔ is equal to the
cone whose base is the circle around the diameter ΓΔ, while its height
is ZH, and that the segment ΓAΔ is equal to the cone whose base is
the same, while its height is ΘZ. So he set himself the task of cutting
the given sphere by a plane, so that the segments of the sphere have to
each other the given ratio, and, making the construction above, he says:
"(1) Therefore the ratio of the cone whose base is the circle around the
diameter ΓΔ, and whose height is ZΘ, to the cone whose base is the
same, while its height is ZH, is given, too;"[458] (2) and indeed, this too
was proved;[459] (3) and cones which are on equal bases are to each other
as the heights;[460] (4) therefore the ratio of ΘZ to ZH is given. (5) And
since it is: as ΘZ to ZA, so EBZ taken together to ZB, (6) dividedly: as
ΘA to AZ, so EB to ZB.[461] (7) And so through the same <arguments>
also: as HB to ZB, so the same line <=EB> to ZA.

So a problem arises like this: with a line, <namely> AB, given in
position, and given two points A, B, and given EB, to cut AB at Z and

---

[455] The following text corresponds to Propositions 7–8 of the Arabic translation of
Diocles' treatise (Toomer [1976] 76–86, who also offers in 178–92 a translation of the
passage in Eutocius with a very valuable discussion, 209–12).

[456] *SC* II.2.

[457] The circle meant is that perpendicular to the "plane of the page," or to the line
AB.

[458] This text is part Diocles' own analysis, part a re-creation of Archimedes' analysis,
now in the terms of Diocles' diagram. Step 1 here corresponds to *SC* II.4 Step 4.

[459] Step 2 probably means: "by proving *SC* II.2, we thereby prove the claim of
step 1."

[460] *Elements* XII.14.   [461] *Elements* V.17.

to add ΘA, BH so that the ratio of ΘZ to ZH will be <the> given, and also, so that it will be: as ΘA to AZ, so the given line to ZB, while as HB to BZ, so the same given line to ZA.

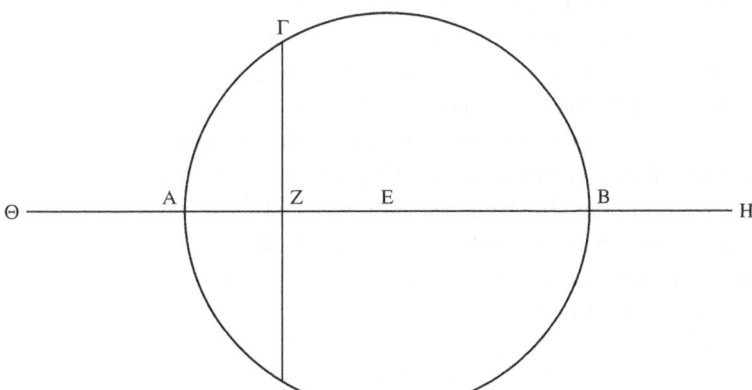

In II.4 Eighth diagram

And this is proved in what follows. For Archimedes, having proved the same thing, rather long-windedly, even so he then reduced it to another problem, which he does not prove in the *Sphere and Cylinder*![462]

Given in position a line AB, and given two points A, B, and the ratio, which Γ has to Δ, to cut AB at E and to add ZA, HB, so that it is: as Γ to Δ, so ZE to EH; and also that it is: as ZA to AE, so a certain given line to BE, and as HB to BE, so the same given line to EA.[463]

(a) Let it come to be, (b) and let ΘAK, ΛBM be drawn at right <angles> to AB, (c) and let each of AK, BM be set equal to the given line. (d) Joining the <lines> KE, ME, let them be produced to Λ, Θ, (e) and let KM be joined, as well, (f) and let ΛN be drawn through Λ, parallel to AB, (g) and let ΞEOΠ <be drawn> through E, <parallel> to NK. (1) Now since it is: as ZA to AE, so MB to BE; (2) for this is assumed; (3) and as MB to BE, so ΘA to AE (4) through the similarity of the triangles,[464] (5) therefore as ZA to AE, so ΘA to AE. (6) Therefore

---

[462] Archimedes transformed the problem of *SC* II.4 into another, more general problem, at a certain remove from the sphere to be cut. To solve that problem, conic sections were required, but since the problem was much more general, and since its solution was removed to an appendix, *Sphere and Cylinder* remained cordoned off from conic sections, preserving a certain elementary aspect. Diocles, on the other hand, applies conic sections to the terms of the problem arising directly from *Sphere and Cylinder*. This makes his approach at the same time more direct, but less elegant.

[463] The "certain given line" remains unlabeled.

[464] The triangles referred to are ΘAE, BEM. That they are similar can be seen through Step b, *Elements* I.27, I.29, I.15 (or I.29, I.32). Step 3 derives from Step 4 through *Elements* VI.4.

ZA is equal to ΘA.[465] (7) So, through the same <arguments>, BH, too, <is equal> to BΛ.[466] (8) And since it is: as ΘAE taken together to MBE taken together, so KAE taken together to ΛBE taken together; (9) for each of the ratios is the same as the <ratio> of AE to EB;[467] (10) therefore the <rectangle contained> by ΘAE taken together and by ΛBE taken together, is equal to the <rectangle contained> by KAE taken together and by MBE taken together;[468] (h) Let each of AP, BΣ be set equal to KA.[469] (11) Now since ΘAE taken together is equal to ZE, (12) while ΛBE taken together is equal to EH, (13) and KAE taken together is equal to PE, (14) and MBE taken together is equal to ΣE, (15) and the <rectangle contained> by ΘAE taken together and by ΛBE taken together was proved to be equal to the <rectangle contained> by KAE taken together and by MBE taken together, (16) therefore the <rectangle contained> by ZEH is equal to the <rectangle contained> by PEΣ. (17) So through this, whenever P falls between the <points> A, Z, then Σ falls outside H, and vice versa.[470] (18) Now since it is: as Γ to Δ, so ZE to EH, (19) and as ZE to EH, so the <rectangle contained> by ZEH to the <square> on EH,[471] (20) therefore: as Γ to Δ, so the <rectangle contained> by ZEH to the <square> on EH. (21) And the <rectangle contained> by ZEH was proved equal to the <rectangle contained> by PEΣ; (22) therefore it is: as Γ to Δ, so the <rectangle

---

[465] *Elements* V.9.

[466] The setting-out and Step a, again, provide the proportion HB:BE::KA:AE and, through the similarity of the triangles KAE, ΛEB the argument is obvious.

[467] By "each of the ratios" Diocles refers to the ratios of the separate lines making up the "taken together" objects. So we have four ratios: ΘA:MB, AE:BE, KA:ΛB, AE:BE (AE:BE occurs twice). All, indeed, are the same as AE:BE, through the similarities of triangles we have already seen. Step 8 follows from Step 9 through successive applications of *Elements* V.18.

[468] *Elements* VI.16. Notice a possible source of confusion. The rectangles are each contained by two lines, and each of these lines is a sum of two lines, denoted by three characters. This is confusing, because often we have a rectangle contained by two lines, and these containing two lines are directly denoted by three characters. Here the summation happens not between the sides of the rectangles, but inside each of the sides.

[469] Thus all lines AP, BΣ, KA, BM are now equal to the unlabeled, given line – this anonymous line is cloned, as it were, all through the diagram.

[470] The "vice versa" means that, conversely to what has been mentioned, also when H falls between B, Σ, then P falls outside Z. ("Outside" here means "away from the center of the diagram" – imagine the diagram as an underground network, and imagine that the lines have two directions, "Inbound" and "Outbound")." This is a remarkable moment. Diocles (or Eutocius?) is aware both of topological considerations, and of a functional relation between variables. But the basic thought is very simple: it is impossible to have two equal rectangles, if the sides of one of the rectangles are both greater than the sides of the other. If one side is greater, the other must be smaller. This is not stated in the *Elements*, but it is implicit in *Elements* VI.16. (That P, Σ, must both be "outside" AB, is implicit in the construction of the points and is learned from the diagram.)

[471] *Elements* VI.1.

contained> by PEΣ to the <square> on EH. (i) Let EO be set equal to BE, (j) and, joining BO, let it be produced to either side, (k) and, drawing ΣT, PY from Σ, P at right <angles to the line AB>, (l) let them meet it <=the line BO, produced> at T, Y. (23) Now since the <line> TY has been drawn through a given <point> B, (24) producing, to a <line> given in position, <namely> to AB, an angle (<namely>, the <angle contained> by EBO), half of a right <angle>,[472] (25) TY is given in position.[473] (26) And the <lines> ΣT, PY, <given> in position, are drawn from given <points,> Σ, P, cutting it <=the line TY, given in position,> at T, Y; (27) therefore T, Y are given;[474] (28) therefore TY is given in position and in magnitude. (29) And since, through the similarity of the triangles EOB, ΣTB,[475] (30) it is: as TB to BO, so ΣB to BE,[476] (31) it is compoundly, also: as TO to OB, so ΣE to EB.[477] (32) But as BO to OY, so BE to EP.[478] (33) Therefore also, through the equality: as TO to OY, so ΣE to EP.[479] (34) But as TO to OY, so the <rectangle contained> by TOY to the <square> on OY, (35) and as ΣE to EP, so the <rectangle contained> by ΣEP to the <square> on EP;[480] (36) therefore also: as the <rectangle contained> by TOY to the <square> on OY, so the <rectangle contained> by ΣEP to the <square> on EP; (37) alternately also: as the <rectangle contained> by TOY to the <rectangle contained> by ΣEP, so the <square> on OY to the <square> on EP. (38) And the <square> on OY is twice the <square> on EP, (39) since the <square> on OB is twice the <square> on BE, too;[481] (40) therefore the <rectangle contained> by TOY, too, is twice the <rectangle contained> by ΣEP. (41) And the <rectangle contained> by ΣEP was proved to have, to the <square> on EH, the ratio which Γ has to Δ; (42) and therefore the <rectangle contained> by TOY has to the <square> on EH the ratio, which twice Γ has to Δ. (43) And the <square> on EH is equal to the <square> on ΞO; (44) for each of the <lines> EH, ΞO is equal to ΛBE taken together;[482] (45) Therefore the <rectangle contained> by TOY has to the <square> on ΞO <the> ratio, which twice Γ has to Δ. (46) And the ratio of twice Γ to Δ is given; (47) therefore the ratio of the <rectangle contained> by TOY to the <square> on ΞO is given as well.

---

[472] From Step i, OE = EB. From Steps b, g, OEB is a right angle. Then the claim of Step 24 is seen through *Elements* I.32.

[473] *Data* 30.   [474] *Data* 25.

[475] Steps b, g, k, *Elements* I.27, 29, 15 (or 32).   [476] *Elements* VI.4.

[477] *Elements* V.18.   [478] Steps b, g, k, *Elements* I.27, VI.2.

[479] *Elements* V.22.   [480] Steps 34–5: both from *Elements* VI.1.

[481] This is through the special case of Pythagoras' theorem (*Elements* I.47) for an isosceles right-angled triangle.

[482] ΞE=ΛB (through Steps b, f, g, *Elements* I.27, 30, 34). EO = EB through Step i. So this settles ΞO = ΛBE. EH = ΛBE can be seen through Step 7.

(48) Therefore if we make: as Δ to twice Γ, so TY to some other <line>, as Φ, and if we draw an ellipse around TY, so that the <lines> drawn down <on the diameter>, inside the angle ΞOB (that is <inside> half a right <angle>), are in square the <rectangles applied> along Φ, falling short by a <figure> similar to the <rectangle contained> by TY, Φ,[483] <the ellipse> shall pass through the point Ξ, (49) through the converse of the twentieth theorem of the first book of Apollonius' *Conic Elements*.[484] (m) Let it be drawn and let it be as YΞT; (50) therefore the point Ξ touches an ellipse given in position. (51) And since ΛK is a diagonal of the parallelogram NM,[485] (52) the <rectangle contained> by NΞΠ is equal to the <rectangle contained> by ABM.[486] (53) Therefore if we draw a hyperbola through the <point> B, around ΘKM <as> asymptotes, it shall pass through Ξ,[487] (54) and it shall be given in position ((55) through the <facts> that the point B, too, is given in position, (56) as well as each of the <lines> AB, BM, (57) and also, through this, the asymptotes ΘKM). (n) Let it be drawn and let it be as ΞB; (58) therefore the point Ξ touches a hyperbola given in position. (59) And it also touched an ellipse given in position; (60) therefore Ξ is given.[488] (61) And ΞE is a perpendicular <drawn> from it; (62) therefore E is given.[489] (63) And since it is: as MB to BE, so ZA to AE, (64) and AE is given, (65) therefore AZ is given, as well.[490] (66) So, through the same <arguments>, HB is given as well.[491]

---

[483] This is the Apollonian way of stating that Φ is the parameter of the ellipse. Imagine that Φ is set at the point T, at right angles to the line YT. You get a configuration similar to that of *Conics* I.13 (see figure. Φ here is transformed into EΘ), for which Apollonius proves that for any point Λ taken on the ellipse EΛΔ, the square on ΛM is equal to the associated rectangle MO.

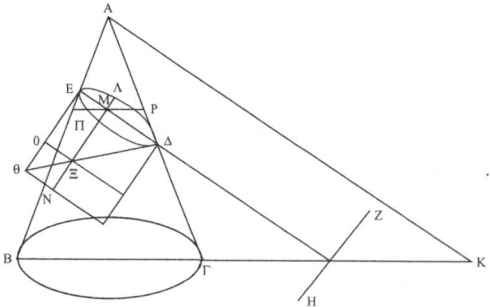

Apollonius *Conics* I.13

[484] What we call *Conics* I.21.   [485] Steps b, c, f, *Elements* I.27, 33.
[486] Based on *Elements* I.43.   [487] Converse of *Conics* II.12.
[488] *Data* 25.   [489] *Data* 30.
[490] With E given, BE is given as well. BM is given from setting-out, Step c, hence BM:BE is given. Step 65 then derives from *Data* 2.
[491] The only difference will be that instead of using the proportion MB:BE::ZA:AE, we use the proportion HB:BE::BM:EA (both from setting-out, Step c).

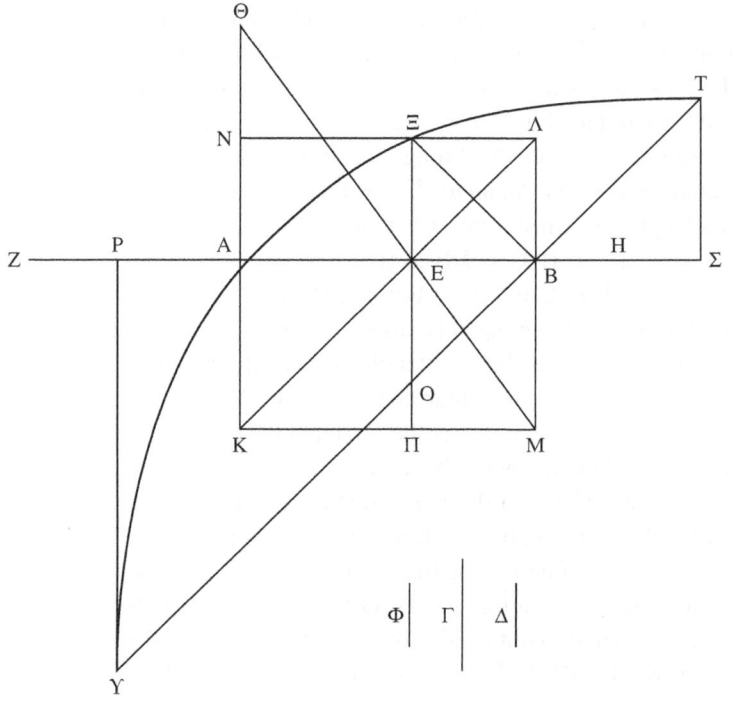

In II.4 Ninth diagram Codex D omits the line ΞB. (Since the line represents a hyperbola, Heiberg draws it as a curved line). Codex D further has the lines KΛ, TY, not parallel, meet at the point K. It also positions the three lines Φ, Γ, Δ at the "slot" beneath BΣ, and has Φ and Δ greater (rather than smaller) than Γ.   Codex A had X instead of K, and has permuted the letters H, N. All are corrected in the codices BG.   Codex D has both X, K where K is required.   Codex H has E instead of Σ.

And it shall be constructed like this:

For (as in the same diagram), let the given line, which it is required to cut, be AB, and let the other given <line> be AK, and <let> the given ratio <be> the <ratio> of Γ to Δ.

(a) Let BM (being equal to AK) be drawn at right <angles> to AB, (b) and let KM be joined, (c) and let AP and BΣ be set equal to KA, (d) and let PY, ΣT be drawn from P, Σ at right <angles to AB>, (e) and let half a right <angle> be constructed at the point B, <namely> the angle <contained> by ABO, (f) and, producing BO to either side, let it cut ΣT, PY at T, Y, (g) and let it come to be: as Δ to twice Γ, so TY to Φ, (h) and let an ellipse be drawn around TY, so that the <lines> drawn down inside half a right angle are in square the <rectangles> applied along Φ, falling short by a figure similar to the <rectangle contained> by TY, Φ, (i) and let a hyperbola be drawn through B, around AK, KM <as> asymptotes, <namely> the <hyperbola> BΞ, (j) cutting the ellipse at Ξ, (k) and let a perpendicular, <namely> ΞE, be drawn from Ξ on AB, (l) and let it be produced to Π, (m) and let ΛΞN be drawn through Ξ parallel to AB, (n) and let KA, MB be produced to Λ, Θ, (o) and, joining ME, let it be produced and let it meet KN at Θ. (1) Now since BΞ is a hyperbola, and ΘK, KM are asymptotes, (2) the <rectangle contained> by NΞΠ is equal to the <rectangle contained> by ABM, (3) through the 8th theorem of the second book of Apollonius' *Conic Elements*,[492] (4) and, through this, ΚΕΛ is a straight <line>.[493] (p) So let AZ be set equal to ΘA, (q) and <let> BH <be set> equal to ΛB. (5) Now since it is: as twice Γ to Δ, so Φ to TY, (6) and as Φ to TY, so the <rectangle contained> by TOY to the <square> on ΞO, (7) through the 20th theorem of the first book of Apollonius' *Conic Elements*,[494] (8) therefore as twice Γ to Δ, so the <rectangle contained> by TOY to the <square> on ΞO. (9) And since it is: as TB to BO, so ΣB to BE,[495] (10) compoundly also: as TO to OB, so ΣE to EB.[496] (11) But as BO to OY, so BE to EP;[497] (12) therefore through the equality, also: as TO to OY, so ΣE to EP.[498] (13) Therefore also: as the <rectangle contained> by TOY to the <square> on OY, so the <rectangle contained> by ΣEP to the <square> on EP;[499] (14) Alternately: as the <rectangle contained> by TOY to the <rectangle contained> by ΣEP, so the <square> on OY to the <square> on EP.[500] (15) But the <square> on OY is twice the <square> on EP (16) through <the fact> that the <square> on BO, too, is <twice> the <square> on BE;[501] ((17) for BE is equal to EO,[502] (18) each of the <angles> at B, O being half right);[503] (19) therefore the <rectangle contained> by TOY, too, is twice the <rectangle contained> by ΣEP.[504] (20) Now since it was proved: as twice Γ to Δ, so the <rectangle contained> by TOY to the <square> on ΞO,[505] (21) also the halves of the antecedents; (22) therefore as Γ to Δ, so the <rectangle contained> by ΣEP to the <square> on ΞO, (23) that is to the <square> on EH; (24) for ΞO is equal to EH, (25) through <the fact> that each of them is equal to ΛBE taken together.[506] (26) Now since it is: as ΘAE taken together to MBE taken together, so KAE taken together to ΛBE taken together; (27) for each of the ratios is the same as the <ratio> of AE to EB;[507] (28) therefore

---

[492] What we call *Conics* II.12.

[493] Converse of *Elements* I.43. (As usual, the assumption that ΝΛΜΚ is a parallelogram is not made explicit. It can be shown on the basis of setting-out, Steps a, m, *Elements* I.30, 33.)

[494] What we call *Conics* I.21.

[495] Steps b, d, *Elements* I.28, 29, and then I.15 (or I.32), and finally VI.4.

[496] *Elements* V.18.   [497] Steps d, k, l, *Elements* I.28, VI.2.

[498] *Elements* V.22.   [499] Successive applications of *Elements* VI.1.

[500] *Elements* V.16.

[501] Step 15 derives from Step 16 through Steps d, k, *Elements* VI.2, then V.17, then VI.22.

[502] Step 16 derives from Step 17 through Step k, *Elements* I.6, 47.

[503] Steps e, k, *Elements* I.32.

[504] Steps d, k, *Elements* VI.2, and then V.18, and then VI.22.   [505] Step 8.

[506] Steps q, 17. Also: Steps a, k, l, m together with *Elements* I.29, 34.

[507] Step 26 derives from 27 through Steps a, *Elements* I.29, VI.2, and then V.18 (that the angle at A is right is an assumption carried over without mention from the setting-out of the analysis).

the <rectangle contained> by ΘAE taken together and by ΛBE taken together is equal to the <rectangle contained> by KAE taken together and by MBE taken together.[508] (29) But ΘAE taken together is equal to ZE,[509] (30) while ΛBE taken together is equal to EH, (31) and KAE taken together is equal to PE, (32) and MBE taken together is equal to EΣ; (33) therefore the <rectangle contained> by ZEH is equal to the <rectangle contained> by PEΣ. (34) But as Γ to Δ, so the <rectangle contained> by PEΣ to the <square> on EH; (35) therefore also: as Γ to Δ, so the <rectangle contained> by ZEH to the <square> on EH. (36) But as the <rectangle contained> by ZEH to the <square> on EH, so ZE to EH;[510] (37) therefore also: as Γ to Δ, so ZE to EH. (38) And since it is: as MB to BE, so ΘA to AE,[511] (39) and ΘA is equal to ZA, (40) therefore as MB to BE, so ZA to AE. (41) And through the same <arguments> also: as KA to AE, so HB to BE.[512]

Therefore given a line, <namely> AB, and another, <namely> AK, and a ratio, <namely that> of Γ to Δ, a chance point has been taken on AB, <namely> E, and lines have been added, <namely> ZA, HB; and ZE was then to EH in the given ratio, and it is also: as the given <line> MB to BE, so ZA to AE, and as the same given <line> KA[513] to AE, so HB to BE; which it was required to do.

These things proved, it is possible to cut the given sphere according to the given ratio, like this:

For let the diameter of the given sphere be AB, and <let> the given ratio, which the segments of the sphere are required to have to each other, be the <ratio> of Γ to Δ; (a) and let E be center of the sphere; (b) and let a point, Z, be taken on AB, (c) and let HA, ΘB be added so that it is: as Γ to Δ, so HZ to ZΘ, and further yet it is: as HA to AZ, so EB, given, to BZ while as ΘB to BZ, so the same given <line,> EA, to AZ; for it has been proved above that it is possible to do this; (d) and let KZΛ be drawn through Z at right <angles> to AB, (e) and let a plane, produced through KΛ, right to AB, cut the sphere. I say that the segments of the sphere have to each other the ratio of Γ to Δ.

---

[508] *Elements* VI.16.

[509] Step p. The original syntactic structure is: "But to ΘAE taken together is equal ZE" (and similarly with the following equalities).

[510] *Elements* VI.1.

[511] Compare the argument for the derivation of Step 26 from Step 27.

[512] Substitute Step q for Step p, in the chain of reasoning, and the argument is indeed the same.

[513] "The same" as MB (equality is taken here for identity: also compare Step c in the synthesis following).

(1) For since it is: as HA to AZ, so EB to BZ,[514] (2) also compoundly;[515] (3) therefore as HZ to ZA, so EB, BZ taken together to BZ; (4) therefore the cone having the circle around the diameter KΛ <as> base, and ZH as height, is equal to the segment of the sphere having the same base and ZA <as> height.[516] (5) Again, since it is: as ΘB to BZ, so EA to AZ, (5) it is also, compoundly: as ΘZ to BZ, so EA, AZ taken together to AZ;[517] (6) therefore the cone having the circle around the diameter KΛ <as> base, and ZΘ as height, is equal to the segment of the sphere having the same base, and BZ <as> height.[518] (7) Now since the said cones, being on the same bases, are to each other as the heights,[519] (8) that is as HZ to ZΘ, (9) that is Γ to Δ, (10) therefore the segments of the sphere, as well, have to each other the given ratio; which it was required to do.

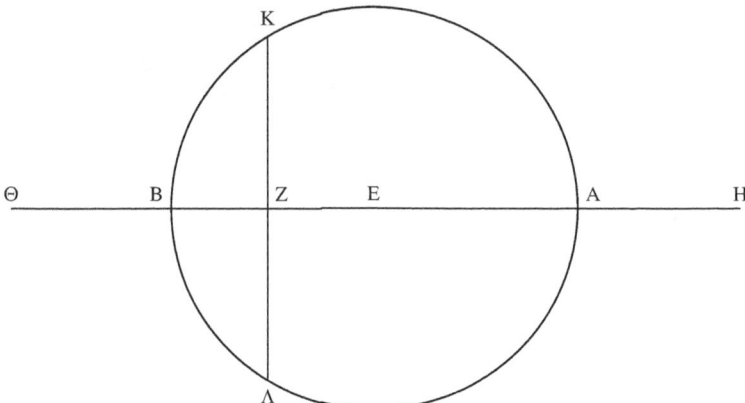

In II.4 Tenth diagram Codex D has AH considerably greater than ΘB.

And we shall prove like this, how one draws a hyperbola through the given point, around the given asymptotes, (as this is not a self-evident outcome of the *Conic Elements*):[520]

Let there be two lines, ΓA, AB, containing a chance angle (that at A), and let some point Δ be given, and let it be put forth: to draw a hyperbola through Δ around ΓA, AB <as> asymptotes.

(a) Let AΔ be joined and produced to E, (b) and let AE be set equal to ΔA, (c) and let ΔZ be drawn through Δ parallel to AB, (d) and let ZΓ be set equal to AZ, (e) and, having joined ΓΔ, let it be produced to B, (f) and let the <rectangle contained> by ΔE, H be equal to the <square> on ΓB,[521] (g) and, producing the <line> AΔ,

---

[514] Step c.   [515] *Elements* V.18.   [516] *SC* II.2.
[517] *Elements* V.18.   [518] *SC* II.2.   [519] *Elements* XII.14.
[520] As noted by Heiberg, at this point we definitely move from Diocles to Eutocius, who refers to Apollonius' *Conics*.
[521] Step f defined the point E.

let a hyperbola be drawn around it,[522] through the <point> Δ, so that the <lines> drawn down <on the axis> are in square <the rectangles applied> along H, exceeding by <a figure> similar to the <rectangle contained> by ΔE, H.[523] I say that ΓA, AB are asymptotes of the drawn hyperbola.

(1) For since ΔZ is parallel to BA, (2) and ΓZ is equal to ZA, (3) therefore ΓΔ, too, is equal to ΔB;[524] (4) so that the <square> on ΓB is four times the <square> on ΓΔ. (5) And the <square> on ΓB is equal to the <rectangle contained> by ΔE, H; (6) therefore each of the <squares> on ΓΔ, ΔB is a fourth part of the figure <contained> by ΔE, H.[525] (7) Therefore ΓA, AB are asymptotes of the hyperbola, (8) through the first theorem of the second book of Apollonius' *Conic Elements*.

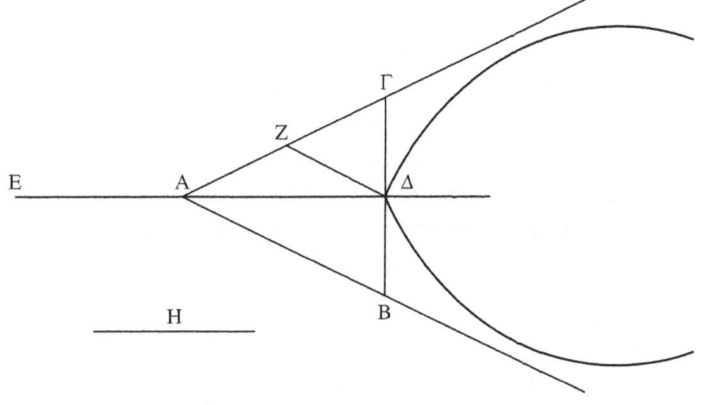

To the synthesis of 4

In II.4 Eleventh diagram Codex D has the hyperbola drawn as a single arc (about a semicircle). (So does the heavily corrected codex B, which also has the asymptotes touch the hyperbola). Codex E does not continue the line EΔ inside the hyperbola itself.   Codex A has omitted the line H with its letter; it is reinserted as a late correction in Codex B, at the position printed.

In the synthesis he produces the diameter of the sphere, ΔB, and sets next to it the line ZB, equal to its half, and he cuts it at Θ by the given ratio, and he takes the <point> X on ΔB in such a way that it is: as

---

[522] This time "around" means "around it as diameter" (and not, as above, "around as asymptote"). If a hyperbola is around one line, it is around a diameter; if around two lines, it is around asymptotes.

[523] The formulaic way of stating that E is the center, and H the parameter of the hyperbola.

[524] *Elements* VI.2.

[525] In Apollonius' *Conics* II, the expression "the fourth of the figure <contained> by . . ." becomes formulaic, hence the word "figure" here, which refers simply, in this case, to the contained rectangle.

Arch. 205   XZ to ΘZ, so the <square> on BΔ to the <square> on ΔX – making
the same construction as before. He then says that; "let it come to be:
as KΔX taken together to ΔX, so PX to XB," and he sets P between
the <points> Θ, Z.

It ought to be proved that this is the case.[526] (1) For since it is: as
KΔX taken together to ΔX, PX to XB, (2) dividedly: as KΔ to ΔX,
PB to XB;[527] (3) alternately: as KΔ to PB, ΔX to BX.[528] (4) But ΔX
is greater than XB; (5) therefore KB, too, is greater than BP[529] (6) that
is ZB <is greater> than BP; (7) so that P falls inside Z. (8) That it also
falls outside Θ shall be proved similarly to the <arguments> in the
analysis (as the entire synthesis of the theorem proceeds < = similarly
to the analysis>).[530] (9) For it is obtained, that it is: as PX to XΛ, BΘ to
ΘZ,[531] (10) so that compoundly, also.[532] And through this the present
proof, too, follows in accordance to what was said above.[533]

Arch. 205   "And through the equality in the perturbed proportion." We learned
in the *Elements* that "a perturbed proportion is, there being three
magnitudes and others equal to them in multitude, when it is: as antecedent to consequent in the first magnitudes, so, in the second magnitudes, antecedent to consequent, while, as consequent to some other
<magnitude> in the first, so, in the second, some other <magnitude>
to antecedent."[534] Now, it has also been proved here that as antecedent

---

[526] I.e. that, given the construction, the position of P is indeed as in the diagram, i.e.
between the points Θ, Z. As Eutocius will make clear, this is essentially the same as his
note to Step 21 of the analysis of this proposition. Nothing in Archimedes' argument
relies on the exact position of the point. This is a commentator's, not a mathematician's
"ought." The force of the "ought" is that this is an interesting point to comment upon,
not that this is a logical lacuna.

[527] *Elements* V.17.   [528] *Elements* V.16.

[529] KΔ, KB are taken to be interchangeable (both radii).

[530] The original grammar is very compressed; perhaps some words have been lost?
The point is clear: the synthesis is the same as the analysis, even with the same labeling
of the diagram, hence precisely the same arguments would apply without any change
including, presumably, Eutocius' comment to Step 21.

[531] Step 16 in the synthesis (inverted).

[532] *Elements* V. 18. The implicit claim is: PΛ:XΛ::BZ:ΘZ.

[533] What Eutocius has done is to show that the construction of Step h in the analysis
holds in the synthesis as well (although it is not made explicit in the synthesis). Having
shown that, he is justified in simply pointing backwards to his argument on that Step h.

[534] Euclid's formulation at *Elements* V Def. 18 (essentially unchanged by Eutocius)
suffers from the difficulty of marking out, without lettering, an object which is arbitrary
and yet fixed. The strange bare nouns, "antecedent in the first magnitudes," "consequent in
the first magnitudes" etc. are just this: one of the terms in the proportion, no matter which,
but the same throughout the definition. Heath's lettered and typographic transcription is
"a *perturbed proportion* is an expression for the case when, there being three magnitudes
a, b, c and three others A, B, C, a is to b as B is to C, b is to c as A is to B."

PΛ to consequent ΛΔ, so antecedent XZ to consequent ZΘ, and as consequent ΔΛ to some other <magnitude>, the <line> ΛX, so some other <magnitude>, the <line> BZ, to antecedent XZ. Therefore, as proved in the fifth book of the *Elements*,[535] it follows through the equality, as well: as PΛ to ΛX, so BZ to ZΘ.[536]

To 5

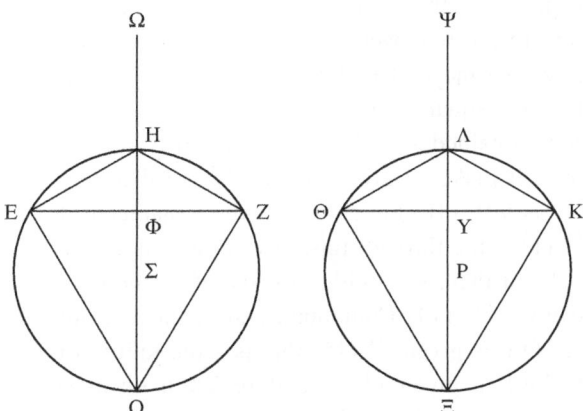

In II.5
Codex D has the lines
EZ, ΘK slanted (E
somewhat higher than
Z, Θ somewhat higher
than K).   Codices
E4 have Z instead of Ξ,
while codices DE4
omit Y. (Codex A
probably had Ξ and, if
it had the Y, it must
have been very
inconspicuous.)

Arch. 209   "And since the segment EZH is similar to the segment ΘKΛ, therefore the cone EZΩ, as well, is similar to the cone ΨΘK." (a) For let the diagrams be imagined set apart,[537] (b) and the <lines> EH, HZ, EO, OZ, ΘΛ, ΛK, ΘΞ, ΞK joined. (1) Now since the segments EZH, ΘKΛ are similar,[538] (2) the angles <contained> by EHZ, ΘΛK are equal, too;[539] (3) so that their halves, too. (4) And the <angles> at Φ, Y are right;[540] (5) therefore the remaining, as well, is equal to the remaining.[541] (6) Therefore the triangle HΦZ is equiangular to the <triangle> ΛYK, (7) and it is: as HΦ to ΦZ, ΛY to YK.[542] (8) So, through the same (the triangles ΦZO, YKΞ being equiangular) (9) as ZΦ to ΦO, KY to YΞ; (10) therefore through the equality: as HΦ to ΦO, ΛY to YΞ.[543]

---

[535] *Elements* V.23.

[536] This is the first time we see a comment whose sole function is to show how Euclid's *Elements directly* validate an Archimedean move. This is extremely interesting in showing how the various components of the *Elements* need not be all equally accessible for all readers. Was Eutocius puzzled by *Elements* V.23?

[537] I.e. we concentrate on just the two circles.   [538] Step a of Archimedes' proof.

[539] *Elements* III Def. 11.   [540] Step c of Archimedes' proof.

[541] *Elements* I.32.   [542] *Elements* VI.4.   [543] *Elements* V.22.

(11) Compoundly also: as HO to OΦ, ΛΞ to ΞY;[544] (12) the halves of the antecedents, too: as ΣO to OΦ, PΞ to ΞY; (13) compoundly also: as ΣOΦ taken together to ΦO, that is ΩΦ to ΦH[545] (14) so PΞY taken together to ΞY[546] (15) that is ΨY to YΛ.[547] (16) But as HΦ to ΦZ, ΛY to YK; (17) therefore through the equality, also: as ΩΦ to ΦZ, ΨY to YK;[548] (18) the doubles of consequents, too; (19) therefore as ΩΦ to EZ, ΨY to ΘK. (20) Therefore the axes and the diameters of the bases of the cones ΩEZ, ΨΘK are proportional; (21) therefore the cones are similar; (22) which it was required to prove.

Arch. 209   "And <the> ratio of ΩΦ to EZ is given." (1) For since the segments of the spheres are given,[549] (2) the diameters of the bases and the heights of the segments are given as well;[550] (3) so that EZ and HΦ are given. (4) Therefore also the half of EZ, <namely> EΦ, shall be given;[551] (5) so that the <square> on it, as well.[552] (6) And it is equal to the <rectangle contained> by HΦO.[553] (7) and if a given is applied along a given, it makes a given breadth;[554] (8) therefore ΦO is given. (9) But so is ΦH; (10) therefore the whole diameter of the sphere is given,[555] (11) and, through this, also its half is given, <namely> ΣO.[556] (12) But then, so is OΦ; (13) therefore the ratio of ΣO to OΦ is also given.[557] (14) Compoundly also: the ratio of ΣOΦ taken together to OΦ is given,[558] (15) that is <the ratio> of ΩΦ to ΦH.[559] (16) And ΦH is given; (17) therefore ΩΦ is given as well.[560] (18) But then, so is EZ; (19) therefore the ratio of ΩΦ to EZ, also, is given.

And the same would be said also in the case of the segment ABΓ, and it shall be obtained that the ratio of XT to AB is given; and through <the fact> that AB is given, XT shall be given as well.

And, while it is quite obvious that when the segments are given, their heights shall be given as well; still, in order that it may be seen that this is obtained following the Elements of the *Data*, this shall be said:

(1) Since the segments are given in position and in magnitude, (2) EZ is given, as well as the angle in the segment;[561] (3) so that its half, too. (4) And if we imagine the <line> EH joined, (4) the angle at Φ (which is right) being given, (5) the remaining <angle> shall be given as well[562] (6) as well as the triangle EHΦ (in figure);[563] (7) so

---

[544] *Elements* V.18.   [545] Step f of Archimedes' proof.   [546] *Elements* V.18.

[547] Step e of Archimedes' proof.   [548] *Elements* V.22.

[549] Step a of Archimedes' proof.

[550] See Eutocius' lemma below.   [551] *Data* 7.   [552] *Data* 52.

[553] *Elements* III.31, VI.8 Cor.   [554] *Data* 57.   [555] *Data* 3.

[556] *Data* 7.   [557] *Data* 1.   [558] *Data* 6.   [559] Step f of Archimedes' proof.

[560] *Data* 2.   [561] *Data* Def. 7.   [562] *Elements* I.32.   [563] *Data* 40.

that the ratio of EΦ to ΦH shall be given as well;[564] (8) and EΦ is given, being half of EZ; (9) therefore ΦH is given as well.

And it is possible to say this in another way, too. (1) Since EZ is given in position, (2) and ΦH <has been drawn> from the given <point> Φ ((3) for it <=Φ> is a bisection of the <line> EZ) (4) at right <angles to a line given> in position, (5) and the circumference of the segment, too, is given in position, (6) therefore H is given;[565] (7) and Φ, too, was given, (8) therefore ΦH is given as well.[566]

Arch. 209   "Since it is: as ΨY to XT, that is the <square> on the <line> BA to the <square> on ΘK, so KΘ to Δ." (1) For since it has just come to be: as ΨY to ΘK, XT to Δ,[567] (2) alternately: as ΨY to XT, KΘ to Δ.[568] (3) But as ΨY to XT, the <square> on AB to the <square> on ΘK[569] ((4) for when cones are equal, the bases are reciprocal to the heights,[570] (5) and as the bases to each other, so the squares on the diameters[571]); (6) therefore also: as the <square> on BA to the <square> on ΘK, ΘK to Δ.

Arch. 210   "And alternately: as AB to ΘK, ς to Δ." (1) Since the ratio of the <square> on the <line> BA to the <square> on ΘK was proved to be the same as the <ratio> of BA to ς[572] and the <ratio> of KΘ to Δ,[573] (2) therefore the ratio of BA to ς, as well, is the same as the <ratio> of KΘ to Δ; (3) so that, alternately, it is: as BA to ΘK, ς to Δ.

To the synthesis of 5

Arch. 211   "Since AB, ΘK, ς, Δ are proportional, it is: as the <square> on AB to the <square> on ΘK, ΘK to Δ." For generally, if there are four proportional lines, it will be: as the <square> on the first to the <square> on the second, the second to the fourth. (1) For since it is, as the first to the second, the third to the fourth, (2) alternately: as the first to the third, the second to the fourth.[574] (3) But as the first to the

---

[564] *Data* 42.   [565] *Data* 29, 25.   [566] *Data* 26.

[567] Step h of Archimedes' proof (immediately before the step on which Eutocius comments right now).

[568] *Elements* V.16.

[569] Step 10 of Archimedes' proof. Interestingly, Eutocius goes on to remind the readers of why this Step 10 is true.

[570] *Elements* XII.15.   [571] *Elements* XII.2.

[572] Step 21 of Archimedes' proof. I had to change Archimedes' word order, "to the ratio X was proved to be the same Y" into "the ratio X was proved to be the same as Y."

[573] Stated as Step 22 of Archimedes' proof (an implicit result of Steps 19–20).

[574] *Elements* V.16.

third, so the <square> on the first to the <square> on the second;[575]
(4) therefore also: as the <square> on the first to the <square> on the
second, the second to the fourth.

To 6

Arch. 214   "And since the KΛM is similar to the segment ABΓ,
therefore it is: as ΛP to PN, BΠ to ΠΘ." (a) For if MN, ΓΘ are joined,
(1) since the segments are similar (3) the angles at B, Λ are equal, as
well. (4) And the angles at M, Γ, also, are right;[576] (5) therefore the
remaining <angle> is equal to the remaining,[577] (6) and the triangles
are equiangular, (7) and it is: as ΘB to ΘΓ, so ΛN to NM.[578] (8) But
as ΘΓ to ΘΠ, so MN to NP (9) through the similarity of the triangles
ΓΘΠ, MNP;[579] (10) therefore through the equality also: as BΘ to ΘΠ,
ΛN to NP;[580] (11) so that dividedly also: as BΠ to ΠΘ, so ΛP to PN.[581]

Arch. 214   "And the ratio of EZ to BΓ is given; for each of the two is given."
(1) For since the segments of the sphere are given, (2) the diameters of
the bases and the heights of the segments are given as well; (3) so that,
since AΓ is given, (4) its half, <namely> ΓΠ, is given as well. (5) And
BΠ is given as well, (6) and they contain a right angle; (7) therefore
BΓ is given as well.[582] (8) So through the same <arguments>, EZ is
given as well; (9) so that the ratio of BΓ to EZ is given as well.[583]

To the synthesis of 6

Arch. 215   "Therefore the segments of circles on KM, AΓ are similar." (a) For
if (as in the analysis[584]) ΓΘ, MN are joined, (1) since the <angles>
at Γ, M are right,[585] (2) and ΓΠ, MP are perpendiculars,[586] (3) they
are mean proportionals between the segments of the base;[587] (4) so
that it is: as the first, BΠ, to the third, ΠΘ, so the <square> on the
first, ΠB, to the <square> on the second, ΠΓ.[588] (5) So through the
same <arguments> also: as ΛP to PN, so the <square> on ΛP to

---

[575] *Elements* VI.20 Cor.   [576] *Elements* III.31.   [577] *Elements* I.32.

[578] *Elements* VI.4.   [579] *Elements* VI.4.   [580] *Elements* V.22.

[581] *Elements* V.17.   [582] *Elements* I.47.   [583] *Data* 1.

[584] I.e. as in Eutocius' immediately preceding comment to the analysis.

[585] *Elements* III.31.   [586] Step f of Archimedes' analysis.

[587] *Elements* VI.8.   [588] *Elements* VI.20 Cor. 2.

the <square> on PM. (6) And it is: as BΠ to ΠΘ, ΛP to PN;[589] (7) therefore also: as the <square> on BΠ to the <square> on ΠΓ, so the <square> on ΛP to the <square> on PM; (8) therefore also: as ΠB to ΠΓ, ΛP to PM. (9) And the sides are proportional around equal angles; (10) therefore the triangles are equiangular.[590] (11) Therefore the angles at B, Λ are equal, (12) and their doubles, the <angles> in the segments; (13) therefore the segments are similar.

To 7

Arch. 219 "Therefore <the> ratio of EΔZ taken together to ΔZ is given."[591] (1) For since EΔ, ΔZ taken together have to ΔZ a given ratio, (2) and if a given magnitude has to some part of itself a given ratio, it shall also have a given ratio to the remaining <part>;[592] (3) so that EΔZ taken together has to EΔ a given ratio. (4) Now since each of EΔ, ΔZ has to EΔZ taken together a given ratio, they also have a given ratio to each other;[593] (5) therefore the ratio of EΔ to ΔZ is given. (6) And EΔ is given; (7) for the diameter is given;[594] (8) therefore ΔZ is given as well.[595] (9) Therefore ZB, remaining, shall be given; (10) so that the <rectangle contained> by ΔZB, too, (11) that is the <square> on AZ,[596] (12) that is AZ shall be given; (13) therefore also the whole AΓ.

And you might argue in another way that AΓ is given. (1) For since the diameter ΔB is given in position,[597] (2) and Z is given as well (as is postulated),[598] (3) and AΓ has been drawn from a given <point> Z at right <angles to ΔB>, (4) therefore AΓ is given in position.[599] (5) But so is the circumference of the circle;[600] (6) therefore A, Γ are given, (7) and AZΓ itself is given.

---

[589] Step 6 of Archimedes' analysis.   [590] *Elements* VI.6.

[591] This is the lemma as it stands in the manuscripts, roughly the same as Archimedes' Step 5 of the analysis. Eutocius now goes on to give an argument deriving Archimedes' Step 8 from this Step 5. Heiberg concluded that Eutocius' text of Archimedes did not contain Archimedes' Steps 6–7, and that the original of Eutocius' contained Steps 5 and 8 as lemma. Heiberg therefore inserted Archimedes' Step 8 to follow Eutocius' Lemma. Heiberg's theory is plausible, but I keep to the manuscripts' reading.

[592] *Data* 5.   [593] *Data* 8.

[594] The sphere is given, and therefore, by definition (*Data* Def. 5) so is its diameter.

[595] *Data* 2.   [596] *Elements* VI.8 Cor.

[597] The sphere and its diameter are supposed given in the analysis.

[598] In the assumption of the analysis.   [599] *Data* 30.

[600] There are infinitely many great circles on the sphere with ΔB as their diameter. However, the argument is invariant to the choice of this circle.

Arch. 219   "And since EΔZ taken together has to ΔZ a greater ratio than EΔB
taken together to ΔB." (1) For since EΔ is greater than half ΔZ,
(2) therefore EΔZ taken together is greater than half as large again
ΔZ. (3) And EΔ, ΔB taken together is half as large again ΔB; (4)
therefore EΔZ has to ΔZ a greater ratio than EΔB to ΔB.

Or also like this: (1) Since ΔB is greater than ΔZ, (2) and <there
is> some other line, EΔ, (3) therefore EΔ has to ΔZ a greater ratio
than EΔ to ΔB;[601] (4) compoundly, EΔZ taken together has to ΔZ a
greater ratio than EΔB taken together to ΔB.[602]

The synthesis of the theorem is clear through the things said here.

To 8

Arch. 222   "ΘZ has to ZH a smaller ratio than duplicate that which the <square>
on BA has to the <square> on AΔ, that is BZ to ZΔ." (1) For since
AZ has been drawn <as> perpendicular in a right-angled triangle,[603]
(2) the triangles next to the perpendicular being similar,[604] (3) as ZB to
BA, AB to BΔ.[605] (4) And as the first to the third, so the <square> on
the first to the <square> on the second[606] (5) and the <square> on the
second to the <square> on the third, (6) as has been proved above;[607]
(7) therefore as ZB to BΔ, the <square> on AB to the <square> on
BΔ. (8) But as BΔ to ΔZ, so the <square> on BΔ to the <square>
on ΔA; (9) for as the first to the third, so the <square> on the first to
the <square> on the second; (10) therefore through the equality also:
as the <square> on BA to the <square> on ΔA, so BZ to ZΔ.[608]

And the same might be obtained also in another way: (1) for since
it is: as BZ to ZΔ, so the <rectangle contained> by ZBΔ to the
<rectangle contained> by BΔZ ((2) BΔ taken as a common height),[609]
(3) and the <square> on BA is equal to the <rectangle contained>
by ΔBZ, (3) while the <square> on ΔA is equal to the <rectangle
contained> by BΔZ,[610] (4) therefore as the <square> on BA to the
<square> on ΔA, so BZ to ZΔ.

Arch. 223   "And since ΘZ has to ZK a smaller ratio than ΘB to BK." For in
general, if there are two unequal magnitudes, and equal <magnitudes>

---

[601] *Elements* V.8.   [602] An extension to inequality of *Elements* V.18.
[603] Setting-out of *SC* II.8, *Elements* III.31.   [604] *Elements* VI.8.
[605] *Elements* VI.4.   [606] *Elements* VI.20 Cor. 2.
[607] Eutocius' commentary to *SC* II.4 (analysis) Step 15.
[608] *Elements* V.22.   [609] *Elements* VI.1.
[610] Steps 2–3 are based on the setting out of *SC* II.8, *Elements* III.31, then *Elements*
VI.8 Cor., 17. Greek word order, "to X is equal Y" had to be changed into the English
word order "Y is equal to X."

are added to them, the greater[611] has to the smaller a greater ratio than the composed to the composed.[612]

For let there be two unequal lines, AB, ΓΔ, and let equal <lines> be added to them, <namely> BE, ΔZ. I say, that AB has to ΓΔ a greater ratio than AE to ΓZ.

(1) For since AB is greater than ΓΔ,[613] (2) therefore AB has to BE a greater ratio than ΓΔ to BE,[614] (3) that is to ΔZ; (4) so that compoundly, also: AE has to EB a greater ratio than ΓZ to ΔZ, (5) through what has been proved already.[615]

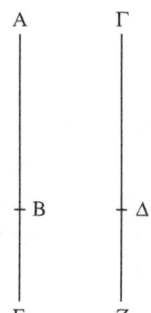

In II.8
In codices BG the line EA is elongated so that the point A becomes higher than the point Γ (so Heiberg).

Arch. 223   "Therefore the <rectangle contained> by ΘZH is smaller than the <square> on ZK." (1) For if there are three continuous[616] lines, as A, B, Γ, so that A has to B a smaller ratio than B to Γ, (2) the <rectangle contained> by the extremes A, Γ is smaller than the <square> on the mean B. (3) For if we produce: as A to B, B to some other <line>, (4) it will be to a <line> greater than Γ ((5) given that it <=the ratio of B to

---

[611] "The greater:" i.e. the greater from among the initial given two magnitudes.

[612] "The composed:" the magnitudes that results from the addition, understood to be in the order of the greater to the smaller. (In the Greek, the words for "addition" and "composition" share a common root.)

[613] This is based on nothing at all, not even the diagram (which apparently just showed two equal and therefore purely conceptual lines). Simply, the choice of greater line is arbitrary, and the convention to prefer to speak of a:b (rather than b:a), when a is greater than b, is sufficiently well understood to merit no comment.

[614] *Elements* V.8.

[615] The reference is to a proof of an extension of *Elements* V.18 to inequalities. Note that this is a remarkably long-range backwards reference, to Eutocius' commentary on *SC* I.2 (!). If this is not a Byzantine interpolation, this shows that Eutocius conceived of his two commentaries as a single unit.

[616] "Continuous:" as used here, this word does not mean anything in strict mathematical sense. The usual meaning of "three continuous lines A, B, Γ" is that the following proportion holds: A:B::B:Γ, but here we are explicitly told that A:B<B:Γ, so that "three continuous lines" means just "three lines in some proportion."

some other line> must be smaller than the ratio of B to Γ[617]). (6) And the <rectangle contained> by A and the <line> greater than Γ will be equal to the square on B;[618] (7) so that the <rectangle contained> by A, Γ is smaller than the <square> on B.

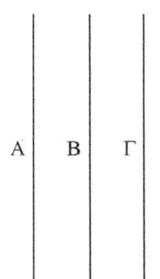

In II.8 Second diagram Codex E has A somewhat bigger than the other lines. Codex G has Γ somewhat bigger than the other lines; B has the same line Γ somewhat smaller than the other lines.

Arch. 223   "Therefore the <rectangle contained> by ΘZH has to the <square> on ZH a smaller ratio than the <square> on KZ to the <square> on ZH". (1) For as ΘZ to ZH, so the <rectangle contained> by ΘZH to the <square> on ZH,[619] (2) and the <rectangle contained> by ΘZH is smaller than the <square> on ZK;[620] (3) and the greater has to the same a greater ratio than the smaller <has>.[621]

Arch. 223   "And since BE is equal to EΔ, therefore the <rectangle contained> by BZΔ is smaller than the <rectangle contained> by BEΔ." (1) For the <rectangle contained> by BEΔ is equal to the <square> on EΔ,[622] (2) while the <rectangle contained> by BZΔ together with the <square> on EZ is equal to the same <square>.[623] (3) And it is clear that, by as much as the <point> Z is removed from the bisection point, it <= the rectangle contained by BZΔ> is smaller than the <rectangle contained> by the equal <lines = the square on EΔ>;[624] (4) for, together with a greater <square>,[625] <namely> that on the <line> between the cuts,[626] it <= the rectangle contained by BZΔ> becomes equal to the <rectangle contained> by the equal <lines>. (5) So that when the line is cut into unequal <segments> at one point, and at another point, the <rectangle contained> by the segments which are

---

[617] *Elements* V.8.   [618] *Elements* VI.17.   [619] *Elements* VI.1.

[620] See previous comment by Eutocius.   [621] *Elements* V.8.

[622] A trivial result of the equality BE=EΔ (or *Elements* VI.1).

[623] *Elements* II.5.

[624] Two things happen here simultaneously. First, Eutocius envisages a functional relation between the position of Z and the magnitude of rect. (BZ, ZΔ). Second, Eutocius explicitly refers to a square as a limiting case of a rectangle.

[625] "Greater" in the dynamic sense: we envisage the point Z gradually moving away from the bisection, with its corresponding square becoming greater in the process.

[626] That is, one cut being the dynamic point Z, the other cut being the static bisection point E.

closer to the bisection point, is greater than the <rectangle contained>
by the more distant segments.

Arch. 223   "Therefore ZB has to BE a smaller ratio than EΔ to ΔZ." For generally, if there are four terms, as A, B, Γ, ΔE, and the <rectangle
contained> by A, ΔE is smaller than the <rectangle contained> by B,
Γ, A has to B a smaller ratio than Γ to ΔE.

For let the <rectangle contained> by B, Γ be equal to the <rectangle
contained> by A, ZE;[627] (1) therefore it is: as A to B, Γ to ZE.[628] (2)
And Γ has to ZE a smaller ratio than <it has> to EΔ;[629] (3) therefore
A, too, has to B a greater ratio than Γ to ΔE.

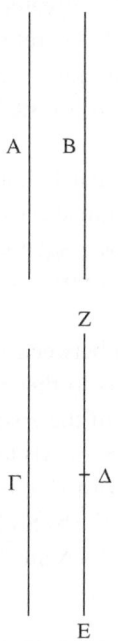

In II.8 Third diagram
Codex A had Λ instead
of Δ (corrected in
codices BD).
Codex E had Ξ instead
of Z, corrected (by the
same hand?).
Codex D has ZE equal
to Γ, and both greater
than A, B (equal to
each other).

Arch. 223   "Therefore it is: as ΘB to BK, the <square> on ΘN to the <square>
on NK." (1) For since the <square> on BN is equal to the <rectangle
contained> by ΘBK,[630] (2) the three lines are proportional: as ΘB to
BN, NB to BK;[631] (3) and as the first to the third (ΘB to BK) (4) so
the <square> on the second to the <square> on the third ((5) that
is the <square> on BN to the <square> on BK) (6) as has been

---

[627] This step determines the point Z (so that ZE> ΔE, since rect. A, ΔE is said in the
enunciation to be smaller than rect. B, Γ , in turn equal to rect. A, ZE). Greek word order
has been altered from "let, to X, Y be equal" into "Let X be equal to Y." Note that I omit
throughout the understood reference to "<terms>."

[628] *Elements* VI.16.   [629] *Elements* V.8.

[630] Step g of *SC* II.8.   [631] *Elements* VI.17.

proved above.[632] (7) Again, since it is: as ΘB to BN, NB to BK, (8) compoundly: as ΘN to NB, KN to KB;[633] (9) alternately: as ΘN to NK, NB to BK;[634] (10) therefore also: as the <square> on ΘN to the <square> on NK, so the <square> on NB to the <square> on BK. (11) But as the <square> on NB to the <square> on BK, so ΘB was shown to be to BK; (12) therefore also: as ΘB to BK, so the <square> on ΘN to the <square> on NK.

Arch. 223   "But the <square> on ΘZ has to the <square> on ZK a greater ratio than the <square> on ΘN to the <square> on NK." (1) For, again, NZ has been added to two unequal <lines>: ΘZ, ZK, (2) and through what is said above,[635] ΘZ has to ZK a greater ratio than ΘN to NK; (3) so that the duplicates, as well.[636] (4) Therefore the <square> on ΘZ has to the <square> on ZK a greater ratio than the <square> on ΘN to the <square> on NK, (5) that is ΘB to BK,[637] (6) that is ΘB to BE,[638] (7) that is KZ to ZH.[639]

Arch. 223   "Therefore ΘZ has to ZH a greater ratio than half as much again the <ratio> of KZ to ZH." For let lines be imagined set separately, as AB, Γ, Δ, so that the <square> on AB has to the <square> on Γ a greater ratio than Γ to Δ. I say that AB has to Δ a ratio greater than half as much again the <ratio> which Γ has to Δ.

(a) For let a mean proportional be taken between Γ, Δ, <namely> E. (1) Now since the <square> on AB has to the <square> on Γ a greater ratio than Γ to Δ, (2) but the ratio of the <square> on AB to the <square> on Γ is duplicate the <ratio> of AB to Γ, (3) while the <ratio> of Γ to Δ is duplicate the <ratio> of Γ to E,[640] (4) therefore AB, too, has a greater ratio to Γ than Γ to E. (b) So let it come to be: as E to Γ, Γ to BZ.[641] (5) And since BZ, Γ, E, Δ are four continuously

---

[632] Eutocius' commentary to *SC* II.4 (analysis) Step 15 (the same reference was made already in Step 6 of the first comment to this proposition).

[633] *Elements* V.18.   [634] *Elements* V.16.   [635] First comment on *SC* II.8.

[636] That is, the same proportion inequality will hold between the duplicate ratios of the ratios mentioned in Step 2. A duplicate ratio can be understood as the ratio between the squares on the lines of an original ratio. The assumption that proportion relations between lines are directly correlated to relations between the squares on those lines is nowhere proved, but it is a simple result of *Elements* VI.1.

[637] Step 34 of *SC* II.8. Here Eutocius begins to go beyond the original step picked up for commentary (Step 35 of *SC* II.8), and to argue for the following steps in the argument (this last step of Eutocius, for instance, explains Step 36 in the extant Archimedean text). Heiberg's interpretation was that Archimedes' original text leapt directly from Step 35 to Step 39, and that Steps 36–8 were added on the basis of Eutocius' commentary.

[638] Step f of *SC* II.8.   [639] Step 18 of *SC* II.8.   [640] *Elements* VI.20 Cor. 2.

[641] That AB>BZ (assumed here in the diagram and used later on in the proof) may be seen through *Elements* V.8.

proportional lines, (6) therefore BZ has a triplicate ratio to Δ than BZ to Γ, (7) that is Γ to E. (8) And also, Γ has to Δ a ratio duplicate of the <ratio> of Γ to E; (9) therefore BZ has to Δ a ratio half as much again the <ratio> which Γ has to Δ; (10) so that AB has to Δ a ratio greater than half as much again the <ratio> of Γ to Δ.[642]

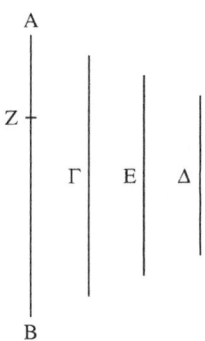

In II.8 Fourth diagram Codex G mirror-inverts the line-arrangement, as in the thumbnail.

Lemma to the following

Let there be four terms, A, Γ, Δ, B. I say that the ratio composed of the <rectangle contained> by A, B to the <square> on Γ, together with the ratio of B to Δ, is the same as the <rectangle contained> by A, B, on B, to the <square> on Γ, on Δ.[643]

(a) Let the <term> K be equal to the <rectangle contained> by A, B, (b) and the <term> Λ equal to the <square> on Γ,[644] (c) and let it come to be: as B to Δ, so Λ to M; (1) therefore the ratio of K to M is composed of K to Λ – that is the <rectangle contained> by A, B to

---

[642] This seems to break free of earlier Greek mathematics: ratios are treated as exponents, to be calculated arithmetically. Did Eutocius assume the general rule for calculation of exponents? Did he stumble upon it, without realizing the general significance of his procedure? Or perhaps (more probably, I think), looking for what sense to give to the expression "a ratio half as much again," he *defined* this as the ratio of the triplicate to the duplicate?

[643] The *on* expression is used here as in the formula "{two dimensional figure} *on* {line}."

[644] Since we are dealing with "terms," two-dimensional objects can be set on the same level as one-dimensional objects: both are single-letter "terms" (i.e. governed by a masculine article). To make the reading slightly less painful, I omit the words "the <term>" from now on (as I usually omit "the <point>" and "the <line>"), but they must be understood.

the <square> on Γ – (2) and Λ to M (3) – that is B to Δ. (d) So let K, having multiplied B, produce N, (e) and let Λ, having multiplied B, produce Ξ, (f) and, having multiplied Δ, <let Λ produce> O.[645] (4) Now since the <rectangle contained> by A, B is K, (5) and K, having had multiplied B, has produced N, (6) therefore N is the <rectangle contained> by A, B, on B. (7) Again, since the <square> on Γ is Λ (8) and Λ, having had multiplied Δ, has produced O, (9) therefore O is the <square term> on Γ, on Δ;[646] (10) so that the ratio of the <rectangle contained> by A, B, on B, to the <square> on Γ, on Δ, is the same as the <ratio> of N to O. (11) Therefore it is required to prove that the ratio of K to M is the same as the <ratio> of N to O.

(12) Now since each of K, Λ, having multiplied B, has produced, respectively, N, Ξ, (13) it is therefore: as K to Λ, so N to Ξ. (14) Again, since Λ, having had multiplied each of B, Δ, has produced, respectively, Ξ, O, (15) it is therefore: as B to Δ, Ξ to O. (16) But as B to Δ, Λ to M; (17) therefore also: as Λ to M, Ξ to O. (18) Therefore K, Λ, M are in the same ratio to N, Ξ, O, taken in pairs; (19) therefore through the equality, it is also: as K to M, so N to O.[647] (20) And the ratio of K to M is the same as the <ratio> composed of the <rectangle contained> by A, B to the <square term> on Γ and of the <ratio> which B has to Δ, (21) and the ratio of N to O is the same as the <rectangle contained> by A, B, on B, to the <square term> on Γ on Δ; (22) therefore the ratio composed of the <rectangle contained> by A, B to the <square term> on Γ and of the <ratio> which B has to Δ, is the same as the <rectangle contained> by A, B, on B, to the <square term> on Γ, on Δ.

And it is also clear that the <rectangle contained> by A, B, on B, is equal to the <square> on B, on A. (23) For since it is: as A to B, so the <rectangle contained> by A, B to the <square> on B (B taken as a common height),[648] (24) and if there are four proportional terms, the <rectangle contained> by the extremes is equal to the <rectangle contained> by the means,[649] (25) therefore the <rectangle contained> by A, B, on B, is equal to the <square> on B on A.

---

[645] Anachronistically (but less anachronistically than elsewhere in Greek mathematics): N=K*B, Ξ=Λ*B, O=Λ*Δ.

[646] The article in the expression "the <square term> on Γ" is masculine (for "term") instead of neuter (for "square"): a remarkable result of the semiotic eclecticism of this text, that keeps veering between general proportion theory, geometry, and calculation terms.

[647] *Elements* V.22.   [648] *Elements* VI.1.

[649] *Elements* VI.16.

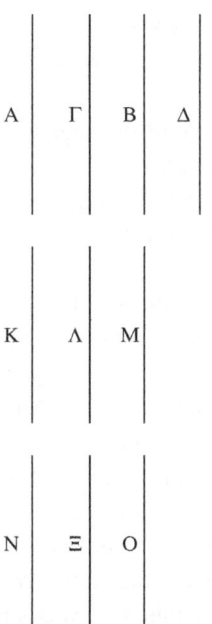

In II.8 Fifth diagram
Codex A had X instead
of K (corrected in
codices BG).

To the alternative of 8

It has been said in the preceding that, if some mean is taken between
two magnitudes, the ratio of the extremes is composed of: the <ratio>
which the first has to the mean, and the mean to the third.[650] So similarly, even when more means are taken, the ratio of the extremes is
composed of the ratios which all the magnitudes have to each other

Arch. 227 in the continuous sequence. Indeed, here he says that "the ratio of the
segment BAΔ to the segment BΓΔ is composed of: the <ratio> which
the segment BAΔ has to the cone whose base is the circle around the
diameter BΔ while <its> vertex is the point A; and the same cone to
the cone having the same base, and the point Γ <as> vertex; and the
said cone to the segment BΓΔ," clearly with the said cones taken as
means between the segment ΔAB and the BΓΔ.

Arch. 227 "But the ratio of the segment BAΔ to the cone BAΔ is the <ratio> of
HΘ to ΘΓ," through the corollary of the second theorem of the second
book; for the segment was said to have to the cone <contained> inside
itself that ratio, which both the radius of the sphere and the height of the
remaining segment, taken together, have to the height of the remaining
segment.

---

[650] Eutocius' comment on *SC* II.4, Step 26.

Arch. 228   "While the <ratio> of the cone BAΔ to the cone BΓΔ is the
<ratio> of AΘ to ΘΓ" for, being on the same base, they are to each
other as the heights.[651]

Arch. 228   "And the <ratio> of the cone BΓΔ to the segment BΓΔ is the
<ratio> of AΘ to ΘZ," through the inversion of the said corollary.[652]

So that the ratio of the segment BAΔ to the segment BΓΔ is composed of the <ratio> of HΘ to ΘΓ and of the <ratio> of AΘ to ΘΓ and
of the <ratio> of AΘ to ΘZ.[653]

Arch. 228   "And the <ratio> composed of the <ratio> of HΘ to ΘΓ, together
with the <ratio> of AΘ to ΘΓ, is the <ratio> of the <rectangle
contained> by HΘA to the <square> on ΘΓ;" for equiangular paral-

Arch. 228 lelograms have the ratio composed of their sides.[654] "And the <ratio>
of the <rectangle contained> by HΘA to the <square> on ΓΘ, together with the <ratio> of AΘ to ΘZ, is the <ratio> of the <rectangle
contained> by HΘA, on ΘA, to the <square> on ΘΓ, on ΘZ," as

Arch. 228 has been proved in the preceding lemma. "And the <ratio> of the
<rectangle contained> by HΘA, on ΘA, is the same <ratio> as <of>
the <square> on AΘ, on ΘH,"[655] for this, too, was simultaneously
proved in the preceding.[656]

Therefore the ratio of the segment to the segment is the same as the
<square> on AΘ, on ΘH, to the <square> on ΓΘ, on ΘZ.[657]

(1) Now since it is required to prove that the segment has to the
segment a smaller ratio than duplicate the ratio of the surface to the
surface, (2) therefore it is required to prove that the <square> on AΘ,
on ΘH, has to the <square> on ΓΘ, on ΘZ, a ratio smaller than
duplicate the <ratio> which the surface of the segment BAΔ has to
the surface of the BΓΔ, (3) that is than the <ratio> which
the <square> on AB has to the <square> on ΓB.[658] (4) But as the
<square> on AB to the <square> on BΓ, so AΘ to ΘΓ; (5) for this has

Arch. 228 been proved in the preceding theorems;[659] (6) therefore it is required

---

[651] *Elements* XII.14.

[652] The corollary spoke of ratio of "segment to cone." Since here we require the ratio
of "cone to segment," Eutocius sees this as relying not on the corollary to *SC* II.2 directly,
but on its inversion (in the sense of proportion theory): *Elements* V.7 Cor.

[653] Eutocius spells out Archimedes' implicit Step B.

[654] *Elements* VI.23.

[655] Remember that here we constantly speak of objects being in ratios to each other
so that it becomes natural, instead of saying that "A is equal to B," to say "the ratio of A
<to some magnitude X> is the same as the ratio of B <to the same magnitude>."

[656] The last paragraph of the preceding lemma.

[657] Again, Eutocius pauses to take stock of what has been implicitly proved so far.

[658] *SC* I.42–3.   [659] *SC* II.3, Step 5, and Eutocius' comment there.

Arch. 228 to prove "that the <square> on AΘ, on ΘH, has to the <square> on ΓΘ, on ΘZ, a ratio smaller than duplicate than the <ratio> of AΘ to ΘΓ."

(1) But duplicate the ratio of AΘ to ΘΓ is the <ratio> of the <square> on AΘ to the <square> on ΘΓ.[660] (2) Therefore that the <square> on AΘ, on ΘH, has to the <square> on ΓΘ, on ΘZ, a smaller ratio than the <square> on AΘ to the <square> on ΓΘ. (3) But as the <square> on AΘ to the <square> on ΘΓ (ΘH taken as a common height), (4) so the <square> on AΘ, on ΘH, to the <square>

Arch. 228 on ΓΘ, on ΘH.[661] (5) Therefore it has to be proved "that the <square> on AΘ, on ΘH, has a smaller ratio to the <square> on ΓΘ, on ΘZ, than (the same) <square> on AΘ, on ΘH, to the <square> on ΓΘ, on ΘH."

But that, to which the same has a smaller ratio, is greater.[662] There-

Arch. 228 fore it is required to prove "that the <square> on ΓΘ, on ZΘ, is greater than the <square> on ΓΘ, on ΘH," that is "that ZΘ is greater than ΘH." And this is obvious; for equals (ZA, ΓH) are added to the unequal (AΘ, ΘΓ).[663]

Saying this, he did not supply the synthesis himself. We shall add it in.

(1) Since ZΘ is greater than ΘH, (2) the <square> on ΓΘ, on ΘZ, is greater than the <square> on ΓΘ, on ΘH;[664] (3) so that the <square> on AΘ, on ΘH, has to the <square> on ΓΘ, on ΘZ, a smaller ratio than the same <magnitude, namely>, the <square> on AΘ, on ΘH, to the <square> on ΓΘ, on ΘH.[665] (4) But as the <square> on AΘ, on ΘH, to the <square> on ΓΘ, on ΘH, the <square> on AΘ to the <square> on ΓΘ;[666] (5) therefore the <square> on AΘ, on ΘH, has to the <square> on ΓΘ, on ΘZ, a smaller ratio than the <ratio> which the <square> on AΘ has to the <square> on ΘΓ. (6) But the ratio of the <square> on AΘ to the <square> on ΘΓ is duplicate the <ratio> of AΘ to ΘΓ; (7) therefore the <square> on AΘ, on ΘH, has to the <square> on ΓΘ, on ΘZ, a smaller than duplicate ratio than the <ratio> of AΘ to ΘΓ. (8) But the ratio of the segments was proved to be the same as the <ratio> which the <square> on AΘ, on ΘH,

---

[660] I print, following Heiberg, as if this sentence is by Eutocius ("But duplicate . . . of ΘΓ"). However this sentence also occurs in the manuscripts for Archimedes, and may therefore be a quotation and not a comment (Eutocius quotes here so extensively, that he may well have chosen to quote even a completely unproblematic assertion, just for the sake of continuity).

[661] An extension of *Elements* VI.1.   [662] *Elements* V.10.

[663] *Elements* I Common Notions 4.   [664] An extension of *Elements* VI.1.

[665] *Elements* V.8.   [666] An extension of *Elements* VI.1.

has to the <square> on ΓΘ, on ΘZ,[667] (9) while the <ratio> of the surfaces, <was proved equal to the ratio> which AΘ has to ΘΓ;[668] (10) therefore the segment has to the segment a ratio smaller than duplicate than the ratio of the surface to the surface.

Arch. 229   Following that, analyzing the other part of the theorem, he adds "So I claim that the greater segment has to the smaller a ratio greater than half as much again the ratio of the surface to the surface. But the <ratio> of the segments was proved to be the same as the <ratio> which the <square> on AΘ, on ΘH, has to the <square> on ΓΘ, on ΘZ, while the <ratio> of the cube on AB to the cube on BΓ is half as much again the ratio of the surface to the surface." (1) For – of the <ratio> of AB to BΓ – the <ratio> of the square on AB to the square on BΓ is duplicate, (2) while the <ratio> of the cube on AB to the cube on BΓ is triplicate. (3) But as the cube on AB to the cube on BΓ, so the cube on AΘ to the cube on ΘB; (4) for as AB to BΓ, so AΘ to ΘB,[669] (5) through the similarity of the triangles ABΓ, ABΘ,[670] ((6) and if there are four proportional lines, the similar and similarly described solids on them are proportional[671]); (7) so that the cube on AΘ has to the cube on ΘB a ratio half as much again as the <ratio> which the square on AB has to the square on BΓ, (8) that is the surface to the surface.

     But as the segment to the segment, so the <square> on AΘ, on ΘH,
Arch. 229   to the <square> on ΓΘ, on ΘZ. "So I claim that the <square> on AΘ, on ΘH, has to the <square> on ΓΘ, on ΘZ, a greater ratio than the cube on AΘ to the cube on ΘB, that is the <ratio> of the <square> on AΘ to the <square> on ΘB and the <ratio> of AΘ to ΘB." For the <ratio> of the <square> on AΘ to the <square> on ΘB, being duplicate the <ratio> of AΘ to ΘB, taking in the <ratio> of AΘ to ΘB, comes to be the same as the <ratio> of the cube on AΘ to the <cube> on ΘB; for each is triplicate the same.[672]

Arch. 229   "But the <ratio> of the <square> on AΘ to the <square> on ΘB, taking in the <ratio> of AΘ to ΘB, is the <ratio> of the <square> on AΘ to the <rectangle contained> by ΓΘB." (1) For since the ratio of AΘ to ΘB is the same as the <ratio> of ΘB to ΘΓ, ((2) ΘB being

---

[667] As noted above by Eutocius (see n. 657), this is an implicit result of Steps 3–9 of *SC* II.8 Alter.

[668] Step 2, *SC* II.8 Alter.

[669] 3 follows from 4 on the assumption of an extension to solids of *Elements* VI.22.

[670] Step 5 is based on *Elements* VI.8. Step 4 follows from Step 5, based on *Elements* VI.4.

[671] Eutocius *asserts* the extension to solids of *Elements* VI.22.

[672] I.e.: both: (1) (sq.) AΘ : (sq.) ΘB, "taking in" AΘ:ΘB, and (2) (cube) AΘ : (cube) ΘB are triplicate (3) AΘ:ΘB.

a mean proportional[673]), (3) the <ratio> of the <square> on AΘ to the <square> on ΘB, together with the <ratio> of AΘ to ΘB, is the same as the <ratio> of the <square> on AΘ to the <square> on ΘB, together with the <ratio> of BΘ to ΘΓ. (4) But the <ratio> of BΘ to ΘΓ is the same as the <ratio> of the <square> on BΘ to the <rectangle contained> by BΘΓ (BΘ taken as a common height);[674] (5) so that the ratio of the <square> on AΘ to the <square> on ΘB, together with the <ratio> of AΘ to ΘB, is the same as the <ratio> of the <square> on AΘ to the <square> on BΘ, together with the <ratio> of the <square> on BΘ to the <rectangle contained> by BΘΓ. (6) But the ratio of the <square> on AΘ to the <rectangle contained> by BΘΓ is the <ratio> composed of the <square> on AΘ to the <square> on BΘ and of the <square> on BΘ to the <rectangle contained> by BΘΓ (the <square> on BΘ taken as a mean); (7) so that the ratio of the <square> on AΘ to the <square> on BΘ, together with the <ratio> of AΘ to ΘB is the same as the <ratio> of the <square> on AΘ to the <rectangle contained> by BΘΓ.

Arch. 229   "And the <ratio> of the <square> on AΘ to the <rectangle contained> by BΘΓ is the same as the <ratio> of the <square> on AΘ, on ΘH, to the <rectangle contained> by BΘΓ, on ΘH" (ΘH

Arch. 229   taken as a common height),[675] "so I claim that the <square> on AΘ, on ΘH, has to the <square> on ΓΘ, on ΘZ, a greater ratio than the <square> on AΘ, on ΘH, to the <rectangle contained> by ΓΘB, on ΘH." And that to which the same has a greater ratio, is smaller;[676]

Arch. 229   "It is to be proved that the <square> on ΓΘ, on ΘZ, is smaller than the <rectangle contained> by BΘΓ, on ΘH, which is the same as proving that the <square> on ΓΘ has to the <rectangle contained> by ΓΘB a smaller ratio than ΘH to ΘZ". For if there are four terms (as, here: the <square> on ΓΘ, and the <rectangle contained> by ΓΘB, and ΘH, and ΘZ), and the <rectangle contained> by the extremes is smaller than the <rectangle contained> by the means, the first has to the second a smaller ratio than the third to the fourth, as has been proved above.[677] Therefore it was validly required to prove that "the <square> on ΓΘ, on ΘZ, is smaller than the <rectangle contained> by ΓΘB, on ΘH, which is the same as proving that the <square> on ΓΘ has to the

[673] *Elements* VI.8. Cor.   [674] *Elements* VI.1.

[675] An extension to solids of *Elements* VI.1.   [676] *Elements* V.10.

[677] The result quoted by Eutocius is an extension to inequality of *Elements* VI.16, proved by himself in his comment to *SC* II.8, Step 30. Here, however, there is a further extension, from areas to solids, which Eutocius glosses over. Notice, related to this, that the formula of "rectangle <contained by>" has now been widened to cover anything we would call "multiplication" – even where the terms involved in the so-called rectangle are not lines.

<rectangle contained> by ΓΘB a smaller ratio than ΘH to ΘZ."[678]
(1) But as the <square> on ΓΘ to the <rectangle contained> by ΓΘB,
ΓΘ to ΘB.[679] (2) Therefore it required to prove that ΓΘ has to ΘB a
smaller ratio than ΘH to ΘZ, (3) that is: HΘ has to ΘZ a greater ratio
than ΓΘ to ΘB.[680]

Arch. 230   "Let EK be drawn from E at right <angles> to EΓ and, from B,
let a perpendicular, <namely> BΛ, be drawn on it < = on EK>; it
remains for us to prove that HΘ has to ΘZ a greater ratio than ΓΘ to
ΘB. And ΘZ is equal to ΘA, KE taken together" for AZ is equal to the

Arch. 230   radius, "therefore it is required to prove that HΘ has to ΘA, KE taken
together a greater ratio than ΓΘ to ΘB; therefore also: subtracting ΓΘ
from ΘH, and EΛ from KE (<EΛ> being equal to BΘ), it shall be
required that it be proved that the remaining ΓH has to the remaining
AΘ, KΛ taken together a greater ratio than ΓΘ to ΘB." (1) For since
it is required that it be proved that HΘ has to ΘA, KE taken together a
greater ratio than ΓΘ to ΘB, (2) also alternately: that HΘ has to ΘΓ a
greater ratio than ΘA, KE taken together to ΘB,[681] (3) that is to ΛE,[682]
(4) also dividedly: HΓ has to ΓΘ a greater ratio than ΘA, KΛ taken
together to ΛE,[683] (5) that is to BΘ,[684] (6) alternately: that HΓ has to
ΘA, KΛ taken together a greater ratio than ΓΘ to ΘB.[685] (7) But as
ΓΘ to ΘB, so BΘ to ΘA,[686] (8) that is ΛE to AΘ;[687] (9) therefore that

Arch. 230   HΓ has to ΘA, KΛ taken together a greater ratio than ΛE to AΘ, "also
alternately: that ΓH, that is KE, has to EΛ a greater ratio than KΛ, ΘA
taken together to ΘA; dividedly: KΛ has to ΛE a greater ratio than the
same KΛ to ΘA, that is that ΛE is smaller than ΘA."

Following that, we shall add in the synthesis. (1) Since ΛE is smaller
than AΘ, (2) therefore KΛ has to ΛE a greater ratio than KΛ to
AΘ;[688] (3) compoundly: KE has to EΛ a greater ratio than KΛ, AΘ
taken together to AΘ.[689] (4) And ΛE is equal to BΘ;[690] (5) therefore
HΓ has to BΘ a greater ratio than KΛ, AΘ taken together to AΘ.[691]
(6) Alternately: therefore HΓ has to KΛ, AΘ taken together a greater

---

[678] Repeating essentially the same quotation. The quotations suddenly become more
than simple lemmata: they are a text to be quoted in support and as an example of a
Eutocian claim. For a brief moment, it is as if instead of Eutocius elucidating Archimedes,
we have Archimedes' text used to show the validity of Eutocius' earlier comments.

[679] *Elements* VI1.   [680] An extension to inequality of *Elements* V.7 Cor.

[681] An extension to inequality of *Elements* V.16.   [682] *Elements* I.34.

[683] An extension to inequality of *Elements* V.17.   [684] *Elements* I.34.

[685] An extension to inequality of *Elements* V.16.   [686] *Elements* VI.8 Cor.

[687] *Elements* I.34.   [688] *Elements* V.8.

[689] An extension to inequality of *Elements* V.18.   [690] *Elements* I.34.

[691] Note that from Step f of Archimedes' proof we have KE=HΓ – which is an implicit
assumption of the move here from Steps 3, 4, to Step 5.

ratio than BΘ to ΘA,[692] (7) that is ΓΘ to ΘB;[693] (8) alternately: HΓ has
to ΓΘ a greater ratio than KΛ, AΘ taken together to ΘB;[694] (9) compoundly: HΘ has to ΘΓ a greater ratio than KΛ, AΘ taken together,
together with ΘB – that is AΘ, KE taken together[695] – (10) to BΘ.[696]
(11) And KE is equal to AZ;[697] (12) therefore HΘ has to ΘΓ a greater
ratio than ZΘ to ΘB; (13) alternately: HΘ has to ΘZ a greater ratio
than ΓΘ to ΘB.[698] (14) But as ΓΘ to ΘB, so the <square> on ΓΘ to
the <rectangle contained> by ΓΘB;[699] (15) therefore HΘ has to ΘZ a
greater ratio than the <square> on ΓΘ to the <rectangle contained>
by ΓΘB. (16) And, through what was said before, the <square> on ΓΘ,
on ΘZ, is smaller than the <rectangle contained> by ΓΘB, on ΘH;[700]
(17) therefore the <square> on AΘ, on ΘH, has to the <square> on
ΓΘ, on ΘZ, a greater ratio than the <square> on AΘ, on ΘH, to the
<rectangle contained> by ΓΘB, on ΘH;[701] (18) <as the latter ratio,>
so is the <square> on AΘ to the <rectangle contained> by ΓΘB;[702]
(19) therefore the <square> on AΘ, on ΘH, has to the <square> on
ΓΘ, on ΘZ, a greater ratio than the <square> on AΘ to the <rectangle
contained> by ΓΘB. (20) But the <ratio> of the <square> on AΘ
to the <rectangle contained> by BΘΓ is composed (the <square> on
BΘ taken as a mean) of the <ratio> which the <square> on AΘ has
to the <square> on ΘB, and of the <square> on BΘ to the <rectangle
contained> by BΘΓ, (21) and the ratio of the <square> on BΘ to the
<rectangle contained> by BΘΓ is the same as the <ratio> of BΘ to
ΘΓ,[703] (22) that is the <ratio> of AΘ to BΘ;[704] (23) therefore the
<square> on AΘ, on ΘH, has to the <square> on ΓΘ, on ΘZ, a
greater ratio than the <square> on AΘ to the <square> on ΘB together with the <ratio> of AΘ to ΘB. (24) But the ratio composed of
the <square> on AΘ to the <square> on ΘB and of the <ratio> of
AΘ tó ΘB is the same as the <ratio> of the cube on AΘ to the cube
on ΘB, (25) that is the cube on AB to the cube on BΓ;[705] (26) therefore
the <square> on AΘ, on ΘH, has to the <square> on ΓΘ, on ΘZ, a
greater ratio than the <ratio> which the cube on AB has to the cube
on BΓ. (27) But the <ratio> of the <square> on AΘ, on ΘH, to the
<square> on ΓΘ, on ΘZ, was proved to be the same as the ratio of

[692] An extension to inequality of *Elements* V.16.   [693] *Elements* VI.8 Cor.
[694] An extension to inequality of *Elements* V.16.   [695] *Elements* I.34.
[696] An extension to inequality of *Elements* V.18.   [697] Step f of Archimedes' proof.
[698] An extension to inequality of *Elements* V.16.   [699] *Elements* VI.1.
[700] See p. 353 above.   [701] *Elements* V.8.
[702] An extension to solids of *Elements* VI.1.   [703] *Elements* VI.1.
[704] *Elements* VI.8 Cor.
[705] *Elements* VI.8, 4, and an extension to solids of *Elements* VI.22.

the segments,[706] (28) while the ratio of the cube on AB to the cube on
BΓ was proved to be half as much again as the ratio of the surfaces;[707]
(29) therefore the segment has to the segment a greater ratio than half
as much again as the <ratio> which the surface has to the surface.

To 9

Arch. 234   "And it is clear that BA is, in square, smaller than double AK, and
greater than double the radius." (a) For, a <line> being joined from B
to the center, (1) with the resulting angle <subtended> by BA being
obtuse,[708] (2) the <square> on AB is greater than the <squares> on
the <lines> containing the obtuse <angle>,[709] (3) which <lines> are
equal;[710] (4) so that it is greater than twice one of them, (5) that is
than <twice> the <square> on the radius. (6) And once again, the
<square> on AB being equal to the <squares> on AK, KB,[711] (7) and
the <square> on AK being greater than the <square> on KB,[712] (8)
the <square> on AB is smaller than twice the <square> on AK. [And
these hold in the case of the figure, on which is the sign ♂, while in the
other figure the opposite may be said correctly.][713]

Arch. 234   "Also, let EN be equal to EΛ, and let there be a cone <set up> from
the circle around the diameter ΘZ, having the point N as vertex; so this
<cone>, too, is equal to the hemisphere at the circumference ΘEZ,"
(1) For since the cylinder having the circle around the diameter ΘZ as
base, and ΛE as height, is three times the cone having the same base
and an equal height,[714] (2) and half as much again as the hemisphere,[715]
(3) the hemisphere is twice the same cone. (4) And the cone having
the circle around the diameter ΘZ as base, and ΛN as height, is also
twice the same cone;[716] (4) therefore the hemisphere, too, is equal to
the cone having the circle around the diameter ΘZ as base, and ΛN as
height.

---

[706] Implicit in Archimedes' Steps 3–9.

[707] This is asserted as Step 16 of Archimedes' proof. The reference however may be
not to Archimedes' assertion but to Eutocius' own comment on that assertion.

[708] Because we assume the case "greater than hemisphere."

[709] *Elements* II.12 (which Eutocius does not quote explicitly). Calling the center X,
we have $(AB)^2 > (AX)^2 + (XB)^2$.

[710] Both are radii.   [711] *Elements* I.47.

[712] Because we assume the case "greater than hemisphere."

[713] Heiberg brackets this last notice because of its reference to an extra figure. For
the textual questions concerning the double figure accompanying this proposition, see
comments on Archimedes' proposition.

[714] *Elements* XII.10.   [715] *SC* I.34 Cor.   [716] *Elements* XII.14.

Arch. 234   "And the <rectangle> contained by ΑΡΓ is greater than the <rectangle> contained by ΑΚΓ (for the reason that it has the smaller side greater than the smaller side of the other)." For it has been said above that if a line is cut into unequal <segments> at one point, and at another point, the <rectangle contained> by segments closer to the bi-section cut is greater than the <rectangle contained> by the segments

Arch. 234   at the more removed <cut>.[717] And it is the same as saying "for the reason that it has the smaller side greater than the smaller <side> of the other;" for by as much as <the side> is smaller,[718] by that much the cut is distant from the bisection.[719]

Arch. 234   "And the <square> on ΑΡ is equal to the <rectangle> contained by ΑΚ, ΓΞ; for it is half the <square> on ΑΒ." (a) For if ΒΓ is joined, (1) <then,> through the fact that ΒΚ was drawn <as> perpendicular from the right <angle> in a right-angled triangle, (2) and <through the fact> that the triangles next to the perpendicular are similar to the whole <triangle>,[720] (3) the <rectangle contained> by ΓΑΚ is then equal to the <square> on ΑΒ;[721] (4) so that the <rectangle contained> by the half of ΓΑ and by ΑΚ, too, (5) that is the <rectangle contained> by ΓΞ, ΑΚ[722] (6) is equal to half the <square> on ΑΒ,[723] (7) that is to the <square> on ΑΡ.

Arch. 234   "Now, both taken together are greater than both taken together, as well." (1) For since the <rectangle contained> by ΑΚ, ΓΞ is equal to the <square> on ΑΡ,[724] (2) while the <rectangle contained> by ΑΡΓ is greater than the <rectangle contained> by ΑΚΓ,[725] (3) and if equals are added to unequals, the wholes are unequal, and that is greater, which was greater from the start, ((4) the <square> on ΑΡ being added to the <rectangle contained> by ΑΡΓ, (5) while the <rectangle contained> by ΑΚ, ΓΞ <is being added> to the <rectangle contained> by ΑΚΓ), (6) the <rectangle contained> by ΑΡΓ together with the <square> on ΑΡ is then greater than the <rectangle contained> by ΑΚΓ together with the <rectangle contained> by ΑΚ, ΓΞ.

     (1) But the <rectangle contained> by ΑΡΓ together with the <square> on ΑΡ comes to be equal to the <rectangle contained> by ΓΑΡ, (2) through the second theorem of the second book of the

---

[717] Eutocius' comment to *SC* II.8, Step 29 (pp. 352–3 above).

[718] The difference in size between the two smaller sides . . .

[719] . . . is also the amount by which one of them is further away the bisection point than the other. Notice Eutocius' careful language, and his avoidance of labeling with letters. He clearly sees the general import of the argument (see my comments to Archimedes' proposition).

[720] *Elements* VI.8.   [721] *Elements* VI.4, 17.

[722] Step a of Archimedes' proposition.   [723] *Elements* VI.1.

[724] Step 13 of Archimedes' proof.   [725] Step 11 of Archimedes' proof.

Elements,[726] (3) and the <rectangle contained> by AKΓ together with the <rectangle contained> by AK, ΓΞ is equal to the <rectangle contained> by AK, KΞ (4) through the first theorem of the same book;[727] (5) so that "the <rectangle contained> by ΓAP is greater than the <rectangle contained> by AKΞ."

Arch. 235 "And the <rectangle contained> by MKΓ is equal to the <rectangle contained> by ΞKA." (1) For it was assumed: as ΞΓ to ΓK, MA to AK;[728] (2) so that compoundly, also: as ΞK to KΓ, so MK to KA.[729] (3) And the <rectangle contained> by the extremes is equal to the <rectangle contained> by the means;[730] (4) therefore the <rectangle contained> by ΞKA is equal to the <rectangle contained> by MKΓ.

But the <rectangle contained> by ΓAP was greater than the <rectangle contained> by ΞKA;[731] therefore the <rectangle contained> by ΓAP is also greater than the <rectangle contained> Arch. 235 by MKΓ.[732] "So that AΓ has to ΓK a greater ratio than MK to AP." (1) For since there are four lines, ΓK, KM, ΓA, AP, (2) and the <rectangle contained> by the first, ΓA, and the fourth, AP, is greater than the <rectangle contained> by the second, MK, and the third, KΓ, (3) the first, ΓA, has to the second, MK, a greater ratio than the third, KΓ, to the fourth, AP;[733] (4) also alternately: ΓA has to KΓ a greater ratio than MK to AP.[734]

Arch. 235 "But the ratio which AΓ has to ΓK, is that which the <square> on AB has to the <square> on BK," (a) for, BΓ joined, (1) through BK's being a perpendicular from the right angle in a right-angled triangle, (2) it is then: as AΓ to ΓB, BΓ to ΓK,[735] (3) and through this, as the first to the third, that is AΓ to ΓK, (4) so the <square> on AΓ to the <square> on ΓB.[736] (5) But as the <square> on AΓ to the <square> on ΓB, so the <square> on AB to the <square> on BK; (6) for the <triangle> ABK is similar to the <triangle> ABΓ;[737] (7) therefore it is also: as AΓ to ΓK, so the <square> on AB to the <square> on BK.

---

[726] *Elements* II.3 in our manuscripts. (Probably, however, our manuscripts are the same as Eutocius' – who simply counted propositions differently.)

[727] This time *Elements* II.1 in our manuscripts, as well.

[728] Step b of Archimedes' proposition.

[729] *Elements* V.18.   [730] *Elements* VI.16.

[731] Step 16 of Archimedes' proof. Original structure of the Greek: "But, of the <rectangle contained> by ΞKA, the <rectangle contained> by ΓAP was greater."

[732] Eutocius asserts explicitly Step 17 of Archimedes' proof, left implicit by Archimedes himself.

[733] Eutocius' comment to *SC* II.8, Step 30.

[734] An extension to inequality of *Elements* V.16.

[735] *Elements* VI.8 Cor.   [736] *Elements* VI. 20 Cor. 2.

[737] *Elements* VI.8. Step 5 follows from Step 6 through *Elements* VI.4, 22.

(8) And AΓ has to ΓK a greater ratio than MK to AP;[738] (9) therefore the <square> on AB, too, has to the <square> on BK a greater ratio than MK to AP; (10) and the halves of the antecedents:[739] (11) the half of the <square> on AB, which is the <square> on AP,[740] (12) has to the <square> on BK a greater ratio than the half of MK to AP, (13) that is MK to twice AP. (14) But the <square> on AP is equal to the <square> on ZΛ,[741] (15) since AB was assumed equal to EZ,[742] (16) while EZ is twice ZΛ, in square;[743] (17) for EΛ is equal to ΛZ;[744] (17) and twice AP is NΛ, (18) since <it=NΛ is> also <twice> ΛZ;[745] (19) so that the <square> on ZΛ "has to the <square> on BK a greater ratio than MK to twice AP, which is equal to ΛN."

Arch. 235

"Therefore the circle around the diameter ΘZ, too, has to the circle around the diameter BΔ a greater ratio than MK to NΛ. So that the cone having the circle around the diameter ZΘ as base, and the point N as vertex, is greater than the cone having the circle around the diameter BΔ as base, and the point M as vertex." (a) For if we make: as the circle around the diameter ZΘ to the <circle> around the diameter BΔ, so KM to some other <line>, (1) it shall be to a <line> smaller than ΛN.[746] (2) And the cone having the circle around the diameter ZΘ as base, and the smaller, <hypothetically> found line as height, is equal to the <cone> MBΔ, ((3) through the bases being reciprocal to the heights),[747] (4) and smaller than the <cone> NΘZ (5) through the <fact> that <cones> which are on the same base are to each other as the heights.[748] "So it is clear that the hemisphere at the circumference EZΘ, too, is greater than the segment at the circumference BAΔ."[749]

Arch. 235

Arch. 235

---

[738] Step 18 of Archimedes' proof.

[739] A brief allusion to the principle that if a:b::c:d, then (half a):b::(half c):d.

[740] This is the hidden definition of the point P.

[741] Original structure of the Greek: "to the <square> on AP, is equal the <square> on ZΛ." The following brief argument takes as its starting-point the hidden definition of AP, namely: (sq. AP) = half (sq. AB).

[742] Cf. Step 4 of Archimedes' proof. So now we may say that (sq. AP) = half (sq. EZ).

[743] So (sq. AP) = (sq. ZΛ), hence AP=ZΛ, the required result.

[744] Step 16 follows from Step 17 through *Elements* I.47.

[745] Step d of Archimedes' proposition.

[746] From Steps 21–2 of Archimedes' proof, with *Elements* XII.2, V.8.

[747] *Elements* XII.15.   [748] *Elements* XII.14.

[749] Heiberg is surprised by Eutocius' text ending in this note, with no comment on the last lemma from Archimedes; suggesting that this lemma may have been imported in Archimedes' text from Eutocius' commentary (and so is not a separate lemma but part of the comment on the preceding lemma). Perhaps: or perhaps this is the most appropriate ending for a commentary on Archimedes, with Archimedes' own triumphant, final words?

[A commentary of Eutocius the Ascalonite to the second book of Archimedes' *On Sphere and Cylinder*, the text collated by our teacher, Isidorus the Milesian mechanical author].[750] [Sweet labor that the wise Eutocius once wrote, frequently censuring the envious.][751]

---

[750] Compare the end of the commentary to *SC I*. Soon after Eutocius had written his commentary, one or more volumes were prepared putting together Archimedes' text and Eutocius' commentary – this being done by Isidorus of Miletus. (The same author mentioned in another interpolation into this commentary, following the alternative proof to Manaechmus' solution of the problem of finding two mean proportionals.)

[751] This Byzantine epigram does not fit in any obvious way the text of Eutocius himself. Perhaps its author read neither Archimedes nor Eutocius and merely entertained himself by attaching, at the end of this volume, an expression of a generalized sentiment, applicable to any work from antiquity.

BIBLIOGRAPHY

Berggren, J. L. 1976. Spurious Theorems in Archimedes' Equilibrium of Planes Book I. *Archive for History of Exact Sciences* 16: 87–103.

Cameron, A. 1990. Isidore of Miletus and Hypatia: on the Editing of Mathematical Texts. *Greek, Roman and Byzantine Studies* 31: 103–27.

Carol, L. 1895. What the Tortoise Said to Achilles, *Mind* 1895: 278–80.

Clagett, M. 1964–84. *Archimedes in the Middle Ages* (5 vols.). Philadelphia.

Dijksterhuis, E. J. 1987. *Archimedes*. Princeton, NJ : Princeton University Press. (First published in 1956, Copenhagen. Original Dutch edition goes back to 1938.)

Fowler, D. H. F. 1999. *The Mathematics of Plato's Academy* (2nd ed.). Oxford.

Heath, T. L. 1897. *The Works of Archimedes*. Cambridge.

Heiberg, J. L. 1879. *Quaestiones Archimedeae*. Copenhagen.

1880–81. *Archimedes / Opera* (3 vols.) (1st ed.). Leipzig.

1907. Eine Neue Archimedesschrift, *Hermes* 42: 234–303.

1910–15. *Archimedes / Opera* (3 vols.) (2nd ed.). Leipzig.

Hoyrup, J. 1994. Platonism or Archimedism: on the Ideology and Self-Imposed Model of Renaissance Mathematicians, 1400 to 1600, in his *Measure, Number and Weight: Studies in Mathematics and Culture* 203–23: Albany NY.

Hultsch, F. 1876–78. *Pappus / Opera* (3 vols.). Berlin.

Jones, A. 1986. *Book 7 of the Collection / Pappus of Alexandria*. New York.

1999. *Astronomical Papyri from Oxyrhynchus*. Philadelphia.

Knorr, W. 1983. Construction as Existence Proof in Ancient Geometry. *Ancient Philosophy*: 125–48.

1986. *The Ancient Tradition of Geometric Problems*. Boston.

1987. Archimedes After Dijksterhuis: a Guide to Recent Studies, in Dijksterhuis 1987, 419–51.

1989. *Textual Studies in Ancient and Medieval Geometry*. Boston.

Lasserre, F. 1966. *Die fragmente des Eudoxus von Knidos*. Berlin.

Lloyd, G. E. R. 1966. *Polarity and Analogy*. Cambridge.

Lorch, P. 1989. *The Arabic Transmission of Archimedes' Sphere and Cylinder and Eutocius' Commentary*. Zeitschrift für Geschichte der Arabische-Islamischen Wissenschaften, 5: 94–114.

Mansfeld, J. 1998. *Prolegomena Mathematica*. Leiden.

Marsden, E. W. 1971. *Greek and Roman Artillery II: Technical Treatises*. Oxford.

Merlan, P. 1960. *Studies in Epicurus and Aristotle*. Klassisch-Philologische Studien 22: 1–112.

Mugler, C. 1970–74. *Archimede/ Oeuvres* (4 vols.). Paris.

Netz, R. 1998. *The First Jewish Scientist?* Scripta Classica Israelica.

1999. *The Shaping of Deduction in Greek Mathematics: a Study in Cognitive History*. Cambridge.

Forthcoming (a). *Issues in the Transmission of Diagrams in the Archimedean Corpus*. Sciamus.

Forthcoming (b) *From Problems to Equations: a Study in the Transformation of Early Mediterranean Mathematics*. Cambridge.

Peyrard, F. 1807. *Archimede / Oeuvres*. Paris.

Rose, V. 1884. Archimedes im Jahr 1269. *Deutsche Literaturzeitung* 5: 210–13.

Rose, P. L. 1974. *The Italian Renaissance of Mathematics: Studies on Humanists and Mathematicians from Petrarch to Galileo*. Geneva.

Saito, K. 1986. *Compounded Ratio in Euclid and Apollonius*. "Historia Scientarium," XXXI, 25–59.

Sesiano, J. 1991. Un Fragment attribué a Archimède. *Museum Helveticum* 48: 21–32.

Toomer, G. J. 1976. *On Burning Mirrors / Diocles*. New York.

Torelli, J. 1792. *Archimedes / Opera*. Oxford.

Ver Eecke, P. 1921. *Archimede / Ouevres*. Paris.

Von Wilamowitz-Moellendorff, U. 1894. Ein Weihgeschenk des Eratosthenes, *Nachrichten der k. Gesselschaft der Wissenschaften zu Göttingen, phil.-hist. Klasse*, 15–35.

Zeuthen, H. G. 1886. *Die Lehere von den Kegelschnitten im Altertum*. Kopenhagen.

INDEX

Alexandria 13

Algebraic tools and conventions 5, 124–125, 143, 162, book1.239n

Almagest com2.297n

Alternative proofs, Use of 59

Ambiguity 66–67, 80, 117, 156, com2.3n

Ammonius 243, com1.16n

Analysis (see also Synthesis) 26, 190–191, 201–202, 207, 217–218, 227, 231, 232, 272, book2.8n, book2.18n, book2.24n, book2.217n

Anthemius com1.128n

Apodeixis, see Proof

Apollonius 33, 61, 278–279, 320, 322, 324, 331, 338, 340, 342, 343, book1.67n, com1.128n, com2.67n, com2.483n

   Conics 34, book1.67n

Arcadius 312

Archimedes, manuscripts

   Archimedes Palimpsest, the 2, 4, 9, 13, 15, 16, 17, 33, 112, 146, 186, com1.128n

   Codex A 8, 14, 15, 16, 17, 18, 33, 56, 146, 182, 186, intro.9n, com1.128n

   Codex B 10, 17, 18, 33, 56

   Codex B 15, 16, 17, 18

   Codex D 10, 16, 18

   Codex E 10, 16, 18

   Codex F 18

   Codex G 10, 16, 18

   Codex H 10, 16, 18

   Codex 4 10, 16, 18

   Codex 13 18

   Marc. Lat. 327 18

Archimedes, works other than *Sphere and Cylinder*

   Assumptions, On 12

   Balances, on 12

Catoptrics 13

Cattle Problem, the (Bov.) 11, 12, 13

Conoids and Spheroids (CS) 11, 12, 13, 14, 15, 19, 24, 186, 187

Construction of the Regular Heptagon 12

Floating Bodies, On (CF) 2, 3, 11, 12, 13, 15, 16

Lemmas, On 12

Length of the Year, On the 13

Measure of a Circle, On the 13

Measurement of the Circle (DC) 11, 12, 13, 14, 15, 19

Mechanics 13

The *Method (Meth.)* 2, 3, 11, 12, 13, 15, 34, book1.8n, com2.153n

Planes in Equilibrium (PE) 11, 12, 13, 14, 15

Plynths and Cylinders, On 13

Polyhedra, On 13

Quadrature of the Parabola (QP) 11, 12, 13, 14, 15, 33, 40

The Sand Reckoner (Arenarius, Aren.) 11, 12, 13, 14, 15

Sphere-Making, On 13

Spiral Lines (SL) 11, 12, 13, 14, 15, 19, 24, 40, 186, 187, intro.15n

Stomachion (Stom.) 2, 3, 11, 12, 13, 15

Surfaces and Irregular Bodies, On 13

Tangent Circles, On 12

Zeuxippus, To 12

Archimedes Palimpsest, the, see Archimedes, manuscripts

Archytas 290–293, 294, 298

Arenarius, see Archimedes, works other than *Sphere and Cylinder*

Aristotle com1.2n, com2.131n

Arithmetical proportion 265, 1.34, 1.44

Assumptions, *On*, see Archimedes, works other than *Sphere and Cylinder*
Authorial voice 57, 83, 102, 117, 138, 183, 201, 242, com2.172n

Backwards-looking arguments 53, 79, 88, 94, 106, 151, 165, 168–169, 178, 206, 212, 216
Balances, *on*, see Archimedes, works other than *Sphere and Cylinder*
Berggren, J.L. 12
Bifurcating structure of proof 66, 88, 183
Byzantium, see Constantinople

Calculation 221, 225, 233, com2.677n
Calculus 23–24
Cameron, A. com1.128n
Carol, L. 49
Catoptrics, see Archimedes, works other than *Sphere and Cylinder*
Cattle Problem, the, see Archimedes, works other than *Sphere and Cylinder*
Catullus 34
"Chapters" of *Sphere and Cylinder I*
    Chapter 1 19, 20, 21, 25, 43–57
    Chapter 2 19, 20, 57–83, 88
    Chapter 3 20, 22, 83–103
    Chapter 4 20, 103–113
    Chapter 5 20, 21, 22–23, 25, 118–153
    Chapter 6 20, 21, 25, 153–184
    Interlude 20, 22, 24, 113–118
    Introduction 19, 20, 23, 43, 71–72, 112–113, book1.111n, com1.5n
    Section 1 20, 24
    Section 2 19, 20–23, 25
Cicero 19
Circle 166–175, I.1, I.3, I.5, I.6, I.21, I.24, I.26, I.29, I.31, I.32, I.35, I.37, I.40, I.43, Segment I.22, I.41
Codex B, see Archimedes, manuscripts
Codex C, see Archimedes, manuscripts: the Archimedes Palimpsest
Codex Vallae, see Archimedes, manuscripts: Codex A
Commandino 17
Composition of ratios 232, 233, 312–316, 355–356, 357, 358, book2.102n, book2.207n, book2.220n, book2.221n
Concavity 35, 36, 42
Conchoids 298–306
Conditionals 111
Coner, A. 9, 217
Conclusion (as part of proposition) 6, 63, 120–121
Cone 131–270, I.7, I.8, I.9, I.10, I.12, I.14, I.15, I.16, I.17, I.18, I.19, I.20, I.25, I.26, I.27, I.31, I.34, I.38, I.44, II.1, II.2, II.7

Conic sections 25, 27, 61–62, book1.2n; see also Apollonius, ellipse, hyperbola, parabola
Conoids and Spheroids, see Archimedes, works other than *Sphere and Cylinder*
Conon 13, 32, 34, 185, 187
Constantinopole 14, 15, 27
Construction (as part of proposition) 6, 7, 8, 70–71, 104, 106, 120–121, 142, 174
Construction by rotation 120, 290–293
Construction of the Regular Heptagon, see Archimedes, works other than *Sphere and Cylinder*
Corollaries, Use of 151–152, 170
Curved-sided polygon 115
Cylinder 185, 260, 270, I.11, I.12, I.13, I.16, I.34, II.1

Definition of goal (as part of proposition) 6, 8, 63, 107, 110, 120–121, 132, 152, 174, 179
Diagrams
    Properties of drawn figures 5, 9, 89, 95, 100–101, 112, 115, 117, 124, 318, book1.108n, book2.258n
    Principles of critical edition 8–10, 104, 112, 124, 130, com1.87n
    Use by the text 6, 42, 46, 104–105, 107, 130, 137, 156, 158, 162, 169, 176–177, 179, 184, 226, 236, 238, 261, 317, com1.40n, com1.54n, com2.397n; see also Letters, referring to diagram
Dijksterhuis, E.J. 1, 4, 18, 226, book1.69n
Diocles 33, 34, 279–281, 283, 286, 318, 334, 343, com2.98n
Dionysodorus 318, 330–334
Diorismos, see Definition of goal, Limits on solubility
Doric dialect 12, 318, com2.248n
Dositheus 13, 27, 31, 34, 185, 187
Doubling the cube, see Two mean proportionals
Dunamis 339, book1.124n
Duplicate ratio, see Exponents

Ekthesis, see Setting out
Ellipse 340
Enunciation (as part of proposition) 6, 8, 14, 42, 63, 104, 110–111, 120–121, 132, 134, 135, 137, 142, 146, 147, 152, 162, 179, 186, 208, 231
Epi locution 228, 229, 231, 233, 320, 323, 326, 328, 355–356
Eratosthenes 13, 294–298, com2.190n
Euclid's *Elements* 24, 40, 42, 43, 44, 56, 60, 71, 80, 99, 114–115, 258, 312, 344, 345, com2.536n
    Book XII 24
Euclid's *Data* 346

Eudemus
Eudoxus 24, 32, 33, 40, 273, 294, 298, book1.8n, com2.98n
Even-sided and equilateral polygon, see under polygon
Exponents 27, 225–226, 232, 233, 358–359, 360, II.8, book2.218n, com2.642n

Figure circumscribed around a sphere 166–269, I.29, I.30, I.31, I.32, I.40, I.41
Figure inscribed in a sphere 22–23, 269, I.24, I.25, I.26, I.27, I.32, I.35, I.37, I.38, I.41
Floating Bodies, on, see Archimedes, works other than *Sphere and Cylinder*
Florence 16
Fontainebleu 16
Formulaic language 76–77, 107–108, 112–113, 132–133, 147, 156, 220; see also verbal abbreviation
Fowler, D.H.F. 221
Fractions 221, book2.102n
Francois the First 16

Galileo 17
Geminus book1.67n
Generality 43, 62, 63, 76, 120, 121, 130, 132, 162–163, 170, 174, 197–198, 208, 237, 239–241, com2.318n, com2.719n
"Given" 201–202
Great circle 22, 185, I.23, I.24, I.25, I.27, I.28, I.30, I.33, I.34, I.36, I.39, I.41

Heath, T.L. 1, 2, 33, II.2
Hemisphere II.9
Heraclius book1.67n
Hero 14, 275–276, 279, 290, com2.67n
Heronas 313
Hippocrates of Chios 294
Huygens 17
Hypatia
Hyperbola 287, 288, 318, 320, 322, 324, 326, 327, 328, 331, 338, 340, 342–343
Hypsicles 33

Identity 59, 127, com2.425n
Imagination 80–81, 94–95, 120, 179, 198, 261, com1.40n, com1.87n, com2.198n; see also Virtual mathematical reality
Implicit argument 100, 152–153, 213, 225, 226, 237–239, book1.258n
Irrationals 33, com2.318n; see also Fractions
Isidorus of Miletus 14, 269, 290, 368, com1.128n, com2.750n

Jacob of Cremona 16, 18
Jones, A. 236, book1.67n

Kataskeue, see Construction
Knorr, W.R. 4, 12, 45, com1.128n, com2.17n, com2.153n, com2.298n

Laurent. XXVIII 4 [Biblioteca Laurenziana, Florence], see Archimedes, manuscripts: codex D
Lemmas, On, see Archimedes, works other than *Sphere and Cylinder*
Length of the Year, On the, see Archimedes, works other than *Sphere and Cylinder*
Leo the Geometer 14
Leonardo 16
Letters, referring to diagram 42, 62–63, 76, 115, 117, 121, 130, 139, 143, 156, 162, 176, 183, 184, 208, book1.243n, book2.84n, com2.133n, com2.167n, com2.204n, com2.468n
Limits on solubility 26, 221, 317, 318, 329, book2.108n, book2.165n
Lloyd, G.E.R. 66
Logical tools and conventions 124–125

Marc. Gr. 305 [Biblioteca Marciana, Venice], see Archimedes, manuscripts: codex E
Mathematical community 186–187
Mathematical existence 45–46; see also Virtual mathematical reality
Maurolico 17
Measure of a Circle, On the, see Archimedes, works other than *Sphere and Cylinder*
Measurement of the circle, see Archimedes, works other than *Sphere and Cylinder*
Mechanical construction 273–279, 281–284, 290–306
Mechanics, see Archimedes, works other than *Sphere and Cylinder*
Menaechmus 286–289, 295, 298
The *Method*, see Archimedes, works other than *Sphere and Cylinder*
Moerbecke, William of 2, 9, 15, 18, com2.102n; see also Archimedes, manuscripts: Codex B
Monac. Gr. 492, see Archimedes, manuscripts: codex 13 6, 8, 42, 107, 110, 120–121, 132, 137–138, 142, 147, 152
Mugler, C. 1

Narrative structure 20–23, 57, 82–83, 147, 155, 201, 241, com2.155n, com2.335n
Nicomachus 313, com1.16n, com1.2n
Nicomedes 298–306
Noein, see imagination
Numbering of propositions 8, 36, 42, 56, 82, 96, 102, 120, 134, 139, 170, 173, 186, 201, 208, book1.116n

'Obvious' 49, 159–160; see also tool-box
Otton. Lat. 1850 [Biblioteca Vaticana], see
    Archimedes, manuscripts: codex B

Pappus 13, 259, 281–284, 286, 312, book1.67n,
    com1.67n, com2.28n, com2.67n, com2.68n,
    com2.78n, com2.187n
Parabola 287, 289, 290, 318, 320, 324, 326, 327,
    328, 331, book1.2n
Parallelogram 98–99
Paris Gr. 2359 [Bibliothèque Nationale Française],
    see Archimedes, manuscripts: codex F
Paris Gr. 2360 [Bibliothèque Nationale Française],
    see Archimedes, manuscripts: codex G
Paris Gr. 2361 [Bibliothèque Nationale Française],
    see Archimedes, manuscripts: codex H
Parts of proposition: see Enunciation, Setting out,
    Definition of goal, Proof, Construction,
    Conclusion
Pedagogic style 56
Pheidias (Archimedes' father) 11
Philo of Byzantium 277–278, 279, com2.67n,
    com2.80n
Plato 273–274, 294, com2.98n
Plynths and Cylinders, On, see Archimedes, works
    other than Sphere and Cylinder
Point-wise construction 279–281
Poliziano 16
Polybius 10
Polygon 258–259, I.1, I.3, I.4, I.5, I.6
    Even-sided and equilateral 22, 253, 260,
    262–263, I.21, I.22, I.23, I.24, I.26, I.28, I.29,
    I.32, I.35, I.36, I.39
Polyhedra, On, see Archimedes, works other than
    Sphere and Cylinder
Problems 25, 45, 80, 190, 201, 207, 208, 232,
    272, book2.109n
Prism I.12
Proclus 44, 186, com1.2n
Proof (as part of proposition) 6, 7, 8, 63, 107,
    120–121, 142, 152
Protasis, see Enunciation
Ptolemy (astronomer) book2.256n
Ptolemy III Euergetes 294, 298
Pyramid 260, I.7, I.8, I.12

Quadrature of the Parabola, see Archimedes,
    works other than Sphere and Cylinder

Ratio and proportion: inequalities 6, 20, 21, 22,
    23, 45, 54, 91, 147, 251–252, 350–351, 352,
    353, 354–355, 359, I.2, I.3, I.4, I.5,
    book1.23n, book1.146n, com1.17n,
    com1.22n, com1.27n, com1.29n,
    com2.615n, com2.636n, com2.677n
Ratio, Types of 250

References between and within
    propositions 89–90, 96–97, 110–111,
    178–179, 207–208, book2.90n
Repetition of text 51–52, 173
Rhombus (solid) 35, I.18, I.19, I.20, book1.189n
Rivault 16

Saito, K. book2.220n
Samos 13, 34
Scholia 49
Sector (see also Circle) I.4, I.6, I.39
Sector (solid) 35, 166–307, I.39, I.40, I.44
Segment of sphere 174–175, 185, I.35, I.37, I.38,
    I.43, II.2, II.3, II.4, II.5, II.6, II.7, II.8, II.9
Setting-out (as part of proposition) 6, 8, 42, 107,
    110, 120–121, 132, 137–138, 142, 147, 152
Socrates 243
Sphere 22–23, 185, I.23, I.25, I.26, I.27, I.28, I.29,
    I.30, I.31, I.32, I.33, I.34, I.35, I.36, I.37,
    I.38, I.39, I.41, II.1, II.3, II.4, II.8
Sphere-Making, On, see Archimedes, works other
    than Sphere and Cylinder
Spiral Lines, see Archimedes, works other than
    Sphere and Cylinder
Sporus 285–286
Stomachion, see Archimedes, works other than
    Sphere and Cylinder
Sumperasma, see Conclusion
Sun symbol 236, 239, 242, 364
Surfaces and Irregular Bodies, On, see
    Archimedes, works other than Sphere and
    Cylinder
Synthesis (see also Analysis) 26, 190, 207,
    217–218, book2.8n, book2.18n, book2.24n,
    book2.217n
Syracuse 10, 13

Tangent Circles, On, see Archimedes, works other
    than Sphere and Cylinder
Tartagla 17
'That is' 127–128
Theon of Alexandria 312
Theorems 25, 45, 80, 207, 208, 227, 232
Titles of books 27, 183, 186
Tool-box 3, 56, 60, 100, 153, com2.536n
Toomer, G.J. com2.47n, com2.98n, com2.455n
Torelli 16, com2.102n
'Toy universe' 132, com2.443n
Triplicate ratio, see Exponents
Two mean proportionals 25, 272–306, II.1

Vatican Gr. Pii II nr. 16 [Biblioteca Vaticana], see
    Archimedes, manuscripts: codex 4
Venice 15, 16
Verbal abbreviation 6–7, 42, 46, 54, 81, 117–118,
    128, 137–138, 139, 147, 169, 170, 173,

183–184, 237, 238; see also Formulaic language
Virtual mathematical reality 52, 71, 117, 169, 170, 179, book2.221n, com2.27n, com2.135n, com2.443n; see also Imagination
Viterbo 15

Wilamowitz-Moellendorf, u. von com2.153n

Zeuthen, H.G. 45
Zeuxippus, To, see Archimedes, works other than *Sphere and Cylinder*
Zig-zag lines 71, 244, 245, book1.13n